An 800,000-Year Paleoclimatic Record from Core OL-92, Owens Lake, Southeast California

Edited by

George I. Smith
and
James L. Bischoff

U.S. Geological Survey
345 Middlefield Road, MS 902 (GIS) and 910 (JLB)
Menlo Park, California 94025

WITHDRAWN
UST
Libraries

SPECIAL PAPER

317

1997

Copyright © 1997, The Geological Society of America, Inc. (GSA). All rights reserved. GSA grants permission to individual scientists to make unlimited photocopies of one or more items from this volume for noncommercial purposes advancing science or education, including classroom use. Permission is granted to individuals to make photocopies of any item in this volume for other noncommercial, nonprofit purposes provided that the appropriate fee ($0.25 per page) is paid directly to the Copyright Clearance Center, 27 Congress Street, Salem, Massachusetts 01970, phone (508) 744-3350 (include title and ISBN when paying). Written permission is required from GSA for all other forms of capture or reproduction of any item in the volume including, but not limited to, all types of electronic or digital scanning or other digital or manual transformation of articles or any portion thereof, such as abstracts, into computer-readable and/or transmittable form for personal or corporate use, either noncommercial or commercial, for-profit or otherwise. Send permission requests to GSA Copyrights.

Copyright is not claimed on any material prepared wholly by government employees within the scope of their employment.

Published by The Geological Society of America, Inc.
3300 Penrose Place, P.O. Box 9140, Boulder, Colorado 80301

Printed in U.S.A.

GSA Books Science Editor Abhijit Basu

Library of Congress Cataloging-in-Publication Data
An 800,000-year paleoclimatic record from Core OL-92, Owens Lake,
  southeast California / edited by George I. Smith and James L.
  Bischoff.
     p.   cm. -- (Special paper ; 317)
   Includes bibliographical references and index.
   ISBN 0-8137-2317-5
   1. Geology, Stratigraphic--Quaternary.   2. Borings--California-
 -Owens Lake (Inyo County)   3. Lake sediments--California--Owens Lake
 (Inyo County)   4. Geology--California--Owens Lake (Inyo County)
 5. Paleoclimatology--California--Owens Lake (Inyo County)
 I. Smith, George I. (George Irving), 1927-   . II. Bischoff, James
 L.   III. Series: Special Papers (Geological Society of America) ;
 317
 QE696.A14   1997
 551.7'9--dc21   96-49549
                                                                       CIP

Cover: Drill site of core OL-92 at Owens Lake, southeast California (view toward west-northwest). *Left:* Laboratory trailer equipped with refrigeration for core storage, a long workbench, sampling equipment, and various instruments; *left center:* drilling-rod storage; *right center:* truck-mounted drilling rig (tower is approximately 10 m high); *right:* water truck. *Foreground:* Lake surface (elevation 1,085 m) is composed of salts that crystallized earlier this century. *Background:* Sierra Nevada; highest peaks in photograph have elevations between 3,600 m and 3,900 m, the southeast-facing lower treeline is near 2,000 m; this part of the range is composed dominantly of Mesozoic plutonic rock.

10 9 8 7 6 5 4 3 2 1

Contents

Preface v

1. **Core OL-92 from Owens Lake: Project rationale, geologic setting, drilling procedures, and summary** 1
   G. I. Smith and J. L. Bischoff

2. **Stratigraphy, lithologies, and sedimentary structures of Owens Lake core OL-92** 9
   G. I. Smith

3. **Climatic signals in clay mineralogy and grain-size variations in Owens Lake core OL-92, southeast California** 25
   K. M. Menking

4. **Responses of sediment geochemistry to climate change in Owens Lake sediment: An 800-k.y. record of saline/fresh cycles in core OL-92** 37
   J. L. Bischoff, J. P. Fitts, and J. A. Fitzpatrick

5. **Movement and diffusion of pore fluids in Owens Lake sediments from core OL-92 as shown by salinity and deuterium-hydrogen ratios** 49
   I. Friedman, J. L. Bischoff, C. A. Johnson, S. W. Tyler, and J. P. Fitts

6. **Paleomagnetism and magnetic susceptibility of Pleistocene sediments from drill hole OL-92, Owens Lake, California** 67
   J. M. Glen and R. S. Coe

7. **Age and correlation of tephra layers, position of the Matuyama-Brunhes chron boundary, and effects of Bishop ash eruption on Owens Lake, as determined from drill hole OL-92, southeast California** 79
   A. M. Sarna-Wojcicki, C. E. Meyer, and E. Wan

8. *A time-depth scale for Owens Lake sediments of core OL-92: Radiocarbon dates and constant mass-accumulation rate* 91
   J. L. Bischoff, T. W. Stafford, Jr., and M. Rubin

9. *A diatom-based paleohydrologic record of climate change for the past 800 k.y. from Owens Lake, California* 99
   J. P. Bradbury

10. *Ostracodes in Owens Lake core OL-92: Alternation of saline and freshwater forms through time* 113
    C. Carter

11. *Paleobiotic and isotopic analysis of mollusks, fish, and plants from core OL-92: Indicators for an open or closed lake system* 121
    J. R. Firby, S. E. Sharpe, J. F. Whelan, G. R. Smith, and W. G. Spaulding

12. *An 800,000-year pollen record from Owens Lake, California: Preliminary analyses* 127
    R. J. Litwin, D. P. Adam, N. O. Frederiksen, and W. B. Woolfenden

13. *Synthesis of the paleoclimatic record from Owens Lake core OL-92* 143
    G. I. Smith, J. L. Bischoff, and J. P. Bradbury

Index 161

Preface

An important goal of paleoclimatic research is to unravel the record of past changes in precipitation in now-arid regions of western North America. Fluctuations in the levels of former lakes are especially informative, but studies of lacustrine sedimentary records older than about 150,000 years are relatively rare because most of them are buried beneath younger sediments.

To improve our knowledge of the changes in precipitation and runoff recorded by lakes during those earlier periods, a core-drilling program was proposed by members of a group attending a U.S. Geological Survey–sponsored paleoclimate workshop in 1991. Owens Lake was identified as one of the promising sites, in part because existing sets of instrumentally collected hydrologic and meteorologic data from this area, some extending back almost a century, would provide a nearly unparalleled resource when converting past changes in regional runoff and lake character into the atmospheric-precipitation element of climate.

The core drilling at Owens Lake, funded by the U.S. Geological Survey's Global Change and Climate History Program, was carried out in the spring of 1992. This volume represents our interpretations of the data collected from the resulting core. A document containing the numerical and other detailed forms of "raw" data collected by this volume's authors was prepared earlier (Core OL-92 from Owens Lake, southeast California, 1993, G. I. Smith and J. L. Bischoff, eds., U.S. Geological Survey Open-File Report 93-683, 398 p.). The interpretations, tables, and diagrams presented herein are derived largely from those data.

George I. Smith
James L. Bischoff

Core OL-92 from Owens Lake: Project rationale, geologic setting, drilling procedures, and summary

George I. Smith and James L. Bischoff
U.S. Geological Survey, 345 Middlefield Road, MS 902 and MS 910, Menlo Park, California 94025

ABSTRACT

Several lines of evidence indicated that Owens Lake, a now-dry lake in southeast California, would probably yield a continuous and climatically informative sedimentary record. Also, the details of modern climate and runoff in the area are exceptionally well known, providing a firm basis for interpreting various types of evidence from a core in terms of past climates. Drilling was carried out in early 1992 to retrieve this record.

The resulting core, OL-92, was taken from the south-central part of the lake (lat 36°22.85′ N, long 117°57.95′ W). Lake surface elevation at the drill site is 1,085 m. The core's length is 322.86 m, recovery was ~80%, and the age of its basal sediments is ~800 ka.

Study of the core has revealed lithologic, chemical, mineralogic, geophysical, and paleontologic evidence that reflects alternating periods of high- and low-volume runoff into Owens Lake. This volume presents these studies and summarizes their paleoclimatic significance.

RATIONALE FOR DRILLING PROJECT

An important goal of paleoclimatic research is to unravel the record of past changes in precipitation within now-arid regions of western North America. More than a century of geologic investigations in this area has produced several lines of evidence indicating that major changes in precipitation and runoff occurred during the Quaternary period. Among the most convincing evidence was that of large changes in the sizes of lakes in the Great Basin. These basins are sometimes termed "nature's rain gauges" because their water levels primarily record annual precipitation amounts within their drainage areas, and their lake-level histories thus almost quantitatively document changes in precipitation.

Reviews of these Pleistocene lake histories by Smith and Street-Perrott (1983) and Benson et al. (1990), based mostly on exposed stratigraphic records and geomorphic criteria, show that our understanding about the fluctuations of these ancient lakes is still inadequate, as is our ability to translate those histories into changes in absolute precipitation amounts. Because changes in precipitation were, and would be in the future, among the most critical to all forms of life, the timing, magnitudes, and causes of past precipitation changes pose important questions to earth and paleoclimate scientists.

Records of lake histories much older than about 150 ka are relatively rare, primarily because of their destruction by erosion or burial by younger deposits. Records of earlier Pleistocene-age lakes can be found, however, in deposits beneath the surfaces of many modern lakes or playas. To study these records, a core drilling program, lasting several years and targeting a number of basins, was envisioned by members of a group attending a U.S. Geological Survey–sponsored paleoclimate workshop in January 1991 (Gardner et al., 1991). Owens Lake was identified as one of the promising sites for a continuous, climatically informative record.

LOCATION AND GEOLOGIC SETTING

Owens Lake is presently a closed basin at the southern end of the Owens Valley in California, on the western edge of the Great Basin and immediately east of the Sierra Nevada (Fig. 1A). The valley is essentially a large graben, with more than 3,300 m of relief, which drains an area of 8,550 km². It is bounded on the west by the Sierra Nevada, which is composed predominantly of Mesozoic plutonic rocks. On the northeast,

---

Smith, G. I., and Bischoff, J. L., 1997, Core OL-92 from Owens Lake: Project rationale, geologic setting, drilling procedures, and summary, *in* Smith, G. I., and Bischoff, J. L., eds., An 800,000-Year Paleoclimatic Record from Core OL-92, Owens Lake, Southeast California: Boulder, Colorado, Geological Society of America Special Paper 317.

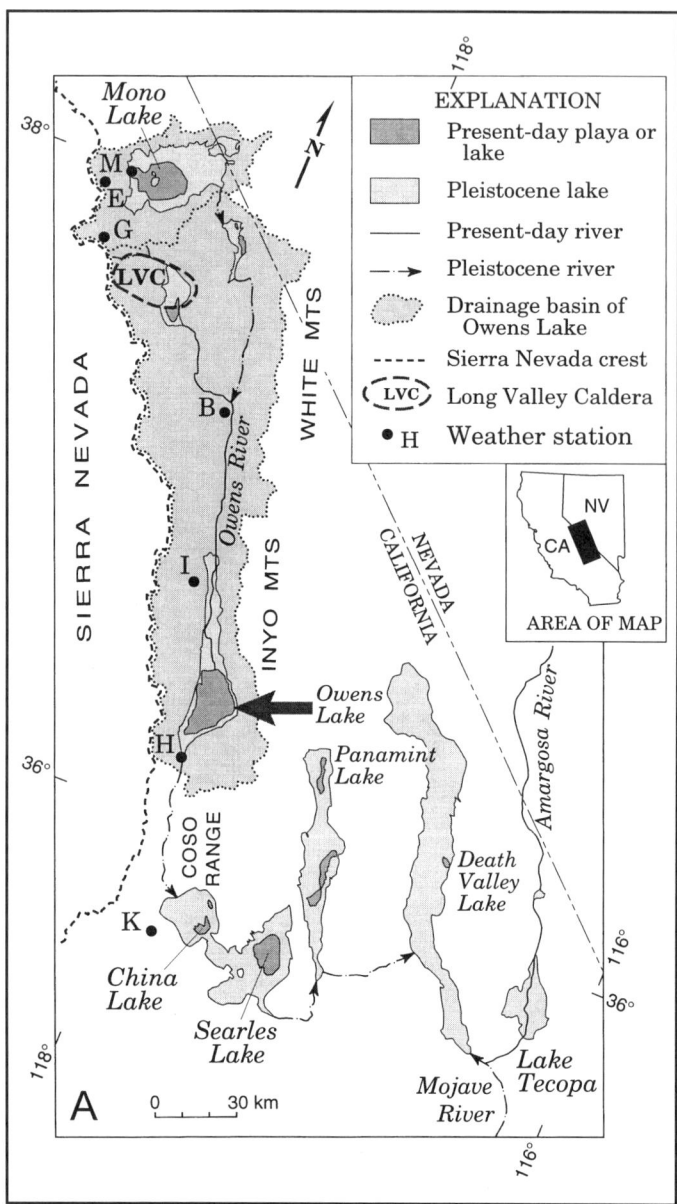

the valley is bounded by the Inyo and White Mountains, which are composed of Paleozoic and Mesozoic sedimentary rocks and Mesozoic plutonic rocks; on the southeast it is bounded by the Coso Range, composed of late Cenozoic volcanic rocks that rest on Mesozoic plutonic rocks. Late Cenozoic volcanic cones and flows are scattered along the east and west edges of the valley, apparently reflecting faults that formed it.

The floor of Owens Valley is semi-arid, receiving about 15 cm of precipitation each year, mostly in the winter. On the east-facing slopes of the Sierra Nevada, precipitation increases with elevation at a linear gradient near 28 cm/1,000 m (Table 1, Fig. 2); on the west-facing slopes of the White and Inyo Mountains, it increases at about half that rate. Temperatures on the valley floor vary seasonally by more than 20 °C; mean monthly temperatures in January are ~3 °C and in July ~25 °C. The lapse rate is near –7.3 °C/1,000 m.

PALEOHYDROLOGY OF OWENS LAKE

Prior to 1913, when the Owens River was diverted into the Los Angeles Aqueduct, the river supplied most of the Owens Lake's water, although not enough to have caused overflow during the past several thousand years. During earlier times, however, the Owens Lake surface stood at its sill depth (~1,145 m) and overflowed. When this occurred, the Owens River drainage area became the main water supply for a chain of lakes that at

Figure 1. A (left), Locations of Owens Lake (arrow) and other lakes hydrologically connected upstream and downstream from it during pluvial periods of the Pleistocene. Names of most modern lakes are same as the host valley except that China Lake is in Indian Wells Valley; names suggested by others for expanded Pleistocene lakes are Lake Russell (for Mono Lake), Lake Gale (in Panamint Valley), and Lake Manly (in Death Valley). Names of adjoining mountain ranges also shown. Meteorological stations listed in Tables 1 and 2 shown as solid circles labeled as follows: M, Mono Lake; E, Ellery Lake; G, Gem Lake; B, Bishop; I, Independence; H, Haiwee; and K, Inyokern. Circular area near headwaters of Owens River labeled LVC is the Long Valley caldera, source of the Bishop ash. B (below), Diagrammatic cross-section of the chain of lakes downstream from the Owens River during pluvial periods of the Pleistocene having maximum intensities. Lake names and elevations of highest lake stands shown above each profiled basin, elevations of present basin floors shown below them.

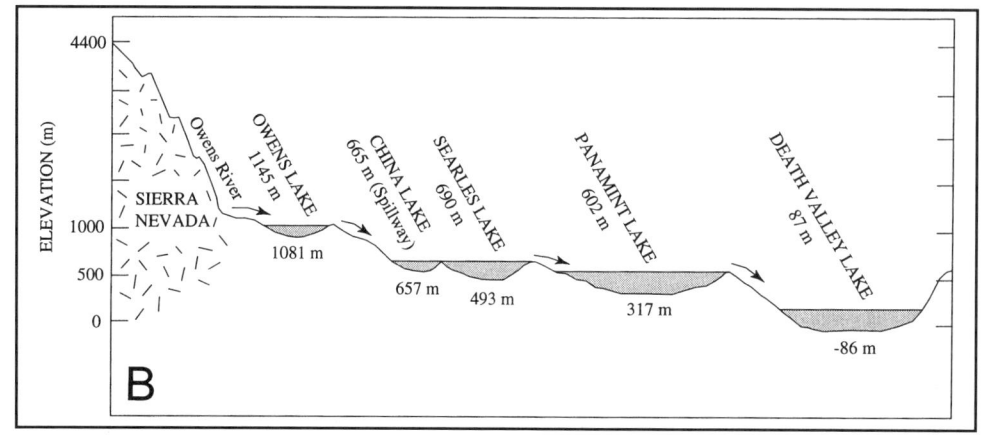

TABLE 1. MEAN PRECIPITATION AMOUNTS AT NATIONAL WEATHER SERVICE AND AFFILIATED STATIONS ON EAST SLOPE OF SIERRA NEVADA AND IN VALLEY AREAS WITHIN 10 KM OF THE RANGE'S EASTERN BOUNDARY

Station[†]	Elevation (m)	Precipitation Summer May 1 to October 31 1951-1980 (cm)	Precipitation Winter November 1 to April 30 1951-1980 (cm)	Annual (cm)	Winter Divided by Annual (W/A)
Mono Lake	1,966	7.90	25.02	32.92	0.760
Ellery Lake	2,940	11.99	46.05	58.04	0.793
Gem Lake	2,734	10.11	37.87	47.98	0.789
Bishop WSO	1,252	2.69	11.56	14.25	0.811
Independence	1,204	2.36	11.33	13.69	0.828
Haiwee	1,166	3.58	13.21	16.79	0.787
Inyokern	743	2.01	8.56	10.57	0.810
Mean		5.81	21.94	27.75	0.797[§,**]

Elevation vs. Summer precipitation: 4.96 cm/1,000 m; $r^2 = 0.96$[**,‡]
Elevation vs. Winter precipitation: 17.60 cm/1,000 m; $r^2 = 0.97$[‡]
Elevation vs. Annual precipitation: 22.57 cm/1,000 m; $r^2 = 0.97$[‡,§§]

*Data from NOAA, 1982.
[†]Stations listed north to south.
[§]Elevation vs. W/A ratio: $r^2 = 0.2$, thus virtually no relation between variables.
[**]r^2 = coefficient of determination, where 1.00 = perfect fit, 0.00 = no fit.
[‡]These r^2 values would be slightly lower if a NWS station (with records between 1951 and 1980) existed near the west edge of the Long Valley caldera (Fig. 1), where precipitation is greater.
[§§]Data for precipitation during various periods at 13 LADWP station in the Owens River drainage area, plus two of the NWS station used in this table, indicate a relation between annual precipitation and elevation of 20.50 cm/1,000 m, $r^2 = 0.94$ (Danskin, in press).

times extended south and east of Owens Lake to Indian Wells, Searles, Panamint, and Death Valleys (Fig. 1B). The floors of these basins are now occupied by playa lakes or salt flats, because the evaporation rate in these basins is high and the inflow from the mountains that surround them is low. Mono Lake, a perennial water body in a closed basin adjoining the north edge of the Owens River catchment area, also overflowed an undetermined number of times into Adobe Valley and then into the upper reaches of what was then the east fork of the Owens River; that fork flowed through areas now known as Benton, Hammil, and Chalfant Valleys. The combined surface area of the perennial lakes in the chain was primarily proportional to the amounts of precipitation falling in their collective drainage areas, including, most importantly, the high eastern slopes of the southern Sierra Nevada which drain into the west (now main) fork of the Owens River. Variations in wind velocities, relative humidity, temperature, and other climatic variables that influence evaporation rates were also factors in determining lake sizes, but changes in them were less important than in precipitation (Smith, 1991).

Studies of lacustrine outcrops, cores, and landforms have allowed reconstruction of the past histories of lakes in Owens Valley and its downstream basins (Gilbert, 1875; Gale, 1914; Smith and Pratt, 1957; Hooke, 1972; R. S. U. Smith, 1975; G. I. Smith, 1979, 1984; Smith et al., 1983). Glacial, geomorphic, and

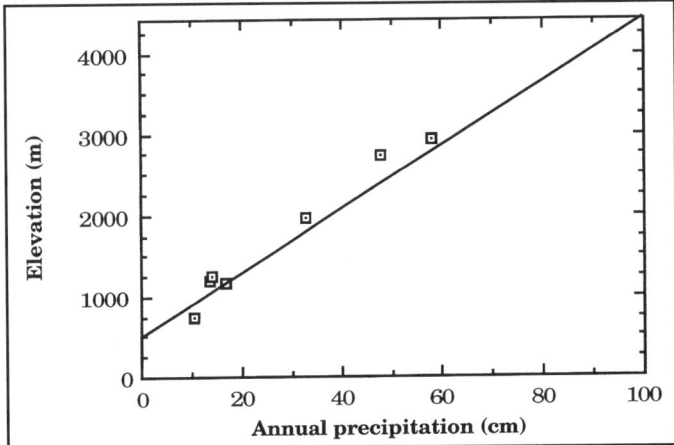

Figure 2. Diagram showing relation between elevation and precipitation for National Weather Service stations listed in Table 1. Stations, small squares, are either on the east slope of the Sierra Nevada or on the adjoining valley floor less than 10 km from base of mountains. Maximum elevation plotted on ordinate (4,417 m) is that of Mt. Whitney, highest peak in the range. Fitted line is a linear regression (y = 486.0 + 44.3x; $r^2 = 0.975$); a second-order polynomial line produces a slightly higher r^2 value but does not project to the maximum elevation in a meteorologically likely manner (r^2 = coefficient of determination, where 1.00 = perfect fit, 0.00 = no fit).

paleobotanical studies in these and adjoining areas provide additional criteria that help determine the sequences and character of past climates (Blackwelder, 1931; Sharp and Birman, 1963; Martin and Mehringer, 1965; Sharp, 1968; Burke and Birkeland, 1979, 1988; Van Devender and Spaulding, 1979; Spaulding et al., 1983; Dorn et al., 1987; Phillips et al., 1990; Bursik and Gillespie, 1993; Koehler and Anderson, 1994). These and other studies provide constraints on the interpretation of the lacustrine record from Owens Lake because all of these areas were part of the same climatic and hydrologic system, and their inferred histories, or some modification of them, must be compatible.

Conditions sought for a core-drilling site

Lacustrine sediments can reflect paleoclimatic changes if (1) climate changes created differences in the character of sedimentation and in the fossil populations; (2) sedimentation rates were relatively constant so that ages can be interpolated or extrapolated from dated horizons with minimal error being introduced; and (3) sedimentation was continuous and in waters deeper than wave base, so that there was little or no loss of the sedimentary record as a result of shallow-water erosion.

These criteria would be nearly impossible to assess prior to receiving *any* core-drilling information. However, the results of coring Owens Lake in 1953 provided evidence that the 278.5 m of sediments that were penetrated, although variable, were all lacustrine (Smith and Pratt, 1957), and that the lacustrine diatom and ostracode abundances and species varied throughout the core (Gardner et al., 1991), partially satisfying criteria 1 and 2.

Several lines of evidence indicate that during most of Pleistocene and early Holocene time, Owens was likely to have been a perennial lake, satisfying criterion 3. Even during late Holocene time—prior to the early twentieth century when the Owens River was diverted into the Los Angeles Aqueduct, causing the lake to desiccate—Owens apparently remained a shallow, moderately saline, perennial lake. The types of evidence indicating this history are as follows.

Evidence based on historical observations. Much of the southwestern United States became more arid between the early 1870s and early 1900s (Knox, 1983). For example, between 1872 and 1905, low runoff and continuing evaporation caused the level of the Great Salt Lake in Utah to fall 4.6 m and nearly double its salinity (Arnow, 1984). Owens Lake was 14.9 m deep in 1872 and 5.8 m deep in late 1905 (Gale, 1914), although in both lakes, the following five years of more normal runoff restored about a third of the water lost during the 33-year drought. Part of the decrease in Owens River flow during this period also reflected the increasing amount of irrigation in Owens Valley. At its lowest point in 1905, the area of Owens Lake had shrunk to about 75% of its 1872 area, but it was still a perennial lake (Gale, 1914), and the planned OL-92 coring site was beneath 2.5 m of water and located more than 3.5 km from the nearest shore. A further decrease in inflow, causing it to shrink to about 50% of its 1872 area, would have been necessary to expose the coring site to subaerial conditions.

Evidence based on earlier core studies. A 278.5-m core (66% recovery), obtained in 1953, recovered sediments that were all interpreted as perennial-lake deposits (Smith and Pratt, 1957; Fig. 3), based on sediment character, the nearly continuous presence of aquatic fossils, and the lack of salts, oolites, mud cracks, soil horizons, or alluvial gravels that would have indicated periods of very shallow water or nonlacustrine deposition. Four shallow cores (Fig. 3) were recovered later in a study of the very late Pleistocene and Holocene deposits and their paleomagnetic variations (Newton, 1991); aquatic fossils were also prominent in them.

Evidence based on past meteorological setting of the site. The elevation of the Sierra Nevada has a major influence today on the precipitation reaching the Owens River drainage area as well as much of the Great Basin. However, uplift of the range

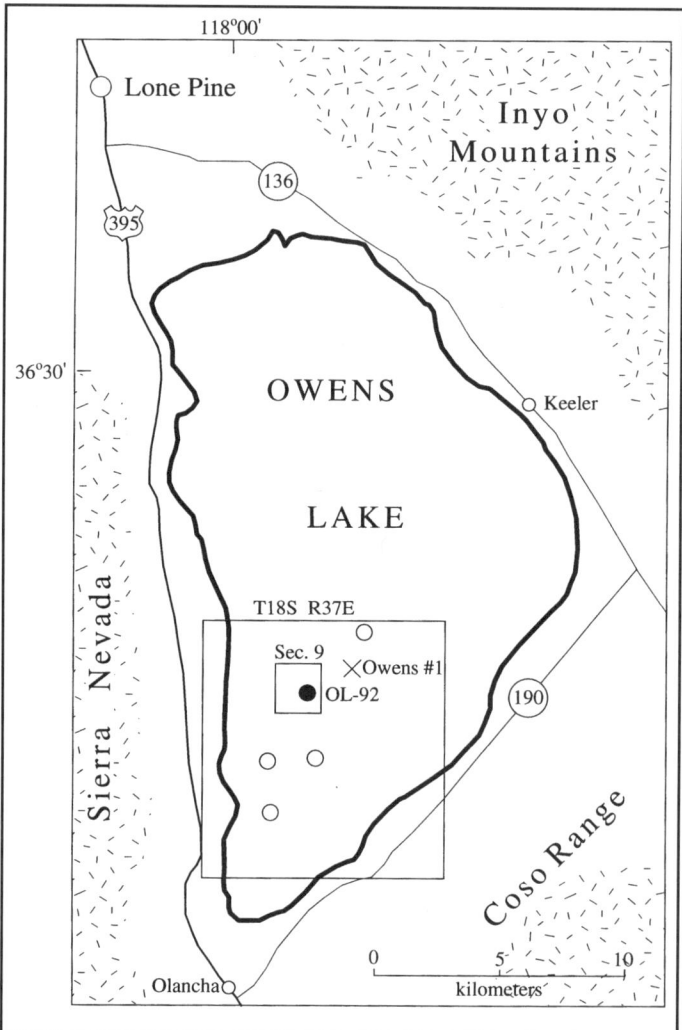

Figure 3. Filled circle represents the location of core OL-92 relative to the boundaries of Owens (dry) Lake and of Section 9 in Township 18 South/Range 37 East; × shows location of Owens #1 described by Smith and Pratt (1957), and open circles show locations of cores described by Newton (1991). Names of mountain ranges, towns, and numbered highways also indicated.

during the 800 k.y. represented by OL-92 was probably meteorologically insignificant. About 3 m.y. ago, the crest of the Sierra Nevada near the headwaters of the Owens River was probably about 1,000 m lower than at present, but undergoing continuing uplift (Huber, 1981). The oldest sediments in core OL-92, therefore, represent only a quarter of that period. If the uplift rate remained constant and affected all segments of the Sierra bounding Owens River drainage area similarly, this part of the range would have been within ~250 m of its present elevation during deposition of all sediments recovered by OL-92.

Precipitation in the Owens River watershed would have been relatively unaffected by this uplift. Modern precipitation gradients in the southern Sierra (Table 2) show that if areas now at high elevations were 250 m lower, annual precipitation on that part of the range would have been reduced by only ~10–20% (Fig. 2). This implies that at 800 ka, runoff was ~80–90% that of present runoff, allowing a little more moisture to remain in the airmass, reducing the "rain shadow" effect as it moved east, but not necessarily condensing much of that moisture within the area draining into the Owens River. As already noted, the historically observed reduction of inflow between 1872 and 1905, which reduced the lake area to about 75% of its 1872 area (Gale, 1914), did not reduce Owens Lake to dryness; thus it is unlikely that earlier periods of comparable drought would have caused the lake to desiccate.

Evidence based on geophysical studies. Seismic, gravity, and aeromagnetic studies show that the deepest and widest part of the bedrock surface beneath southern Owens Valley lies under Owens Lake, and that this area is underlain by more than 1.8 km of low-density sediments (Pakiser et al., 1964). A core of almost any feasible length would probably be dominated by basin-center—and therefore mostly lacustrine—sediments.

TABLE 2. MEAN SUMMER, WINTER, AND ANNUAL TEMPERATURE AT NATIONAL WEATHER SERVICE AND AFFILIATED STATIONS, ALONG EAST EDGE OF SIERRA NEVADA AND IN VALLEY AREAS WITHIN 10 KM OF THE RANGE'S EASTERN BOUNDARY

Station	Elevation (m)	Temperatures*		
		Summer (May 1 to October 31 1951-1980) (°C)	Winter (November 1 April 30 1951-1980) (°C)	Annual (°C)
Mono Lake	1,966	15.14	2.31	8.72
Bishop WSO	1,252	20.08	6.62	13.33
Haiwee	1,166	21.96	7.99	15.00
Inyokern	743	24.37	10.41	17.39
Mean		20.4	6.8	13.6
Lapse rate (deg/1,000 m)		-7.78	-6.73	-7.27
Coefficient of determination (r^2)		0.98	0.99	0.99

*Data from NOAA, 1982.

Modern data relevant to paleoclimatic reconstruction

The existing network of hydrologic and meteorologic data points, mostly operated by the Los Angeles Department of Water and Power and the Southern California Edison Company, serves as a nearly unparalleled resource for the investigation of past changes in the hydrologic element of climate in this area.

Hydrologic and meteorologic data. Owens Valley has been the focus of almost a century of measurements by scientists and engineers concerned with the water supply for the City of Los Angeles (W. T. Lee, 1906; C. H. Lee, 1912; City of Los Angeles, 1916; Hollett et al., 1991). Well over a half century of instrumentally obtained data on seasonal and annual precipitation, runoff, and evaporation are on record. Knowing the present evaporation rate and its probable variation during the Pleistocene, one can translate the history of Owens Lake and its downstream chain of lakes into the upstream requirements for precipitation and runoff.

The Owens River's water primarily comes from creeks originating in the high-elevation, east-draining slopes of the Sierra Nevada. About 80% of the precipitation falls during the six months that include winter (Table 1), most of it at high elevations as snow. The crest of this part of the range has an average elevation near 3,500 m, and it includes 34 peaks whose elevations exceed 4,000 m. The lowest elevation along the crest is at Minarette Summit (~2,800 m), immediately west of the Long Valley caldera (Fig. 1). Stable-isotope studies of the snow pack, as well as the vegetation character east of this point, show that the topographic low allows significantly more precipitation to reach an area extending about 15–20 km east of the pass (Friedman and Smith, 1970). This region also would have accounted for about 40% of the water that flowed into Owens Lake, if the river had not been diverted upstream into the Owens Valley Aqueduct (Hollett et al., 1991, p. B38, Table 2). The remaining, higher parts of the bounding section of the Sierra Nevada produce virtually all of the balance.

Stable-isotope data from Sierran snow packs also show that storms predominantly bring moisture to these high, east-slope elevations via trajectories that pass over the crest of the range, with much of the precipitation condensing in the vicinity of the crest and drifting east (Friedman and Smith, 1970, 1972). Major increases in past precipitation, therefore, required barometric gradients that were steeper than today, forcing more moist-air masses over this 3,500-m barrier.

Studies of precipitation falling on the lower levels of Owens Valley, however, reveal that while many storms reaching those lower levels include isotopically light moisture that arrived via a high-elevation trajectory, some storms bring isotopically heavy precipitation which arrived via a low-elevation trajectory south of the Sierra Nevada (Smith et al., 1979). The existence of the alternative lower-elevation route, which enables an air mass to retain more of its moisture, appears to account for the ~20% higher annual precipitation at Haiwee, south of Owens Lake, relative to Independence and Bishop to its north (Table 1).

An isohyet map of Owens Valley (Hollett et al., 1991) shows

maximum annual precipitation in the Sierra, west of Owens Valley, to be between ~75 cm (~30 in) and ~100 cm (~40 in). A line fitted to data representing precipitation between 1951 and 1980 at seven east-slope stations (Fig. 2) confirms this range of maximum values at the elevation of the range's highest peak.

CORE-HOLE LOCATION AND CORING PROCEDURES

Transporting drilling and core-processing equipment to an area near the probable depositional center of Owens Lake—at a reasonable cost—was an important consideration. Most of the lake's central surface is too soft for heavy-vehicle travel, but fortunately, an existing heavy-duty road, built for development of the lake's soda ash potential, led to an area near the longitudinal center of the lake surface. The accessible area is about 4 km south of the latitudinal center of the lake, but this was considered to be an advantage because it might have minimized past variations in the clastic contributions from the Owens River, which entered the lake ~65 km north of the proposed drill site when the lake was at its spillway level. The drill site is about 2 km south of the lowest point in the underlying bedrock.

The drill site was located about 140 m west and 420 m north of the southeast corner of Sec. 9, T. 18 S., R. 37 E., Mount Diablo Base and Meridian. Its coordinates are lat 36°22.85′N, long 117°57.95′W. Elevation of this part of the dry-lake floor is about 1,085 m; for reference, the lake's water surface in 1872 was at 1,096 m.

Core OL-92 reached a total depth of 322.86 m (1,059.25 ft). "Core OL-92" is actually composed of three subcores from adjacent sites, designated by Smith (1993) as cores OL-92-1 (5.49–61.37 m), OL-92-2 (61.26–322.86 m), and OL-92-3 (0.00–7.16 m) The two deeper cores were obtained by use of a rotary rig equipped with a core barrel that recovered 7.6-cm (3-in) diameter cores using a split-spoon liner. The shallow core was recovered using first a small rotary rig to make a hole in the hard upper part of the oolite layer, through which we then pushed a 5.5-m (20-ft) long, 7.6-cm-diameter, polyvinyl chloride (PVC) tube into the lower part of the soft oolite bed and the underlying mud; a cap was cemented on top. The tube was inserted and retrieved using a large backhoe (Smith, 1993).

Core drilling commenced on April 23 and finished on June 8, 1992. In the field, the core was split, logged, and sampled; it was then stored in a refrigerated compartment at the drill site. Overall core recovery was about 80%. Some of the recovered sediment at the top of a run had clearly caved into the hole after retrieval of the core barrel from the previous run. We generally discarded the most conspicuously soft sediments, but segments of uncertain origin were saved and noted in the core log as being questionable (Smith, 1993); three samples from these intervals were confirmed by ^{14}C dating as caved-in material (Bischoff et al., this volume chapter 8). Most caving took place during the drilling of core OL-92-1 because the uppermost sediments are wetter, less compacted, and therefore weaker than the deeper sediments. We eventually sealed off that core hole at a depth of 61.37 m, moved to a new site about 3 m away, drilled using a solid bit to 61.26 m, set 61 m of casing in the new hole, and began recovery of core OL-92-2.

SUMMARY OF PROJECT RESULTS

Most of the data collected by various investigators during the core-drilling period, plus the year and a half following completion of that phase, are presented in 20 chapters compiled by Smith and Bischoff (1993). This present volume summarizes those results, reports work that was unfinished at the time that volume was completed, and offers our paleoclimate interpretations of them. Brief overviews of the results of these studies follow. They represent a multidisciplinary approach, and each study contributes either directly or indirectly to our paleoclimatic goals.

• Core OL-92, from Owens (dry) Lake, southeast California, represents virtually all of the past 800 k.y. of lacustrine deposition. During the wettest parts of Pleistocene time, Owens was connected by overflow to downstream lakes in China, Searles, Panamint, and Death Valley basins.

• The lowest 37% of the core's clastic section is relatively sandy, representing shallow-water deposition; it contains about 30% sand-sized material and 70% clay and silt. The upper 63% of that section is predominantly clay and silt, representing deep-water deposition. The relatively coarse clastic sediments in the clay and silt deposits have mean sizes ~15 μm, and the relatively fine materials have mean sizes ~5 μm. Chemical sediments overlie the clastic sediments; 5 m of oolites are overlain by more than 2 m of anthropogenic salts deposited early in the twentieth century.

• Changes in sediment chemistry and mineralogy of the OL-92 core reflect past climate cycles. The most sensitive chemical criteria are the relative contents of $CaCO_3$ and organic C, reflections of the residence time of the lake water. Clay minerals are dominantly illite and smectite, and their ratios, relative to the clay-sized fraction, are inversely related: illite was deposited during wet periods and smectite during dry periods. Generally, the climatic cycles appear dominated by the ~100-k.y. periodicities that characterize marine records of high-latitude glaciation. Four of the last five marine isotope terminations are clearly shown in the Owens Lake record.

• Pore-water composition and isotopic make-up show that throughout most of the core, diffusion has smoothed initial variations in their character. Near the bottom of the core, however, abrupt changes in these properties document the post-depositional introduction of water into the sediments, making the pore waters much younger than the host sediment, and possibly indicating a migration of the original fluids into a lower-elevation valley 40 km to the east.

• An apparently constant sedimentation rate allows construction of a well controlled time-depth curve. Radiocarbon ages provide age control for the upper 23 m of core, and show that during that interval, only the lacustrine sediments deposited between ca. 8.3 ka and ca. 5.1 ka were removed by erosion. The Bishop ash and Matuyama-Brunhes paleomagnetic reversal provide age control near the bottom of the core. Calculations of mass accumulation rate (MAR), based on laboratory measurements of the sediments within and below the radiocarbon-dated section, suggest that the MAR was virtually constant at 51 g/cm^2/k.y. down to the 760 ka Bishop ash near the base of the core. This allowed correction for compaction and construction of an age-depth curve.

• Paleomagnetic studies identified the boundary between the Matuyama and Brunhes paleomagnetic epochs (780 ka) near the base of the core. Ten magnetic-field excursions of approximately known ages were located in the sediments that support the constant MAR time-depth curve. Paleomagnetic-susceptibility intensities also reflect climate cycles.

• Tephra recovered from the core include 5+ m of airfall and reworked Bishop ash (760 ka), the Dibekulewe ash bed (470–610 ka), and other tephra that have poorly constrained ages.

• Oolites, the lake's natural depositional record during the past 5 k.y., and the lack of comparable deposits at greater depths imply that the late Holocene period has been drier in the Owens Lake drainage area—and presumably a large region around it—than at any time during the preceding 800 k.y. The absence of gaylussite or gypsum in earlier sediments indicates that lake salinity never exceeded ~15 wt %, a limit that requires overflow every 10 k.y. or less.

• Correlation of the OL-92 record with the downstream Searles Lake record shows that similar climatic trends are recorded in both basins. However, sedimentation in the Searles Lake area, which is significantly more arid than the Owens Lake area, responded to an approaching period of increasing aridity ~40 k.y. before the Owens sedimentary record showed a response to this trend. Conversely, the Owens Lake sedimentation record reflected an approaching period of increased moisture ~60 k.y. before Searles began to receive significantly greater volumes of water from Owens. We tentatively interpret these "delays" as evidence that middle-latitude precipitation changes during late Pleistocene time proceeded gradually.

• Fossil diatoms and ostracodes, found throughout most of the core, add significant details of the lake's paleolimnology, such as fluctuations of its temperature, salinity, ionic make-up, and concentration of dissolved components, as well as variations in the seasonal distribution of precipitation and in lake depth.

• Fossil mollusca and fish further reveal the lake's history. Several mollusk taxa represent significantly fresh water regimes; other taxa are less restrictive but rule out the presence of very brackish water. Fossil fish remains include those of sucker and tui chub (now present at low elevations in the Owens Lake drainage system) plus whitefish and trout (now found only in very fresh Sierran lakes at higher and cooler elevations). Analyses for ^{13}C and ^{18}O in mollusk shells help confirm the lake's fresh-water character as well as rapid fluctuations in it.

• Fossil pollen, indicative of variations in the flora of the region, suggest that the southern Sierra Nevada region underwent nine late Pleistocene cool-to-warm vegetational shifts plus the Pleistocene-Holocene transition. These pollen zones are correlated with marine isotope records.

• Paleoclimate reconstructions show that Owens Lake also underwent long term cycles that are reflected by overflow volumes. Five of the last six ~100-k.y. cycles in the isotopic-paleotemperature record from Devils Hole (Nevada) and in the deep-sea reconstructions of polar-ice volumes are identified in the OL-92 record. The ages of individual climatic fluctuations in these two records, however, apparently differ from the OL-92 record by as much as ~30 k.y., with about half the OL-92 wet and dry periods preceding and half following the correlated inflections in the isotopic-paleotemperature and the polar-ice-volume records (a few of which differ from each other by as much as ~20 k.y.).

• We suggest that Pleistocene changes in precipitation in the headwaters of Owens Lake, in air-temperature levels at Devils Hole, and in polar ice volumes—each reflecting different elements of climate—responded at different rates to successive atmospheric-circulation reorganizations because each element of climate had its own inherent rate of change and mechanics of reaction and interaction.

ACKNOWLEDGMENTS

This drilling program was made possible by funding from the U.S. Geological Survey's "Global Change and Climate History Program," coordinated by Milan J. Pavich and Richard Z. Poore. The drill site was on land leased from the State of California by the Lake Minerals Corp., a subsidiary of Cominco American. Our efforts were assisted and greatly expedited by William C. McClung and Paul A. Lamos of Lake Minerals' Lone Pine office and by James and Bruce Pischel at the company's Owens Lake facility. Arthur F. Nitsche of the [California] State Lands Commission, Mineral Resources Management Division, helped us obtain the state's permission to drill at this site.

Geologists at the drill site included Bischoff, Fitts, Fitzpatrick, Glen, Menking, and Smith, all authors of one or more chapters in this volume, plus Mary L. McGann and Robert J. Rosenbauer. The drilling crew, supervised by Arthur C. Clark, was composed of Todd E. Hunter, Stephen J. Grant, Daniel J. Sweeney, and John C. Palmer; we greatly appreciated their professional competence and cooperative spirit. David P. Adam, Mary L. McGann, and Michael E. Torresan provided much useful advice during the planning and core-archiving stages. Joseph P. Smoot and E. Wesley Hildreth helped us interpret some sedimentary structures in the core. Thomas E. Chase generously made a photographic record of the entire core. All of the above were members of the U.S. Geological Survey while working on the project.

We also wish to acknowledge the authors of papers in this volume of which we are the editors. Much effort and time was required to complete their studies and prepare their contributions to this volume. Reviewers of these papers include D. P. Adam, L. Benson, F. H. Brown, D. E. Champion, O. K. Davis, W. E. Dean, R. M. Forester, J. A. Gardner, J. M. Glen, R. L. Hay, D. B. Herbst, R. Hershler, B. Hausback, B. F. Jones, T. Ku, M. L.

McGann, D. Z. Piper, J. G. Rosenbaum, A. M. Sarna-Wojcicki, J. P. Smoot, R. S. Thompson, S. W. Tyler, and P. E. Wigand. To all of the above, our thanks.

REFERENCES CITED

Arnow, T., 1984, Water-level and water-quality changes in Great Salt Lake, Utah, 1847-1983: U.S. Geological Survey Circular 913, 22 p.

Benson, L. V., Currey, D. R., Dorn, R. I., Lajoie, K. R., Oviatt, C. G., Robinson, S. W., Smith, G. I., and Stine, S., 1990, Chronology of expansion and contraction of four Great Basin lake systems during the past 35,000 years: Palaeogeography, Palaeoclimatology, and Palaeoecology, v. 78, p. 241–286.

Blackwelder, E., 1931, Pleistocene glaciation in the Sierra Nevada and Basin Ranges: Geological Society of America Bulletin, v. 42, p. 865–922.

Burke, R. M., and Birkeland P. W., 1979, Reevaluation of multiparameter relative dating techniques and their application to the glacial sequence along the eastern escarpment of the Sierra Nevada, California: Quaternary Research, v. 11, p. 21–51.

Burke, R. M., and Birkeland P. W., 1988, Soil catena chronosequences on eastern Sierra Nevada moraines, California, U.S.A.: Arctic and Alpine Research, v. 20, p. 473–484.

Bursik, M. I., and Gillespie, A. R., 1993, Late Pleistocene glaciation of Mono Basin, California: Quaternary Research, v. 39, p. 24–35.

City of Los Angeles, 1916, Complete report on construction of the Los Angeles Aqueduct: Board of Public Service Commissioners of the City of Los Angeles, 330 p.

Danskin, W., in press, Evaluation of the hydrologic system and selected water-management alternatives in Owens Valley, California: U.S. Geological Survey Water-Supply Paper 2370-H.

Dorn, R. I, Turrin, B. D., Jull, A. J. T., Linick, T. W., and Donahue, D. J., 1987, Radiocarbon and cation-ratio ages for rock varnish on Tioga and Tahoe morainal boulders of Pine Creek, eastern Sierra Nevada, California, and their paleoclimatic implications: Quaternary Research, v. 28, p. 38–49.

Friedman, I., and Smith, G. I., 1970, Deuterium content of snow cores from Sierra Nevada area: Science, v. 169, p. 467–470

Friedman, I., and Smith, G. I., 1972, Deuterium content of snow as an index of winter climate in the Sierra Nevada area: Science, v. 176, p. 790–793.

Gale, H. S., 1914, Salines in the Owens, Searles, and Panamint basins, southeastern California: U.S. Geological Survey Bulletin 580-L, p. 251–323.

Gardner, J. V., Sarna-Wojcicki, A. M., Adam, D. P., Dean W. E., Bradbury, J. P., and Rieck, H. J., 1991, Report of a workshop on the correlation of marine and terrestrial records of climate changes in the western U.S.: U.S. Geological Survey Open-File Report 91-140, 48 p.

Gilbert, G. K., 1875, The glacial epoch: Exploration and surveys west of the 100th meridian, (Wheeler) Report, v. 3, p. 86–104.

Hollett, K. J., Danskin, W. R., McCaffrey, W. F., and Walti, C. L., 1991, Geology and water resources of Owens Valley, California: U.S. Geological Survey Water-Supply Paper 2370-B, 77 p.

Hooke, R. LeB., 1972, Geomorphic evidence for late Wisconsin and Holocene tectonic deformation, Death Valley, California: Geological Society of America Bulletin, v. 83, p. 2073–98.

Huber, N. K., 1981, Amount and timing of late Cenozoic uplift and tilt of the central Sierra Nevada, California—Evidence from the upper San Joaquin River basin: U.S. Geological Survey Professional Paper 1197, p. 1–28.

Knox, J. C., 1983, Response of river systems to Holocene climates, *in* Wright, H. E., ed., Late Quaternary Environments of the United States, Volume 2: Minneapolis, University of Minnesota Press, p. 26–41.

Koehler, P. A., and Anderson, R. S., 1994, Full-glacial shoreline vegetation during the maximum highstand at Owens Lake, California: Great Basin Naturalist, v. 54, p. 142–149.

Lee, C. H., 1912, An intensive study of the water resources of a part of Owens Valley, California: U.S. Geological Survey Water-Supply Paper 294, 135 p.

Lee, W. T., 1906, Geology and water resources of Owens Valley, California: U.S. Geological Survey Water-Supply Paper 181, 28 p.

Martin, P. S., and Mehringer, P. J., Jr., 1965, Pleistocene pollen analysis and biogeography of the southwest, *in* Wright, H. E., Jr., and Frey, D. G., The Quaternary of the United States: Princeton, New Jersey, International Association for Quaternary Research, Princeton University Press, p. 433–451.

Newton, M. S., 1991, Holocene stratigraphy and magnetostratigraphy of Mono and Owens Lakes, eastern California [Ph.D. thesis]: Los Angeles, University of Southern California, 330 p.

NOAA (National Oceanic and Atmospheric Administration), 1982, Monthly normals of temperature, precipitation, and heating and cooling days, 1951-80, California: Nashville, North Carolina, National Oceanic and Atmospheric Administration, National Climatic Center, Climatology of the United States No. 81, 38 p.

Pakiser, L. C., Kane, M. F., and Jackson, W. H., 1964, Structural geology and volcanism of Owens Valley region, California—A geophysical study: U.S. Geological Survey Professional Paper 438, 65 p.

Phillips, F. M., Zreda, M. G., Smith, S. S., Elmore, D., Kubic, P. W., and Sharma, P., 1990, Cosmogenic chlorine-36 chronology for glacial deposits at Bloody Canyon, eastern Sierra Nevada: Science, v. 248, p. 1529–1532.

Sharp, R. P., 1968, Sherwin-Bishop Tuff geological relationships, Sierra Nevada, California: Geological Society of America Bulletin, v. 79, p. 351–364.

Sharp, R. P., and Birman, J. H., 1963, Additions to classical sequence of Pleistocene glaciations, Sierra Nevada, California: Geological Society of America Bulletin, v. 74, p. 1079–1086.

Smith, G. I., 1979, Subsurface stratigraphy and geochemistry of late Quaternary evaporites, Searles Lake, California: U.S. Geological Survey Professional Paper 1043, p. 1–130.

Smith, G. I., 1984, Paleohydrologic regimes in the southwestern Great Basin, 0–3.2 my ago, compared with other long records of "global" climate: Quaternary Research, v. 22, p. 1–17.

Smith, G. I., 1991, Continental paleoclimatic records and their significance, in Chapter 2, Quaternary paleoclimates (Smiley, T. L., et al.), *in* Morrison, R. B., ed., Quaternary nonglacial geology: Conterminous U.S., Boulder, Colorado, The Geological Society of America, The Geology of North America, v. K-2, p. 35–41.

Smith, G. I., 1993, Field log of Core OL-92, *in* Smith, G. I., and Bischoff, J. L., eds., Core OL-92 from Owens Lake, southeast California: U.S. Geological Survey Open-File Report 93-683, p. 4–57.

Smith, G. I., and Bischoff, J. L., eds., 1993, Core OL-92 from Owens Lake, southeast California: U.S. Geological Survey Open-File Report 93-683, 398 p.

Smith, G. I., and Pratt, W. P., 1957, Core logs from Owens, China, Searles, and Panamint Basins, California: U.S. Geological Survey Bulletin 1045-A, p. 1–62.

Smith, G. I., and Street-Perrott, F. A., 1983, Pluvial lakes of the western United States, *in* Porter, S. C., ed., Late Quaternary of the United States: Minneapolis, University of Minnesota Press, p. 190–212.

Smith, G. I., Friedman, I., Klieforth, H., and Hardcastle, K., 1979, Areal distribution of deuterium in eastern California precipitation, 1968–1969: Journal of Applied Meteorology, v. 18, no. 2, p. 172–188.

Smith, G. I., Barczak, V. J., Moulton, G. F., and Liddicoat, J. C., 1983, Core KM-3, a surface-to-bedrock record of late Cenozoic sedimentation in Searles Valley, California: U.S. Geological Survey Professional Paper 1256, 24 p.

Smith, R. S. U., 1975, Late Quaternary pluvial and tectonic history of Panamint Valley, Inyo and San Bernardino Counties, California [Ph.D. thesis]: Pasadena, California Institute of Technology, 330 p.

Spaulding, W. G., Leopold, E. B., and Van Devender, T. R., 1983, Late Wisconsin paleoecology of American Southwest, *in* Porter, S. C., ed., The late Pleistocene: Volume 1, Late-Quaternary Environments of the United States (Wright, H. E., series ed.): Minneapolis, University of Minnesota Press, Minneapolis, p. 259–293.

Van Devender, T. R., and Spaulding, W. G., 1979, Development of vegetation and climate in the southwestern United States: Science, v. 204, p. 701–710.

MANUSCRIPT ACCEPTED BY THE SOCIETY JUNE 17, 1996

Geological Society of America
Special Paper 317
1997

Stratigraphy, lithologies, and sedimentary structures of Owens Lake core OL-92

George I. Smith
U.S. Geological Survey, 345 Middlefield Road, MS 902, Menlo Park, California 94025

ABSTRACT

Owens Lake, a now-dry lake in southeastern California immediately east of the southern Sierra Nevada, was the site of a coring project designed to obtain a long paleoclimatic record. During the ensuing study, lacustrine deposits were recovered by the 323 m long core designated "OL-92." The presence of the Bishop ash (ca. 760 ka) and the Matuyama-Brunhes paleomagnetic reversal (ca. 780 ka) near the base of core OL-92 shows that this core represents about 800 k.y. of deposition in Owens Lake.

The sediments are dominantly lacustrine clay, silt, and fine sand, although some intervals contain as much as 40 wt % $CaCO_3$. The lowest ~57 m of recovered sediments is mostly silt or clay, but several sand beds are present; the overlying ~60 m of sediment is similar, but its sand content is more dispersed. Together, these two units are composed of ~70 wt % silt and clay and ~30 wt % sand, suggesting deposition in lakes that fluctuated between moderately deep and shallow. Overlying them is ~201 m of sediments that were mostly deposited in deep water; they consist predominantly of silt and clay but include two thin, coarse-sand beds. An oolite bed forms the upper ~4 m of natural deposits, and an anthropogenic salt bed, >2 m thick, forms much of the present surface. In addition to the Bishop ash, several much thinner tephra layers are also present.

About 70% of the clastic-sediment units are massive, some clearly because of bioturbation; other units display a thin bedding defined by changes in color or grain size. Rhythmic bedding, observed in numerous segments <1 m thick, seems to represent cyclical events ~100 yr long. Thin color bands caused by the chemical alteration of sediments on each side of hairline fractures create irregular subvertical veins. Clastic dikes, as much as ~2 cm wide and ~75 cm long, characterize some zones. Bioturbation structures, sand pods, ice-rafted(?) granules, small faults, minor discontinuities, and possible turbidity-current structures are also present.

Lithologic variations, in combination with other evidence, indicate that from ca. 810–645 ka, Owens was most commonly a moderately deep fresh-water lake; from ca. 645–450 ka, it was more commonly a shallow—but still fresh-water—lake; from ca. 450–5 ka, it was almost continuously a deep, mostly fresh-water lake; and after ca. 5 ka, it was a shallow, moderately saline lake. Other variations in the sediments and their contents, however, indicate additional cycles of average lake-overflow volumes that are not reflected by sediment-size changes.

SETTING OF OWENS LAKE AND CORE OL-92

Owens Lake is at the south end of the Owens Valley, on the west edge of the Great Basin and just east of the Sierra Nevada in California. Details of its location and its geologic, physiographic, and climatic settings are given in the introductory chapter by Smith and Bischoff (this volume), who also describe how, during much of the Pleistocene, Owens was the first in a chain of lakes that periodically extended southward and eastward into Indian Wells, Searles, Panamint, and Death Valleys, the floors of which

Smith, G. I., 1997, Stratigraphy, lithologies, and sedimentary structures of Owens Lake core OL-92, *in* Smith, G. I., and Bischoff, J. L., eds., An 800,000-Year Paleoclimatic Record from Core OL-92, Owens Lake, Southeast California: Boulder, Colorado, Geological Society of America Special Paper 317.

are now occupied by playa lakes or salt flats. The number and cumulative water-surface areas of those lakes primarily reflected the amounts of precipitation falling in their collective drainage areas, the most important of which were the high eastern slopes of the southern Sierra Nevada that drain into the Owens River.

Core OL-92 was recovered from a site in the south-central part of Owens Lake (Smith and Bischoff, this volume). Core recovery was ~80%. The sediments in OL-92 are assigned ages according to the time-depth relations developed by Bischoff et al. (this volume chapter 8), which appear to be verified by paleomagnetic studies (Glen and Coe, this volume) and tephrochronologic studies (Sarna-Wojcicki et al., this volume). The average sedimentation rate for the entire core is ~40 cm/k.y.

LITHOLOGIES

Graphic and descriptive logs of core OL-92 (Fig. 1; Table 1) are generalized from the more detailed field descriptions of individual beds (Smith, 1993) that were made mostly by visual inspection, using a hand lens and a binocular or petrographic microscope where necessary. Additional laboratory

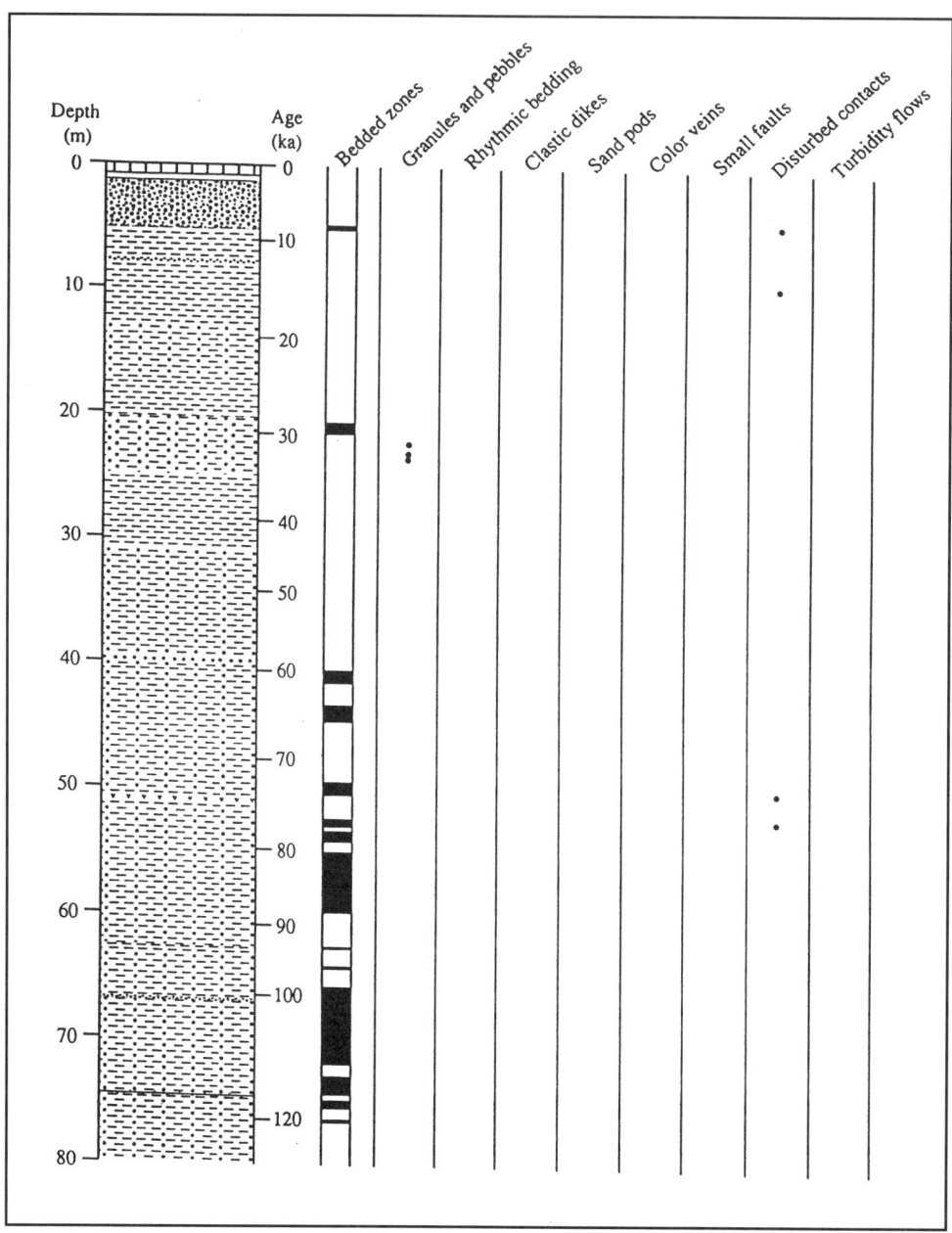

Figure 1 (on this and following four pages). Graphic log of core OL-92, showing depths at which bedded zones (black segments) and various sedimentary structures and uncommon lithologies (black dots in labeled columns) were observed. Graphics by M. D. Medrano.

study included thin-section examination, X-ray diffraction analyses, petrographic study of smear slides, and microscopic study of selected sediment samples.

Bedding and other structures defined by color were conspicuous when first observed after core recovery. Within hours, however, the medium-dark to dark Fe-bearing minerals (greigite, hydrotroilite, or mackinawite?), which were stable phases in the sediments under subsurface reducing conditions and for a short while after recovery of the core, began to oxidize. Apparently, oxidation altered the Fe minerals that helped define bedding and other structures to a new, less diverse suite of minerals, which in the process lost much of their original color contrast. As a result, details of bedding and other structures, easily visible when the cores were fresh, are now more difficult to recognize in the archived core.

All sediments recovered by core OL-92 appear to be lacustrine. Possibly the most convincing evidence of this interpretation is the virtual ubiquity of diatoms (Bradbury, this volume), ostracodes (Carter, this volume), and other aquatic fossils (Firby et al., this volume), which most workers consider indicative of a lacustrine environment, although spring marshes or slow-moving streams might be hosts to similar assemblages. The bedding and sorting characteristics of the clastic materials also resemble those of sediments commonly found in cores from modern lakes and in outcrops of moderately deep-water to deep-water lacustrine deposits (Muessig et al., 1957; Smith

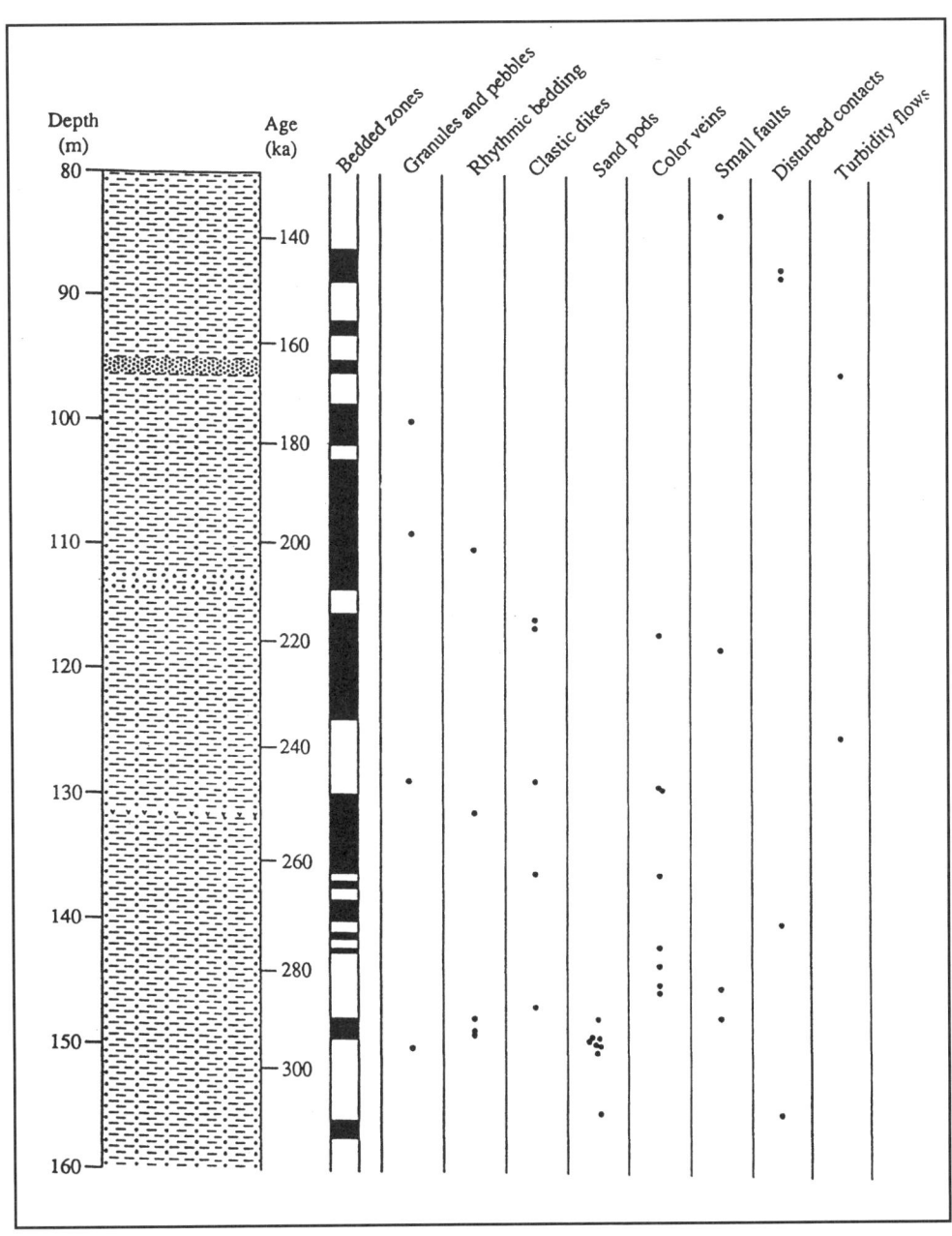

and Pratt, 1957; Bassett et al., 1959; Haines, 1959; Jones and Bowser, 1978; Sly, 1978; Smith, 1979, 1987).

Although shallow-water lacustrine deposits generally are distinguishable from riverine deposits when observing large outcrops, where discontinuous- and cross-bedding, scour structures, and zones of poor sorting are identifiable, that distinction is difficult where the "outcrop" is a split core only 7.6 cm wide. However, comparison of the log of core OL-92 core log (Smith, 1993) with that of the core obtained from an area 2.2 km to the northeast in 1953 (Smith and Pratt, 1957) provides some of the perspective offered by a large outcrop: It shows that at least eight horizons in both cores lie at similar depths (Table 2), although the correlated horizons in the more centrally located 1953 core are consistently 7–14 m deeper than those in the 1992 core. Consistently deeper correlated horizons throughout much of the time coarse sediment was being deposited suggests that if the depositing shallow-water body were a river, it was flowing the "wrong" way, from southwest to northeast, and had to follow a course that consistently flowed over both recently cored sites for nearly 370 k.y. A lake having minimum depths between ~8 m and ~15 m seems to be the more likely environment of deposition.

Among the distinctive lithologic properties of these lacustrine deposits are the following: (1) Although the cored sediments are poorly sorted according to laboratory measurements of the silt- and clay-rich sections, hand-lens inspection shows that it is much better sorted than alluvial or fluvial deposits. (2) Many clay and silt beds are thin (<2 cm thick), and a few are finely lam-

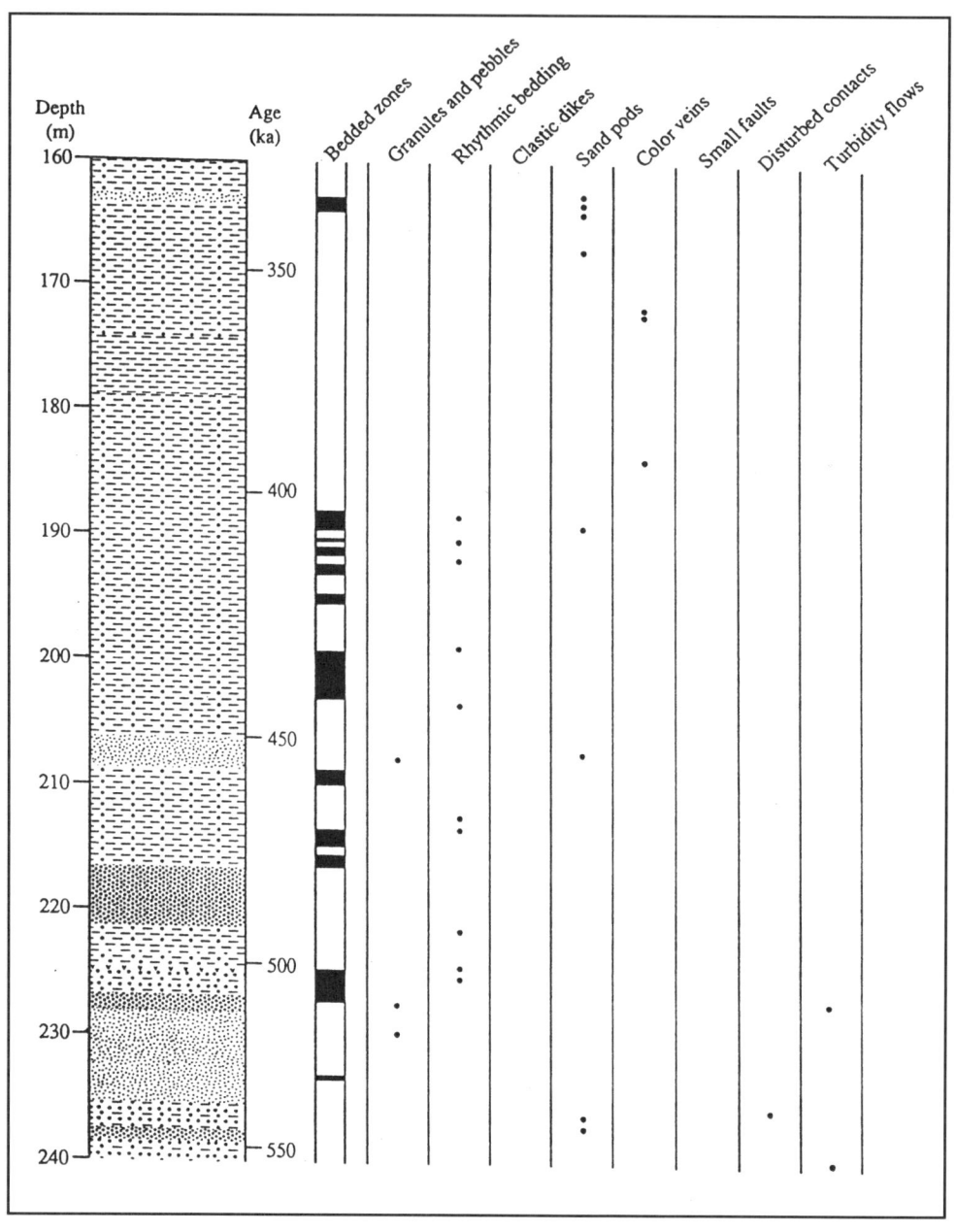

inated (<2 mm). (3) Beds are mostly horizontal. (4) Beds defined only by color (with no grain-size change) are common. (5) Many massive beds show tubelike structures commonly attributed to subbottom bioturbation. (6) The sediments exposed on freshly split core faces commonly formed bubbles filled with a gas that was determined to be either H_2S or NH_3 (identified by odor) or CH_4 (identified by flammability). Also, as noted above, the surfaces of nearly all fresh core segments composed mostly of silt or clay contained medium- to dark-gray minerals that oxidized to lighter, greener colors within hours, a common indication of sediments that have been removed from a reducing environment.

In addition to these criteria indicating that the sediments are lacustrine, some core samples contain nearly 50 wt % nonclastic $CaCO_3$, whereas other samples contain <1 wt % (Bischoff et al., this volume, chapter 4). Such variation is characteristic of sediments deposited in lakes that undergo fluctuations in the evaporative concentration of CO_3-bearing waters.

Clastic sediments

Virtually all the epiclastic components of the 317.7 m of sediments underlying the salt and oolite beds (0–5-m depth) in core OL-92 (Fig. 1) are clay, silt, and sand. Colors of the sediments vary within moderately narrow limits. Using the "Rock-

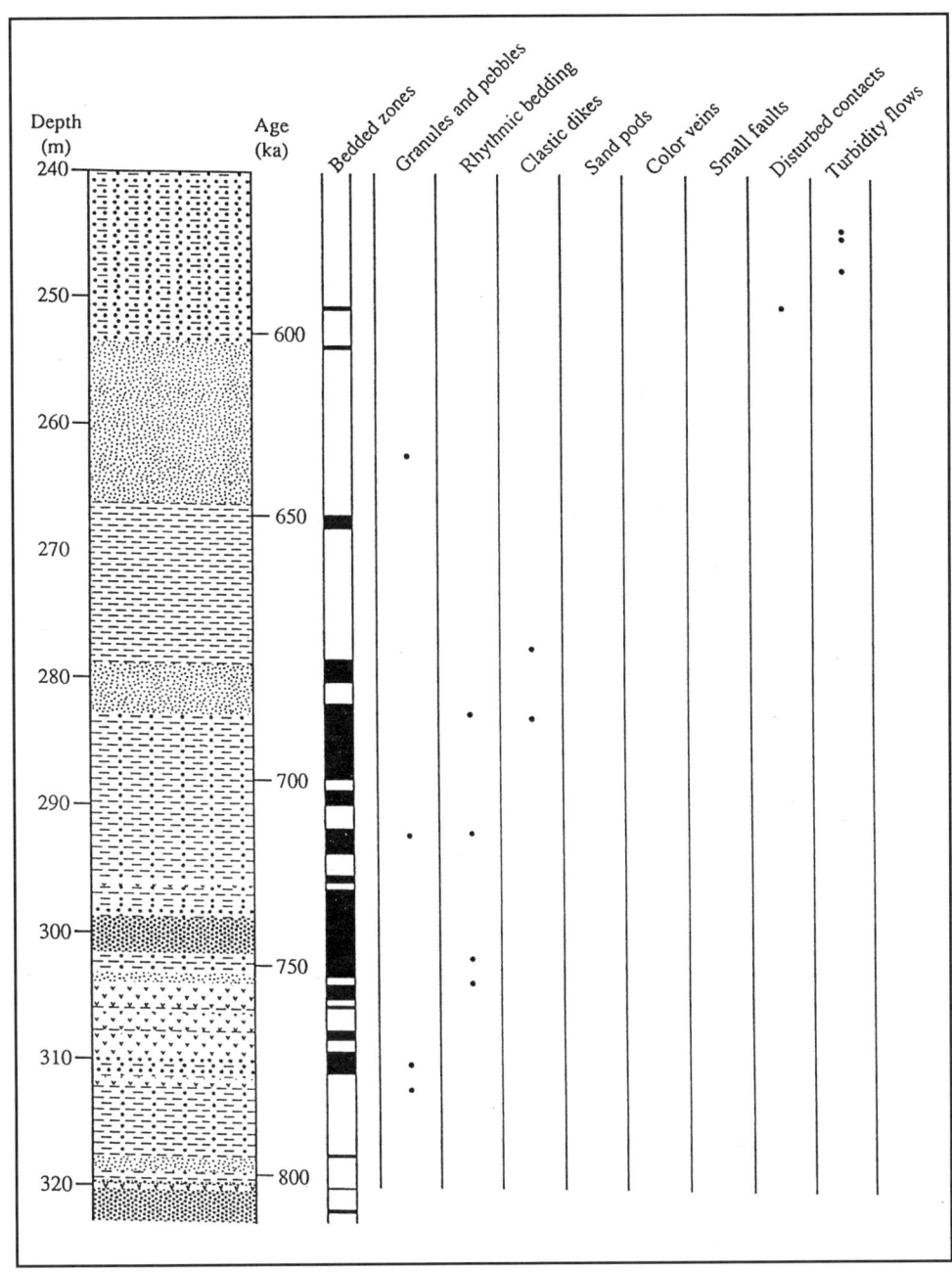

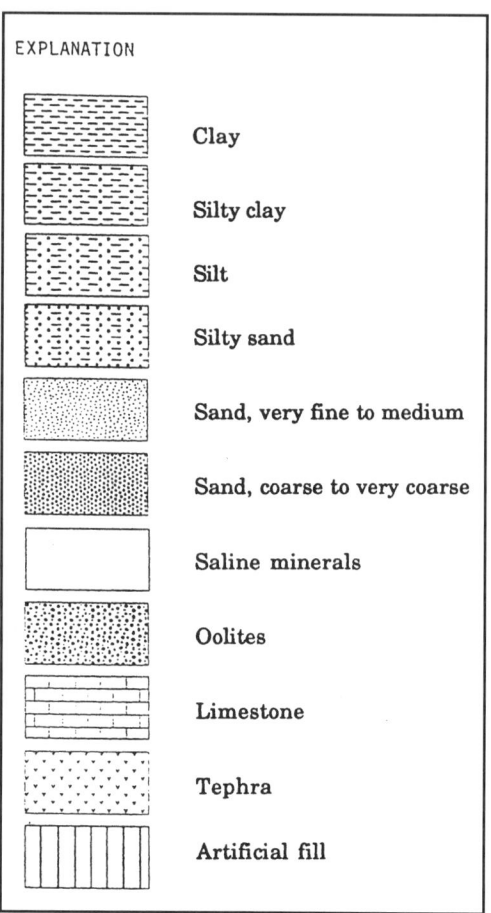

Color Chart" classification system (Geological Society of America, 1991), hues range mostly from yellowish green (5Y to 10Y) through green (5GY and 5G), with rare thin beds of gray (N3 to N5). When the cores were fresh, moist, and unoxidized, most lightness values ranged from 3 to 5 (on an increasing scale of 1 to 9), and most chroma (color saturation) values from 1 to 4 (on an increasing scale of 1 to 6). Silt and clay beds throughout the core have about the same colors, as do sand beds that contain significant amounts of silt or clay. Sand beds that are free of finer material, however, commonly have hues that range from yellowish brown (10YR) to middle-gray (N4 to N7). Although the cores were wrapped in airtight flexible plastic immediately after logging and sampling, and then refrigerated at 5–7 °C, oxidation of the sediments over the next few days and weeks caused their hues to become more yellow (for example, from 5GY or 10Y to 5Y) and their color-saturation values to increase by one to two steps (for example, from 1 to 2 or from 2 to 4). Lightness values remained about the same.

Sediment-size analyses (Menking, this volume) show that almost all of the clastic sediments are, on average, poorly sorted in the sense defined by Folk and Ward (1957), reflecting minimal influence of beach and shallow-water winnowing and sorting. In fact, much of the finest sediment may have been carried directly to its depositional site as suspended matter, sorted only by the earlier loss of coarser grains by settling. The upper 201 m of clastic sediments is notably enriched in the finest grains, as indicated by the positive skewness documented by grain-size analyses. Mean grain sizes in this part of the core range mostly from 2–20 µm. Although the underlying 117 m of more sandy silt and clay has a nearly normal grain-size distribution, mean grain sizes range mostly from 10 µm to as much as 50 µm or even 120 µm (Menking, this volume).

Granules and small pebbles, most commonly in units composed of silt or clay, were noted at 14 horizons (Fig. 1; Tables 1 and 3). In some places, such clasts were noted in closely spaced horizons, suggesting that they are dropstones recording a substantial period when ice commonly formed on the lake surface, rafting coarse, near-shore sediments to offshore parts of the lake. Today, mean air temperatures in the study area during the coldest month (January) are ≥3 °C, implying that a freshwater lake today would not freeze solid during most years.

Cores from the upper half of the clastic section were generally soft to slightly firm, and those from the lower half were notably firm. A few short segments and several spherical concretions in the lower half of the clastic section had rocklike properties and fractured conchoidally.

The clastic section is here divided into three sequences: a lower sequence that extends from a depth of 323 m to 266 m, a middle sequence that extends from 266 m to 206 m, and an upper sequence that extends from 206 m to the base of the oolite bed at a depth of 5 m (Smith, 1993).

Interval from 322.86 m to 266.22 m. Sediments in the lower sequence, which represent 18 vol % of the clastic section, are composed of about 40 wt % granules, sand, and impure to pure tephra; the rest is silt and clay. Within this interval, of which 89% was recovered, units logged as dominantly sand or granules make up 29 vol %, as silt and clay 59 vol %, and as tephra 12 vol % (Smith, 1993). Grain-size analyses performed later in the laboratory, which did not distinguish between epiclastic and pyroclastic materials, indicate that from 25 wt % (spot samples) to 30 wt % (channel samples) of the sediments are sand size or coarser (Menking et al., 1993, Tables 2 and 4).

Much of the sand-size sediment in the lower sequence is in five discrete beds that are rich in medium to very coarse sand (Table 1). The beds, >0.5 m to <4 m thick, are centered at about 320 m, 318 m, 303 m, 300 m, and 280 m depth. Although most of the clay and silt beds also contain some sand, three clay or silt beds centered at about 314 m, 290 m, and 260 m depth are thick (6–13 m; Fig. 1; Table 1). These beds are likely to have been the major source of the 35 wt % clay-size sediment measured in channel samples from this segment of the core (Menking et al., 1993, Table 4). The standard deviation of those analytical results is high (22.3%), indicating a large variation in clay percentages in the 19 successive channel samples of core. Substantial changes in lake-water depths seem most likely to be responsible for the changes in bed lithologies.

Interval from 266.22 m to 206.17 m depth. The depth of the upper limit of the middle sequence is debatable. It is here placed at 206.17 m, the base of an overlying unit that is com-

TABLE 1. LITHOLOGIC LOG OF CORE OL-92*

From (m)	To (m)	Thickness (m)	Description
0.00	0.94	0.94	**Artifical fill;** top of this unit is substituted for "lake-surface level" at the drill site (see text).
0.94	1.32	0.38	**Saline minerals,** mostly well bedded halite, trona, and burkeite, light olive gray (5YR4-6/1); maximum thickness of salts elsewhere on lake surface about 2.75 m.
1.32	5.16	3.84	**Oolites,** granule to coarse-sand size, some silt, massive to faintly bedded, light olive gray (5Y4-8/1-2); oolites are rounded to subrounded, composed of calcite and aragonite; fairly well bedded; upper half of unit very hard.
5.16	7.72	2.56	**Clay,** some silt, massive to mottled, bioturbated(?); contains disseminated grains of sand.
7.72	7.73	0.01	**Sand,** medium to very coarse, silt matrix, moderate olive brown (5Y4/4).
7.73	12.83	5.10	**Clay,** some silt, mottled (bioturbated?), olive gray (5Y2-6/1).
12.83	17.56	4.73	**Clay and silt,** mottled (bioturbated?), medium gray (N4-N6).
17.56	20.19	2.63	**Clay,** some silt, massive, medium dark gray (N4) to dusky yellow (5Y4-6/2).
20.19	24.77	4.58	**Silt,** mostly massive, faint or discontinuous bedding near top; contains disseminated round pebbles in zone just below middle of unit; dusky yellow (5Y4-6/2).
24.77	30.86	6.09	**Clay,** mostly massive, mottled, dusky yellow (5Y4-6/2).
30.86	39.49	8.63	**Silty clay,** mottled to massive, borings(?) make conspicuous tubemarks as much as 10 mm wide; grayish green (5G-5Y3-4/1-4), with irregular subhorizontal areas of light olive gray (5Y6/1).
39.49	40.18	0.69	**Silt and sand,** very fine, mixed; at top, very fine sand, pure, well bedded, mottled, mostly light olive gray (5Y6/2), some greenish black (5GY2/1).
40.18	50.64	10.46	**Silty clay,** mostly massive; some zones of faint horizontal color beds, dark greenish gray (5Y-5G2-6/1-2); thin beds of very fine sand in lower third.
50.64	50.74	0.10	**Tephra,** impure, very fine sand size, massive in upper half, laminated in lower half, laminae 3 to 10 mm thick, medium gray (N4-7).
50.74	62.47	11.73	**Silty clay,** very faint to conspicuous color bedding, most beds 0.2 to 0.5 cm thick, light olive gray (5Y3-5/2-4).
62.47	62.64	0.17	**Clay,** faintly bedded, yellowish gray (5Y7/2).
62.64	66.74	4.10	**Silty clay,** conspicuous beds in upper part, faint beds in lower; olive gray (5Y2-4/1-2).
66.74	66.99	0.25	**Sand,** fine to coarse, many grains coated with carbonate, faintly bedded in lower half; dusky yellow (5Y4-6/4).
66.99	74.36	7.37	**Silty clay,** bedding is faint to conspicuous, expressed by changes in both color and clast size, beds commonly 5-10 mm thick, some appear rhythmic; moderate olive brown (5Y-5GY2-5/1-4).
74.36	74.47	0.11	**Limestone,** impure, highly indurated, weakly bedded, light olive gray (5Y6/1).
74.47	76.56	2.09	**Silty clay,** faint color beds, 2 to 10 mm thick to massive, mostly dark greenish gray (5GY2-5/1); basal 0.29 m of unit is medium bluish gray (5B5/1).
76.56	85.64	9.08	**Silty clay,** massive, greenish gray (5GY4-5/1); several 1 to 2 cm thick beds of diatom-rich marl; normal fault that dips about 70° offsets sand bed more than 20 cm.
85.64	95.24	9.60	**Silty clay,** many beds contain disseminated fine quartz(?) grains; most of zone is variegated (bioturbated?), some zones weakly color bedded, dark greenish gray (5Y-5G2-5/1); basal 0.30 m contains some sand.
95.24	96.51	1.27	**Sand,** mostly medium to very coarse, grains angular to subrounded, fairly well to well sorted, silt matrix, granitic source, some clay fragments; unit contained methane (and other gases?) under high pressure.
96.51	98.52	2.01	**Silty clay,** mottled (bioturbated?), tube structures as large as 10 by 70 mm, greenish black (5Y-GY2/1).
98.52	112.58	14.06	**Silty clay,** contains disseminated sand grains and a few thin (<3 cm thick) beds of fine to medium sand; massive to faint color bedding, dark greenish gray (5G-5Y2-6/1-2).
112.58	113.79	1.21	**Silt and sand,** very fine, faint bedding defined by color and grain-size change; dark greenish gray (5G-5GY3-5/1); some clay in lower third.
113.79	119.98	6.19	**Silty clay,** scattered very thin beds of silt, faint bedding also defined by color changes; olive gray (5Y-5GY4-6/1).
119.98	122.30	2.32	**Silty clay and sand,** very fine, massive, olive gray (5Y-5GY4-5/1).
122.30	131.17	8.87	**Silty clay,** scattered zones characterized by beds composed of silt and very fine sand, <2 mm thick, commonly about 10-20 mm apart; olive gray (5Y-5GY4-6/1-4

TABLE 1. LITHOLOGIC LOG OF CORE OL-92* (continued - page 2)

From (m)	To (m)	Thickness (m)	Description	From (m)	To (m)	Thickness (m)	Description
131.17	131.18	0.01	**Tephra and sand,** very fine to fine, coarser in upper half, well sorted and bedded, very hard, medium dark gray (N4); tephra grains are small, clear, and commonly triangular, rectangular, or diamond shaped.	224.17	226.37	2.20	**Silt and sand,** very fine, some clay, massive greenish gray (5Y-5GY4-5/1).
131.18	163.30	32.12	**Silty clay,** some thin beds of silt or fine sand, bedding mostly defined by variation in clast sizes; a few color beds, dark greenish gray (5Y-5GY3-5/1-4); many color veins, some clastic dikes, dispersed granules and pebbles near 151 m.	226.37	227.77	1.40	**Sand,** fine to very coarse, massive to faintly bedded; beds defined by changes in clast size, greenish gray (5Y-5GY4-6/1); at base, a 0.26 m thick, grayish green (10G4/2) layer composed of granules and a 20 mm long pebble.
163.11	163.30	0.19	**Sand,** fine to very fine, very hard, light olive gray (5Y4-5/2-4).	227.77	235.11	7.34	**Sand,** mostly medium, well sorted, massive to faintly bedded, dark greenish gray (5G-5GY3-5/1).
163.30	174.16	10.86	**Silty clay,** some beds of silt and sand, sand in beds (1 to 10 mm thick) and pods (1 to 3 mm across), massive except for sand beds; many zones of disseminated sand grains; dark greenish gray (5G-5Y3-5/1).	235.11	237.18	2.07	**Silt and sand,** fine to very fine, massive, but lithologic variations define many subunits <0.6 m thick, dark greenish gray (5G-5GY4-5/1).
				237.18	238.08	0.90	**Sand,** medium to coarse, quartz rich, well sorted, subangular, massive, dark greenish gray (5GY5/1).
174.16	178.74	4.58	**Clay,** with some silt or very fine sand, mostly massive; some color beds, greenish gray (5GY5/1).	238.08	253.47	15.39	**Sand and silt,** a few silt beds interbedded with fine to very coarse sand layers; most beds are individually massive but <0.3 m thick; sand moderately well to very well sorted, dark greenish gray (5Y-5GY4-6/1).
178.74	184.82	6.08	**Silty clay,** contains disseminated grains of fine to very fine sand, massive; greenish gray (5GY5/1).				
184.82	190.63	5.81	**Clay, silt, and sand,** thin beds of sand, 1-5 cm thick, mostly very fine, with some medium to coarse sand; also contains disseminated grains of sand or granules, massive; greenish gray (5G-5Y4-6/6/1).	253.47	266.22	12.75	**Sand,** mostly very fine to medium, some coarse, well sorted, massive, greenish gray (5GY3-5/1).
185.94	206.17	15.54	**Clay and silt,** similar to overlying unit but fewer sand grains and coarse fragments, massive, greenish gray (5GY4-6/1-2).	266.22	278.86	12.64	**Clay,** some silt and very fine sand, mottled to faintly bedded; many units contain ~2 vol % dispersed grains of quartz(?) and biotite, greenish gray (5Y-GY4-6/1); some grayish black (N2) clay beds break with conchoidal fracture.
206.17	208.61	2.44	**Sand,** fine to very coarse, fairly well sorted, subangular to subrounded, granitic source, massive, dark greenish gray (5Y-5GY4-5/1).				
208.61	216.54	7.93	**Silt and clay,** numerous thin beds of silt or very fine sand, moderately well bedded, dark greenish gray (5GY4-5/1).	278.86	282.74	3.88	**Sand,** fine to very coarse, top 3 cm very well indurated, massive to thinly bedded, poorly sorted, greenish gray (5Y-5GY4-6/1); from granitic terrain.
216.54	221.25	4.71	**Sand,** coarse to very coarse, some silt, massive to faintly bedded; some sand medium gray (N2-N6), but most dark greenish gray (5GY4-5/1).	282.74	295.95	13.21	**Clay and silt,** intermixed fine and very fine sand; faint bedding caused by color and grain-size change; some faint laminar bedding, mostly light olive (5Y-5GY4-5/1), but 1.9 m thick dark yellowish brown (10YR3-5/1-4) zone present ~4 m below top of unit.
221.25	223.08	1.83	**Clay and silt,** very little fine sand as partings, massive, dark greenish gray (5GY4-5/1).				
223.08	224.14	1.06	**Silt and sand,** very fine, grading down to silt and clay, massive, greenish gray (5Y-5GY5-6/1-2).	295.95	296.06	0.11	**Tephra,** upper half is impure, lower half is pure, massive; sharp basal contact, light olive gray (5Y6/1).
224.14	224.17	0.03	**Tephra,** very fine grained, mottled, medium light gray (N5-N7), darker near base; identified as the Dibekulewe ash (Davis, 1978; Sarna-Wojcicki et al., this volume).	296.06	297.42	1.36	**Clay and silt,** some pods and streaks of fine to medium sand, faintly color bedded to massive, dark yellowish brown (10YR-5Y3-4/2).
				297.42	298.25	1.03	**Silt and sand,** very fine, some clay, mostly well bedded by color change or clast-size variation, moderate brown (5YR2-5/1-5).

TABLE 1. LITHOLOGIC LOG OF CORE OL-92* (continued - page 3)

From (m)	To (m)	Thickness (m)	Description	From (m)	To (m)	Thickness (m)	Description
298.25	301.79	3.54	**Sand,** medium to coarse, zones of very coarse sand, grains subangular to subrounded, fairly well sorted, quartzose, very faintly bedded, medium light gray (N6) to light olive brown (5Y5/5-6).	309.14	311.01	1.87	**Silt and sand,** fine to medium, some clay near base; massive to well bedded by clast-size and color variations, various shades of olive (5Y-5GY3-6/1-6), some pure sand is medium gray (N4-5).
301.79	302.94	1.15	**Clayey silt,** some sand, massive; some faint beds caused by clast-size changes, olive gray (5Y-5GY2-6/1) grading down to dark yellowish brown (10YR3-5/2-4).	311.01	311.69	0.68	**Tephra,** medium-sand-size to very coarse sand-size tephra grading downward to very fine sand size, mixed with epiclastic minerals, some beds are poorly sorted, bedding good; defined by color and grain-size change, medium light gray (N5-7) to light olive gray (5Y6/1).
302.94	303.53	0.59	**Sand and granules,** upper quarter and lower half are fine to medium sand; rest is very coarse sand and granules that are faintly bedded; coarser fractions very poorly sorted; mostly moderate yellowish brown (5Y-10YR5-6/1-2).	311.69	317.71	6.02	**Clay and silt,** some very fine to very coarse sand, massive; olive gray (5Y-5GY4-6/1); several pebbles were noted near top (largest, 17 mm).
303.53	305.39	1.86	**Tephra,** numerous 1 cm thick beds of green impure (silt and clay) ash, faintly bedded to laminated; fragments are fine to medium sand size, light olive gray (5Y4-8/1) to medium light gray (N6).	317.71	319.02	1.31	**Sand,** fine to medium, some very well sorted; grains angular, massive, mostly dark greenish gray (5G-5GY3-4/1).
				319.02	319.84	0.82	**Clay and silt,** massive, mostly dark greenish gray (5G4/1), but some thin zones are more yellow or brown.
305.39	305.71	0.32	**Silt, clay, and sand,** very fine, with numerous beds (1 to 2 mm thick) of **tephra,** well bedded; light olive gray (5Y5-6/1), tephra is light gray (N6-8).	319.84	319.96	0.12	**Granules and sand,** fine to coarse, subangular, very poorly sorted, massive to faintly bedded, olive gray (5Y3/2) to dark gray (N3).
305.71	307.06	1.35	**Tephra,** mostly as granule-size, pebble-size, and very coarse sand-size fragments, rounded; one fragment 30 mm long, most <10 mm grading downward to medium-sand-size to very-fine-sand-size fragments of same material, massive; lapelli are grayish pink (5R8/2) and matrix is very light gray (N8) to light brownish gray (5Y-5YR7/1).	319.96	320.04	0.08	**Tephra,** medium-sand-size fragments, angular, thin bedded, beds distinguished by grain-size change; medium dark gray (N4).
				320.04	322.86	2.82	**Sand,** medium, grading downward to coarse and very coarse, some fine, mostly angular and poorly sorted; scattered granules, micaceous, massive, medium gray (N3-5).
307.06	307.44	0.38	**Clay and silt,** some very fine sand, massive, light olive gray (5Y7/1).				
307.44	309.14	1.70	**Tephra,** upper two-third is like that between 305.71 and 307.06 m depth; lower third is purer, composed of fine- to medium-sand-size fragments.				

Bottom of Core OL-92

*Color names and alphanumeric designations are from the Geological Society of America Rock-Color Chart, 1991. Core descriptions generalized from Smith, 1993.

posed almost entirely of clay and silt; alternatively, the depth could be placed at 224.14 m, the top of a zone composed almost entirely of sand. About half the beds between these two depths are sand, and the other half are silt.

According to laboratory analyses, the sediments in this middle sequence, which represents 19 vol % of the clastic section, are composed of 30 wt % (spot samples) to 34 wt % (channel samples) sand and granules; the rest is silt and clay (Menking et al., 1993, Tables 2 and 4). Within this interval, units logged as dominantly sand or granules make up 35 vol %, and these dominantly silt and clay make up 65 vol % (Smith, 1993); however, only 62% of this part of the lake deposits was recovered as core, indicating that the actual composition of this interval is poorly constrained.

Although the overall sand and granule content of the middle sequence is only ~5 wt % greater than that of the lower sequence, the distribution of these components differs considerably. According to the core log (Table 1), 84 vol % of the sediment within the middle sequence has sand as a major component, whereas in the lower sequence, most of the sand is concentrated in five distinct beds that constitute 29 vol % of that unit. The clay-size fraction makes up 32 vol % of the middle sequence, slightly less than in the lower sequence. However, the standard deviation of the 12 grain-size analyses from this middle sequence

TABLE 2. POSSIBLE CORRELATIONS BETWEEN SIMILAR HORIZONS IN THE 1992 CORE (OL-92) AND THE 1953 CORE, BOTH FROM OWENS LAKE, CALIFORNIA*

Depths 1953 Core (m)	Depths 1992 Core (m)	Difference in Depth† (m)	Correlated Feature and Age§
109.30	95.52	13.78	Isolated sand-rich bed**; ca. 164 ka
127.59	119.96	7.63	Tephra; 226 ka
142.52	131.17	11.35	Tephra; 251 ka
200.86	186.77‡	14.09	Uppermost mollusks; 398 ka
199.43	187.87	11.56	Uppermost sand bed (below 110 m); 403 ka
228.60	216.54	12.06	Top of thick sand-rich zone; 477 ka
230.49	224.14	6.32	Tephra; 500 ka
270.11	262.30	7.81	Base of thick sand-rich zone; 635 ka
Mean value		10.58	
Standard deviation		2.95	

*Data on 1953 core from Smith and Pratt, 1957. The 1953 core site is about 550 m west and 650 m north of SE corner of Sec. 3, T. 18 S., R. 37 E.; lake-floor elevation is ~0.6 m lower than 1992 core site. Data on 1992 core from Smith, 1993.
†Calculation: 1953 core depth minus 1992 core depth.
§Age based on model of Bischoff et al., this volume, chapter 8.
**Thickness of sand-rich bed in 1953 core, 0.12 m; in 1992 core, 0.28 m.
‡Data from Firby et al., this volume.

TABLE 3. LOCATIONS AND DESCRIPTIONS OF GRANULES AND PEBBLES NOTED IN CORE OL-92

Depth (m)	Long Dimension (cm)	Lithology	Source*
22.41	0.5	?	...
23.16	0.4	?	...
23.42	1.0	?	...
100.10	1.0	Sandstone	E
109.18 ± 0.30†	2.0	Volcanic(?)	E?
109.18 ± 0.30†	1.0	Volcanic	E
109.18 ± 0.30†	0.7	Volcanic	E
128.68	0.7	Quartz	W
150.24	1.2	?	...
185.39 ± 0.55	2.4	?	...
208.02	0.3	?	...
227.61	2.0	Basalt cinder	E
229.92	<0.8	Granitic	W
262.07 ± 0.4	0.4	Granitic	W
292.52	0.5	Granitic	W
310.46§	0.8	?	...
312.51 ± 0.02	1.7	?	...

*E = east side of Owens Valley (White Mountains, Inyo and Coso Ranges); W = west side of valley (Sierra Nevada).
†Core lost; described fragments were retained by core catcher.
§Not reported by Smith, 1993.

is 13.0 vol %, little more than half that in the lower sequence, indicating that clay is more evenly distributed throughout the middle sequence. This result is interpreted to mean that water depths during deposition of the middle sequence varied less than during deposition of the lower sequence and that depths which promoted sand deposition were most often the norm.

Interval from 206.17 m to 5.16 m depth. Sediments in the upper sequence, which represents 63 vol % of the clastic section, is mostly composed of silt- and clay-size material. About 87% of this part of the lake deposits was recovered as core. Newton (1991) concluded that some of the material near the top of this sequence is dominated by glacial flour. Menking (this volume) reports that no samples of the clay-size fraction throughout core OL-92 contain more than about 30 wt % non-clay minerals (predominantly K-feldspar, plagioclase, and quartz). Glacial flour, however, especially the quartz fraction, is commonly found to be silt size (Krinsley and Doornkamp, 1973), and study of ~100 smear slides made by using the silt-size fraction of samples from the upper 92 m of core OL-92 (0 to ca. 155 ka) reveals numerous conchoidal fragments of quartz, much like the samples of glacial flour illustrated by Krinsley and Doornkamp (1973), as well as many highly angular fragments of feldspar, biotite, and hornblende.

Two beds of medium to very coarse sand occur within the upper sequence, one (0.25 m thick) at ~67 m depth and the other (1.27 m thick) at ~95 m depth. The deeper of these sand beds contained a substantial volume of gas under high pressure, some or all of which was nearly odorless but flammable (CH_4?); it vented for nearly a day.

Chemical sediments

A layer, 0.38 m thick, composed of saline minerals was recovered beneath the 0.94 m thick drill-rig pad; an unknown additional thickness was removed during construction of the pad. These saline minerals are mostly trona ($Na_2CO_3 \cdot NaHCO_3 \cdot 2H_2O$), halite (NaCl), and burkeite ($Na_2CO_3 \cdot 2Na_2SO_4$); detailed descriptions of the stratigraphy and mineralogy, and of the seasonal variations observed in these deposits, are given elsewhere (Dub, 1947; Smith et al., 1987). These salts were deposited within a few years after 1913, when the Owens River was first diverted into the Los Angeles Aqueduct. In other parts of Owens Lake, this saline layer is as much as 2.75 m thick. Therefore, the drill-pad thickness has been added to the preserved-salt thickness in determining depths to all deeper levels, making the total "saline layer" 1.32 m thick. Elsewhere on the lake, depths measured using this correction would be closer to those measured from the surface of an undisturbed part of the saline layer.

Beneath the saline layer, as well as outside its areal limits, calcite and aragonite oolites form a layer that locally is very hard; thin-section study shows that the hard zones are also cemented by these minerals. Oolite sizes range from fine to coarse sand. Thin-section and X-ray diffraction studies show that in addition to the (primary?) aragonite and (secondary?) calcite that constitute the oolites and their cement, these deposits contain significant amounts of the saline minerals gaylussite ($Na_2CO_3 \cdot CaCO_3 \cdot 5H_2O$) and halite (NaCl), both of which coat

TABLE 4. SEQUENCES OF CYCLIC BED PAIRS DEFINED BY VARIATION IN GRAIN SIZE*

Interval		Thickness of Coarser Beds		Thickness of Finer Beds		Lithology† (fine/coarse)	Number of Bed Pairs
From (m)	To (m)	Mean (cm)	Median (cm)	Mean (cm)	Median (cm)		
110.38	110.77	~0.7	....	6.5	4	Slt-cly/vfs	6
131.48	132.26	0.5	0.5	11.1	8	Slt-cly/vfs+fs	7
147.93	149.01	0.5	0.1	4.8	3	Slt-cly/vfs+fs	21
148.88	149.04	~0.2	....	2.1	1	Slt-cly/vfs+fs	7
149.23	149.47	~0.1	....	2.3	5	Slt-cly/vfs+fs	10
188.70	189.07	~0.3	....	6.2	2	Cly+slt/vfs	6
190.65	191.27	0.9	0.6	3.8	3	Slt-cly/vfs	13
192.18	193.05	0.2	0.1	9.4	9	Slt-cly/fs-vcs	9
199.16	201.65	~0.2	....	7.5	4	Slt-cly/fs+ms	32
203.71	204.20	~0.1	....	3.2	2	Slt-cly/vfs-fs	15
212.67	213.14	~0.1	....	9.4	6	Slt+cly/vfs	5
213.59	214.19	~0.2	....	12.0	4	Slt+cly/vfs	5
221.75	222.52	~0.1	....	5.0	5	Cly+slt/slt+vfs	15
224.73	225.30	~0.1	....	5.2	3	Slt/vfs	11
225.57	225.96	~0.1	....	7.8	6	Slt/vfs	5
282.90	284.16	~0.1	....	4.6	4	Slt+vfs/vfs+fs	27
292.29	293.42	~0.2	....	6.8	4	Cly-slt/vfs+ms	16
302.11	302.89	~0.1	....	7.7	7	Cly-slt/vfs-ms	10
304.02	305.00	~1.0	....	9.8	6	Slt/vfs+fs	9
Mean		0.30	....	6.59	4.53		
Standard deviation		0.28	....	2.88	2.09		
Median		0.1		6.8	4		

*Listed only when five or more pairs of beds were recorded. Products of thicknesses of bed pairs times number of pairs do not always match size of interval because of rounding.
†Cly = clay; cly-slt = clayey silt; slt-cly = silty clay; slt = silt; vfs = very fine sand; fs = fine sand; ms = medium sand; cs = coarse sand; vcs = very coarse sand. Terminology follows that of Wentworth, 1922.

the spherical oolite grains and fill cavities within them, acting as an additional cement. The gaylussite is apparently a byproduct of an intermediate stage in the desiccation of Owens Lake after 1913, when the brine became sufficiently dense to displace the underlying interstitial fluid and replace it with a fluid with a composition (Na-carbonate rich) and salinity (15+ wt %) that could crystallize gaylussite (Bury and Redd, 1933). The halite probably also crystallized from that brine. Primary gaylussite formed again in Owens Lake in early 1970 after the salt body was flooded and partly dissolved, and while the dissolved salts were being recrystallized (Smith et al., 1987).

The oolites are interpreted here as products of shallow, moderately saline, alkaline lakes. The lake waters must have been alkaline enough to precipitate much of the $CaCO_3$ dissolved in the inflowing water, and shallow enough to allow wind energy to rework the lake-bottom sediments so that nuclei could accrete $CaCO_3$ coatings. Inasmuch as the twentieth-century saline deposits rest directly on ~4 m of oolites (Smith, 1993), oolites were probably being deposited in Owens Lake before 1913, when it was a fluctuating perennial body of water with a maximum recorded depth of ~15 m and recorded salinities that ranged from about 6–9 wt % (Gale, 1914). The major components of the lake water at that time (recalculated to 100 wt %) were 36–38 wt % Na, 22–26% Cl, 22–24% CO_3, and 10–15% SO_4. Calculations using these values indicate that the brine at that time had a pH of ~10 (Bischoff et al., 1993).

Tephra

Eruption of the ca. 760 ka Bishop ash from Long Valley Caldera (Izett et al., 1970) led to the subsequent deposition of ~5 m of pure to impure tephra in Owens Lake (Fig. 1; Table 1), possibly over a geologically instantaneous period of time. The base of the Bishop ash bed in core OL-92 is at 309.14 m (Sarna-Wojcicki et al., this volume), and the age of the clastic sediment immediately beneath that horizon could be anywhere from ca. 760 ka (if no significant time period is represented by the overlying mixture of reworked ash and sediment) to ca. 772.5 ka (if the average deposition rate for that ~5 m of ash-sediment mixture is the same as the overall rate in this core, 0.4 m/k.y.).

A 0.68-m-thick tephra bed was also recovered from a slightly deeper (311 m) layer in core OL-92 (Table 1), and several thin tephra beds were identified at intermediate levels, one of which (at 224 m depth) is correlated with the Dibekulewe ash (Sarna-Wojcicki et al., this volume).

Most tephra beds appear to be composed dominantly of

sand-size or larger fragments (Table 1). Many identifiable segments of the reworked Bishop ash are poorly sorted, and some contain granule- to small-pebble-size lapilli that are rounded and light pink or yellow. Bedding ranges from laminar to faint or massive.

LAKE-WATER DEPTHS INFERRED FROM LITHOLOGIES

The 810 k.y. period represented by the lacustrine deposits in core OL-92 can be divided into four intervals, using lithologic criteria to estimate lake-water depths. I emphasize that depth reconstructions are not necessarily reliable as paleoclimatic tools because depths are also affected by sedimentation rates and tectonic vagaries. This section illustrates the importance of recognizing this effect, as well as of using multiple criteria of climate change.

(1) The oldest interval, bounded by the dates of ca. 810 ka and ca. 645 ka, is represented by the sediments from the depths of 323 m and 266 m. These sediments indicate a fluctuating series of lakes that were too deep for coarse-sediment deposition about 85% of the time, judging from the thickness of silt and clay, and much shallower about 15% of the time, when five beds of fine to very coarse sand were deposited, each apparently representing a period of ~1–4 k.y.

(2) The next younger interval, bounded by dates of ca. 645 ka and ca. 450 ka, is represented by the sediments from the depths of 266 m and 206 m. These sediments consist mostly of silty sand to very coarse sand, although finer sediments occur near the top of this interval. The thickness of sand suggests that the series of lakes were relatively shallow ~80% of the time. Evidence from associated fossil assemblages supports this inference of frequent shallow-water stages during this period but also indicates that some or most of the shallow lakes were fresh (Bradbury, this volume; Carter, this volume; Firby et al., this volume), a seeming paradox that is discussed in a following section.

(3) The next still-younger interval, bounded by dates of ca. 450 ka and ca. 5 ka, is represented by the sediments between 206-m depth to the base of the oolite bed at 5-m depth. These sediments indicate a succession of lakes that were mostly deep, fresh, and much of the time, at least, overflowing. Brief, possibly shallow periods at ca. 165 ka and ca. 101 ka are represented by coarse-sand beds at 95 m and 67 m depth. However, chemical and other data (Bischoff et al., this volume, chapter 4; Glen and Coe, this volume; Menking, this volume) reveal several climatically caused changes in lake hydrology that, with two exceptions, did not reduce lake levels enough for sand beds to be deposited.

(4) The youngest interval, represented by oolites deposited between ca. 5 ka and A.D. >1913, appears to indicate a series of lakes that were continuously shallow, nonoverflowing, and moderately saline.

The lithologic, chemical, and paleontologic evidence from segments between the depths of 266 m and 206 m in core OL-92 indicate that Owens Lake was simultaneously shallow, fresh, and most of the time depositing little CO_3. This indicates either that (1) the shallow periods were so brief that evaporation had insufficient time to concentrate dissolved solids in the inflowing Owens River waters (now ~240 mg/L) to levels that deposited much CO_3 or left a record of detectably increased salinity, or that (2) the lake frequently overflowed because its spillway and lake floor at the site of core OL-92, now separated by 60 m, were within a few meters of the same elevations.

The first possibility seems unsupported by the lithologic and fossil records. For example, Owens Lake in 1872 was 15 m deep and had concentrated the inflowing river water to a salinity of as much as 9 wt %. Bischoff et al. (this volume, chapter 8) calculate that concentrating river water to a salinity of 15%, which is the low-salinity edge of the stability field of gaylussite (Bury and Redd, 1933), would have required evaporation of Owens Lake water for a period of ~8 k.y. if the lake stood at its 1872 level and a period of ~55 k.y. if the lake stood just below its spillway level. Gaylussite crystals or pseudomorphs, which have distinctive shapes, were not observed in the pre-Holocene sediments, tentatively eliminating the possibility of such periods.

A shallow-but-overflowing freshwater lake, therefore, seems more likely to have been the result of (1) an increased rate of erosion that lowered the spillway nearly to the level of the lake floor at the site of core OL-92, (2) a decreased tectonic-depression rate that allowed normal sedimentation to nearly fill the basin, or (3) an increased lake-sedimentation rate that nearly exceeded the tectonic-depression rate.

Although either possibility 1 or 2 could have caused Owens Lake to be fresh and shallow, possibility 3 seems more likely for the following reasons: The Sherwin glaciation in the Sierra Nevada reached its climax before 760 ka, when the Bishop Tuff was deposited on an outcrop of its till that extended east of the Sierra Nevada; the till was exposed long enough to develop a mature soil before being covered by tuff (Sharp, 1968). That glaciation has been interpreted to be correlative with a persistent pluvial period that lasted from ca. 1.3–1.0 Ma, inferred on the basis of subsurface evidence from Searles Lake (Smith et al., 1983), implying that the Sherwin glaciation lasted ~300 k.y. A long, intense glacial period is also suggested by both the thickness and areal extent of Sherwin till near its type area (Blackwelder, 1931; Sharp, 1968, 1972), an area ~30 km northwest of Bishop. Outwash from such a long glacial period could have caused a marked and prolonged increase in the amount of clastic debris reaching Owens Lake, nearly filling the basin with sediment.

On a lithologic basis alone, I conclude that the sediments deposited during the period between ca. 810 ka and ca. 450 ka represent moderately shallow to shallow-lake periods when fluctuations in lake level, probably caused by climate changes, led to the deposition of a wide range of clastic-sediment sizes. This interval, however, can be divided into two periods of approximately equal length. The earlier period (ca. 810 ka to ca. 645 ka) probably represents a wetter time, as it is represented by discrete beds of moderately deep-water deposits alternated with beds of shallow-water deposits. During the later period (ca. 645 ka to ca. 450 ka), deep-water deposits are relatively minor. The next youngest interval, which records sedimentation between ca. 450 ka and ca. 5 ka, represents an overall deep-lake period when fluctuations in its surface level caused by climatic changes were generally insufficient to affect deep-water, clastic-sediment characteristics.

TABLE 5. DIMENSIONS OF CLASTIC DIKES IN CORE OL-92

Depth (m)	Length (cm)	Width (mm)	Dip (°)
115.98	12	3–5	70–80
146.92	76	5–15	60–90
277.76	4	>20	~90
283.14	6	3–5(?)	~90

SEDIMENTARY STRUCTURES

Bedding

About 70% of the units in core OL-92 are logged as massive (Fig. 1), and some are recognizably bioturbated. The rest of the core exhibits faint to conspicuous bedding, ranging in thickness from a few millimeters to a fraction of a meter, that are defined either by variations in grain size or by color differences that were most visible when the core was fresh. Only five units were described in the field as thinly laminated (Smith, 1993), a likely indication of anoxic conditions. Three of these laminated units are in tephra-rich layers that represent periods when deposition was probably accelerated because of the availability of unconsolidated air-fall ash, allowing differentiation of layers representing annual or even shorter periods of time.

Intervals characterized by rhythmic sequences of beds, defined by alternating grain sizes, were noted and measured in the field (Fig. 1; Table 4). Most of these intervals make up <1 m of section. The most commonly observed pattern is that of a finer layer that is ~5–15 times thicker than the coarser layer of the pair. Most of the coarser layers, consisting predominantly of fine to very fine sand, are <1 cm thick, and many are only one or two grains thick. These sequences appear to record a regularly recurring succession of events: Conditions that produced the finer layers represented the norm, and periodic brief changes in those conditions produced the coarser layers.

The mean thickness of most fine-plus-coarse-bed pairs within a rhythmic-bedding interval is greater than the median thicknesses within the same interval (Table 4), indicating that many measured sequences included a few bed pairs that were substantially thicker than the median thickness and few, if any, that were thinner. I consider the median, rather than the average, thickness to be a better measure of a cyclic event's most common period. The median thickness for all the finer layers is 4 cm; if the thin bed of coarser material was deposited instantly, and the sedimentation rate of the finer material was 40 cm/k.y., the recurrence interval was ~100 yr.

Clastic dikes

Clastic dikes, mostly silt intrusions into beds of silty clay or clay, were observed in two parts of the core: two dikes between 116–147 m in depth, and two dikes between 277 m and 284 m in depth (Fig. 1; Table 5). The dikes' lower ends either taper to a point or terminate at a bed of the dike-filling material (Fig. 2). The upper ends of all these dikes taper to a point, and none have the upward-widening, truncated-top characteristic of desiccation cracks that were filled later from above. The dikes could be syneresis cracks, although evidence of desiccation, which normally initiates this process, is absent. The walls of the clay dike at a depth of ~147 m have beds that were offset 5–15 mm along the fracture containing the dike, suggesting that it, and possibly other dikes, resulted from earthquake shaking that opened cracks in the lake-floor sediments that were almost instantly filled with liquified sediment from below.

Sand pods

Approximately equidimensional pockets or pods of sand, mostly 1–3 cm in diameter, that are suspended in a matrix of finer sediment were noted in several parts of the core (Fig. 1). A total of 12 of the 16 recorded sand pods occur within a few meters of one or more of the others, suggesting that water depth or some other influence on deposition may have been optimal at those times. Most sand in the pods was logged as clean or well sorted, suggesting that the water depth was shallow, where sorting processes are most effective, although 10 of the 16 pods occur in fine-grained sediment that implies deep waters. The origin of these pods is not currently understood.

Color veins

Thin, sharply defined, subvertical curving bands that cut across bedding and are slightly lighter, darker, or of a different hue than the host sediment are herein called color veins (Figs. 1 and 2; Table 6). The grain size is identical in both the color-vein sediment and its host. Most veins straddle hairline cracks that apparently served as conduits for fluids with a different chemistry from that of the pore waters of the host sediment; that is, these color veins are alteration zones. Many of the color veins are in close stratigraphic proximity to other veins (Fig. 1).

As the cores oxidized after being brought to the surface, the veins became difficult to see because the color contrast was reduced. This observation suggests that the origin of the color veins is related to differences in the redox potential of the interstitial brine occupying this part of the core versus the brine that flowed through the hairline crack and diffused into the adjoining sediment, a difference that was destroyed by uniform oxidation after exposure.

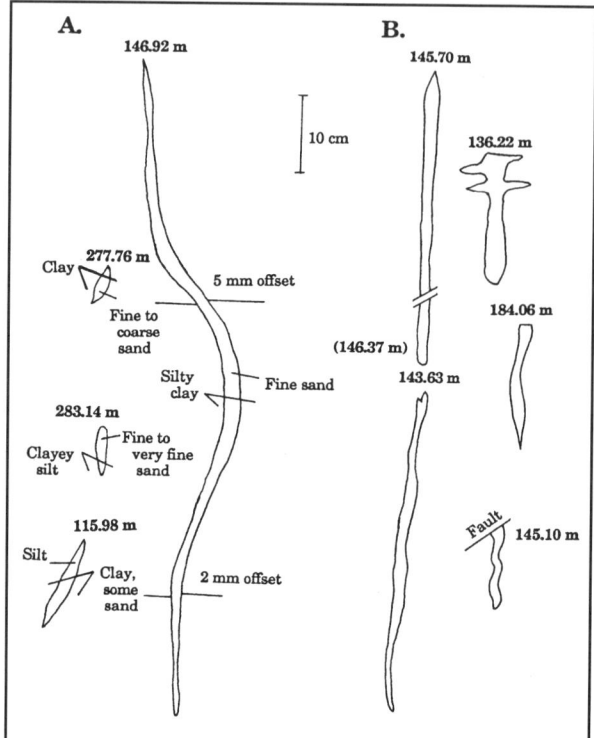

Figure 2. Field sketches of typical clastic dikes (A) and color veins (B) from core OL-92. Lithologies of clastic dikes and host sediment are indicated in A; colors of color veins and host sediments are listed in Table 6. Approximate vertical dimensions are indicated by scale; fault offsets, dips, and horizontal dimensions are not to scale (see Tables 5 and 6). Depths are to top of each structure.

TABLE 6. DEPTHS AND DESCRIPTIONS OF COLOR VEINS IN CORE OL-92

Depth (m)	Length (cm)	Width (mm)	Colors* (vein : host)
117.07	38	2-3	5GY5/1 : 5GY4/1
129.85	15	2	5Y5/1 : 5GY4-6/1
129.28	25	2	5Y6/1 : 5GY4-6/1
136.22(2)	19, 18	2-10	5Y6/2 : 5Y3-5/2
142.09	47	1-15	5G3/1 : 5G4/1
143.63	44	1-3	5Y4/4 : 10Y5/2
145.10	15	3	N5-6 : 5Y-5GY4/1-4
145.70	67	1-2	5G4/4 : 5Y-5GY5/1
171.88	25	1	5GY3/1 : 5G4/1
172.41	13	1	5GY3/1 : 5G4/1
184.06(2)	17, 15	2	5G4/1 : 5GY5/1

*Geological Society of America Rock-Color Chart, 1991.

Bioturbation structures

Structures interpreted as evidence of bioturbation are common in core OL-92. About half the horizons logged as massive contain subtle to conspicuous structures that resemble worm tubes or burrowings. Those structures parallel to the split-core surface are generally 2–10 cm long and 0.05 cm to nearly 1 cm wide; borings normal to that surface are circular. Their orientations relative to the lake floor appear to be random. The sediment filling the tubular shape generally differs slightly in color, and some has a different grain size or texture.

These bioturbation structures indicate periods when the lake water was well oxygenated and mixed down to the sediment/water interface, at least during part of the year, allowing burrowing organisms to occupy sediment that was characterized at other times by more intensely reducing conditions.

Other structures

Small faults with moderately steep dips and offsets of several centimeters were noted at four horizons (Fig. 1). Faults are to be expected in these lake-floor sediments. The historically active Owens Valley fault lies a few kilometers west of the drill site, and a Holocene, northwest-trending fault was inferred by Carver (1970), on the basis of low-sun-angle aerial photographs, along a line that passes ~0.3 km west of the drill site. Dips on the faults observed in the core range from 45° to 70°. The offset along one fault was only 1.5 cm, but offsets along three other faults exceeded the length of the fracture exposed in the core, confirming that these fractures were not caused by stresses within the core barrel during drilling, which would have trapped both halves of any fracture that was initiated within it.

Contacts between beds that suggest erosion or disturbed bedding planes were noted at several horizons (Fig. 1), although no evidence of subaerial exposure or desiccation accompanied them. None of these disturbed contacts was at the top or bottom of a cored unit, and a few centimeters above or below the disturbed contacts, we observed undisturbed bedding, tending to eliminate stresses generated during coring as a cause. Some of these erosional or disturbed contacts dip ≥15°, whereas others have wavy or small-scour shapes. These features might have resulted from times when the lake depth increased, compacting and deforming the underlying soft sediment, although no consistent relation is apparent between the presence or absence of these features and lake depth, as inferred from chemical, mineralogical, and paleontological criteria (Bischoff et al., this volume, chapter 4; Bradbury, this volume; Carter, this volume; Firby et al., this volume; Menking, this volume).

Graded bedding, commonly interpreted as evidence of turbidity currents, was observed at seven horizons (Fig. 1). Although the three closely spaced beds near 245-m depth (ca. 580 ka) might indicate a period when turbidity currents were more frequent, the deepest of these three beds has a layer of clean sand near the middle, making it atypical of normal graded-bedding sequences.

SUMMARY

About 800 k.y. of lacustrine deposition is represented by the OL-92 core. The first ~350 k.y. of this period was characterized by deposition of sand-rich (30 wt %) clay and silt; this interval is divided into two subunits on the basis of whether the sand is concentrated in discrete beds, as is found in the lower subunit, or is dispersed throughout it. Subsequent deposits, representing ~450 k.y., were nearly sand free except for two thin beds in the upper part. Oolites were first deposited at ca. 5 ka, and salts crystallized on the lake surface after diversion of the Owens River in the early 1900s. The oolite bed indicates that the late Holocene has been more arid in this area than at any time during the preceding 800 k.y.

The clastic materials are moderately well sorted, range in color (when fresh) from green to yellowish green, are soft in the upper part of the core and well indurated toward the base, and contain small amounts of CH_4, NH_3, and H_2S. About 70% of the sediments are massive; the rest display faint to distinct thin beds. Some segments <1 m long display rhythmic bedding, suggesting cyclical (climatic?) events ~100 yr long. The average size of grains in the sand-rich sections ranges from 10 µm to as much as 120 µm, and the sand-poor sections from 2 µm to 20 µm.

The 760 ka air-fall and reworked Bishop ash forms a >5 m thick bed near the base of the core, and several much thinner volcanic-ash beds are present above it.

Besides bedding, sedimentary structures observed in core OL-92 core include clastic dikes, sand pods, color veins, bioturbation structures, small faults, and graded or disturbed bedding.

REFERENCES CITED

Bassett, A. M., Kupfer, D. H., and Barstow, F. C., 1959, Core logs from Bristol, Cadiz, and Danby Dry Lakes, San Bernardino County, California: U.S. Geological Survey Bulletin 1045-D, p. 97–138.

Bischoff, J. L., Fitts, J. P., and Menking, K., 1993, Sediment pore-waters of Owens Lake Drill Hole OL-92, in Smith, G. I., and Bischoff, J. L., eds., Core OL-92 from Owens Lake, southeast California: U.S. Geological Survey Open-File Report 93-683, p. 100–105.

Blackwelder, E., 1931, Pleistocene glaciation in the Sierra Nevada and Basin Ranges: Geological Society of America Bulletin, v. 42, p 865–922.

Bury, C. R., and Redd, R., 1933, The system sodium carbonate-calcium carbonate-water: Chemical Society [London] Journal, p. 1160–1162.

Carver, G. A., 1970, Quaternary tectonism and surface faulting in the Owens Lake basin, California: Reno, Nevada, Mackay School of Mines, Technical Report AT-2, 103 p. + Appendix.

Davis, J. O., 1978, Quaternary tephrochronology of the Lake Lahontan area, Nevada and California: Nevada Archeological Survey Research Paper 7, 137 p.

Dub, G. D., 1947, Owens Lake [California]—Source of sodium minerals: American Institute of Mining, Metallurgical, and Petroleum Engineers Technical Publication 2235, p. 1–13.

Folk, R. L., and Ward, W. C., 1957, Brazos River bar, a study in the significance of grain-size parameters: Journal of Sedimentary Petrology, v. 39, p. 781–786.

Gale, H. S., 1914, Salines in the Owens, Searles, and Panamint basins, southeast California: U.S. Geological Survey Bulletin 580-L, p. 251–323.

Geological Society of America, 1991, Rock-Color Chart: Boulder, Colorado, The Geological Society of America, 10 p.

Haines, D. V., 1959, Core logs from Searles Lake, San Bernardino County, California: U.S. Geological Survey Bulletin 1045-E, p. 139–317.

Izett, G. A., Wilcox, R. E., Powers, H. A., and Desborough, G. A., 1970, The Bishop ash bed, a Pleistocene marker bed in the western United States: Quaternary Research, v. 1, p. 121–132.

Jones, B. F., and Bowser, C. J., 1978, The mineralogy and related chemistry of lake sediments, in Lerman, A., ed., Lakes: Chemistry, geology, physics: New York, Springer-Verlag, p. 179–235.

Krinsley, D. H., and Doornkamp, J. C., 1973, An atlas of quartz sand surface textures: Cambridge, U.K., Cambridge University Press, 91 p.

Menking, K., Hannah, M. M., Fitts, J. P., Bischoff, J. L., and Anderson, R. S., 1993, Sediment size analyses of the Owens Lake core, in Smith, G. I., and Bischoff, J. L., eds., Core OL-92 from Owens Lake, southeast California: U.S. Geological Survey Open-File Report 93-683, p. 58–74.

Muessig, S., White, G. N., and Byers, F. M., Jr., 1957, Core logs from Soda Lake, San Bernardino County: U.S. Geological Survey Bulletin 1045-C, p. 81–96.

Newton, M. S., 1991, Holocene stratigraphy and magnetostratigraphy of Owens and Mono Lake, eastern California [Ph.D. thesis]: Los Angeles, University of Southern California, 330 p.

Sharp, R. P., 1968, Sherwin Till–Bishop Tuff geological relationships, Sierra Nevada, California: Geological Society of America Bulletin, v. 79, p. 351–364.

Sharp, R. P., 1972, Pleistocene glaciation, Bridgeport Basin, California: Geological Society of America Bulletin, v. 83, p. 2233–2260.

Sly, P. G., 1978, Sedimentary processes in lakes, in Lerman, A., ed., Lakes: Chemistry, geology, physics: New York, Springer-Verlag, p. 65–89.

Smith, G. I., 1979, Subsurface stratigraphy and geochemistry of late Quaternary evaporites, Searles Lake, California: U.S. Geological Survey Professional Paper 1043, 130 p.

Smith, G. I., 1987, Searles Valley, California: Outcrop evidence of a Pleistocene lake and its fluctuations, limnology, and climatic significance, in Hill, M. L., ed., Centennial field guide Volume 1, Cordilleran Section of the Geological Society of America: Boulder, Colorado, Geological Society of America, p. 137–142.

Smith, G. I., 1993, Field log of Core OL-92, in Smith, G. I., and Bischoff, J. L., eds., Core OL-92 from Owens Lake, southeast California: U.S. Geological Survey Open-File Report 93-683, p. 4–57.

Smith, G. I., and Pratt, W. P., 1957, Core logs from Owens, China, Searles, and Panamint Basins, California: U.S. Geological Survey Bulletin 1045-A, p. 1–62.

Smith, G. I., and Street-Perrott, F. A., 1983, Pluvial lakes of the western United States, in Porter, S. C., ed., Late-Quaternary environments of the United States, Volume 1: The late Pleistocene: Minneapolis, University of Minnesota Press, p. 190–212.

Smith, G. I., Barczak, V. J., Moulton, G. F., and Liddicoat, J. C., 1983, Core KM-3, a surface-to-bedrock record of late Cenozoic sedimentation in Searles Valley, California: U.S. Geological Survey Professional Paper 1256, 24 p.

Smith, G. I., Friedman, I., and McLaughlin, R. J., 1987, Studies of Quaternary saline lakes—III. Mineral, chemical, and isotopic evidence of salt solution and crystallization processes in Owens Lake, California, 1969-1971: Geochimica et Cosmochimica Acta, v. 51, p. 811–827.

Wentworth, C. K., 1922, A scale of grade and class terms for clastic sediments: Journal of Geology, v. 30, no. 5, p. 377–392.

MANUSCRIPT ACCEPTED BY THE SOCIETY JUNE 17, 1996

Geological Society of America
Special Paper 317
1997

Climatic signals in clay mineralogy and grain-size variations in Owens Lake core OL-92, southeast California

Kirsten M. Menking*
Earth Sciences Department, Earth and Marine Sciences Building, University of California, Santa Cruz, California 95064

ABSTRACT

Mean grain size and clay mineralogy of sediments in Owens Lake core OL-92 reflect climatic conditions in the Sierra Nevada region. Variations in mean grain size of mud within the core reflect climatic oscillations in the region, with relatively coarse sediments (mean grain size ~15 µm in diameter) probably deposited during Owens Lake lowstands and relatively fine sediments (mean grain size ~5 µm in diameter) deposited during highstands. The mineralogy of the clay-size fraction, as determined by X-ray diffraction (XRD), also reflects the climatic history of cyclic glaciation. Mineral assemblages display two end-member XRD scans: Those having large illite, quartz, and feldspar peaks but a small smectite peak (interpreted to be glacial rock flour), and those with small illite, quartz, and feldspar peaks and a large smectite peak (interglacial sediments). On the basis of diffraction peak areas, illite and smectite probably constitute 70% or more of the clay-size minerals and are inversely abundant. In addition, core sections high in smectite correlate well with high mean grain size and carbonate content in Owens Lake sediments, whereas sediments rich in illite, quartz, and feldspar (rock flour) correlate well with finer mean grain size and low carbonate content. Variations in clay mineralogy and grain size, which are strikingly similar to variations in deep-sea oxygen isotopic composition ($\delta^{18}O$), indicate that the lake-level variations and nature of sediments delivered to the lake vary in concert with global climate changes.

INTRODUCTION

This study is part of an ongoing U.S. Geological Survey (USGS) project to examine sediments cored from Owens Lake in southeastern California. The purpose of the USGS study is to construct a more complete record of Pleistocene and Holocene climatic variations in the area than has been possible with the discontinuous and moderately age-constrained moraine record in the Sierra Nevada (Blackwelder, 1931; Sharp, 1972; Birman, 1964; Gillespie, 1982; Phillips et al., 1990) and the "filtered" climatic record in the Searles Lake core (Smith, 1984). Core OL-92, taken from Owens Lake in the spring of 1992 by a USGS drilling crew, measures 7.5 cm in diameter and was drilled to 323 m.

Recovery was 80%. Sediments vary from an evaporite package at the top of the core to lacustrine clay-, silt-, sand-, and granule-size clastic materials plus carbonate muds. Clastic sediments at the top of the core are predominantly clay to silt size, and become interbedded silts and sands below about 200 m depth. Ages are constrained via magnetostratigraphy (Glen and Coe, this volume), ^{14}C dating (Bischoff et al., 1993b; this volume, chapter 8), and tephrochronology (Sarna-Wojcicki et al., 1993; this volume). Bischoff (1993; this volume, chapter 8) constructed an age-versus-depth model based upon an assumption of a constant mass-accumulation rate for the core. This model agrees closely with the magnetostratigraphic record of Glen et al. (1993; this volume) and yields an average mass-accumulation rate of 51.4 g/cm²/k.y. and an average age-depth relation of 40.1 cm/k.y.

In this chapter, I report the results of grain-size analyses and X-ray diffraction (XRD) analyses of clay minerals in the core

---

*Present address: Department of Geology and Geography, Box 255, Vassar College, Poughkeepsie, New York 12601.

Menking, K. M., 1997, Climatic signals in clay mineralogy and grain-size variations in Owens Lake core OL-92, southeast California, *in* Smith, G. I., and Bischoff, J. L., eds., An 800,000-Year Paleoclimatic Record from Core OL-92, Owens Lake, Southeast California: Boulder, Colorado, Geological Society of America Special Paper 317.

and compare the relation of variations in grain size and clay mineralogy to other climate proxy records in the core. In addition, I use the age-depth model of Bischoff (1993; this volume, chapter 8) to compare the Owens Lake record with the $\delta^{18}O$ record derived from deep-sea sediments (Imbrie et al., 1984).

PREVIOUS WORK

Lacustrine grain size and its relation to climate change

Many workers have noted a relation between distance to the shoreline and grain size of materials deposited in lakes. Sarmiento and Kirby (1962) found that sediments from the center of Lake Maracaibo, Venezuela, are substantially finer grained than those near the shores of the lake. Furthermore, a series of short cores (2–5.5 m long) reveal a coarsening-upward trend in one of the stratigraphic units, which possibly records a shallowing of the lake. Picard and High (1972, 1981) reported on the lithofacies of several lakes throughout the world. In general, they noted a progression from coarse sand and gravel at the shores of these lakes to fine muds near the centers. They also noted that sediments of particular sizes fell along concentric belts, generally parallel to the shores of the lakes. They attributed this pattern of deposition to the location of wave base and wave agitation in the lakes. Sediments deposited below wave base are typically finer than those deposited above wave base, where wave energy is strong and fine grains are winnowed away.

Waitt (1980, 1985) interpreted a package of fining-upward rhythmites in bottom sediments from glacial Lake Missoula as evidence for periodic filling and catastrophic flooding of that lake. Failure of the ice dam confining Lake Missoula resulted in the instantaneous deposition of silt, which slowly fined upward to clay as the lake refilled. At least 40 filling and flood cycles are recorded by these bottom sediments. In each case, as the lake enlarged, deposits at any given site fined upward as the distance to the shorelines increased and as wave base moved away from the center of the lake.

In addition to the influence of wave-base location on sediment distribution, Boggs (1987) included river inflow that can send a plume of fine sediment far into the lake, long-shore currents, convective overturn of lake water caused by thermal stratification, and density underflows. Smith (this volume) also identified possibly ice-rafted pebbles and sand lenses as an important transport mechanism for coarse-grained material.

Clay minerals in the Owens Lake system

Droste (1961a, 1961b) first reported on the nature and relative abundances of clay minerals in the Owens chain of lakes. He identified montmorillonite (a smectite mineral), illite, chlorite, and kaolinite in about 20 samples each from 300–1,000-ft cores of Owens, China, Searles, and Panamint Lakes taken in the 1950s. On the basis of mineralogic assemblages, Droste concluded that clays found in China and Searles Lakes had, to some extent, originated in the Owens Lake drainage basin and had been transported by overspill into basins down the chain during pluvial-lake highstands. Lacustrine clay assemblages from pluvial lake lowstands, on the other hand, reflected their unique drainage basins (Droste, 1961b). He further noted that variations in clay mineralogy in a core from Owens Lake dominantly reflect variations in the proportion of montmorillonite to illite (Droste, 1961b) and that clay minerals constitute at least 70% of the clay-size fraction (Droste, 1961a). Looking at samples of core sediments from Searles Lake, Hay and Moiola (1963) and Hay et al. (1991) identified illite, montmorillonite, chlorite, and kaolinite in the clay-size fraction; the silt and sand-size fractions were dominated by microcline, orthoclase, albite, andesine, and quartz. Like Droste (1961b), they noted a variation in dominance of illite and montmorillonite in Searles Lake sediments, and Hay and Moiola (1963) suggested that the absence of montmorillonite in several samples was caused by its diagenetic destruction. Feth et al. (1964) identified montmorillonite, kaolinite, and "micaceous clay minerals" (i.e., illite) in the <4-µm fraction of sediments (size convention of Wentworth, 1922) from various locations in the Sierra Nevada. Newton (1991) conducted XRD analyses on sediments from shallow cores (<30 m) from Owens and Mono Lakes. He interpreted the presence of quartz and feldspar in the <4-µm fraction as indicative of glacial abrasion in the Sierra Nevada. Furthermore, he detected only trace amounts of clay minerals and concluded that the clay-size fraction was wholly rock flour.

ORIGINAL HYPOTHESES

Because of the proximity of Owens Lake to the Sierra Nevada, and the fact that Owens Lake receives most of the runoff from the eastern side of the range, the lake's sediments should contain a record of the cyclic glaciation that occurred in the mountain range. From the results of the previous lacustrine grain-size studies (Picard and High, 1972, 1981; Sarmiento and Kirby, 1962; Waitt, 1980, 1985), I hypothesized that variations in lake level might manifest themselves by variations in grain size of sediments deposited at the core OL-92 site. In particular, I expected fine-grained deposition to characterize Owens Lake highstands, coarser grained materials to characterize lowstands, and grain-size variations that would generally reflect variations in the distance between the core site and the lake shore. Furthermore, I hypothesized that the lake sediments might alternate in mineralogy between glacial and nonglacial mineralogic end members. Following the lead of Newton (1991), I expected sediments characteristic of glacial periods to contain abundant quartz and feldspar in the clay-size fraction because of the abrasive action of valley glaciers that produce vast amounts of rock flour. Interglacial sediments were expected to reflect less physical weathering and a concomitant increase in chemical weathering. As a result, smectite and other clay minerals were expected to outweigh the amounts of quartz and feldspar deposited during these periods. To test these hypotheses, I conducted grain-size and X-ray diffraction studies of core sediments.

GRAIN-SIZE ANALYSES

Sample types

Two types of samples were taken from the core for grain-size analysis. Point samples, which represent about 2–3 cm of core length and which were cut out of the center of the core with a knife, were collected at the drill site at 1–2-m spacings (~2,500–5,000 yr between samples). These samples were taken to coincide with visually obvious changes in lithology (Smith, 1993; this volume), and each comprised about 60 g of sediment. Because of the time required for analyses, a selection of these samples (one sample about every 5 m) was chosen for analyses of grain size (this chapter), pore-water chemistry (Friedman et al., 1993; this volume), and selected solid elements (Bischoff et al., 1993a; this volume, chapter 4). The remainder of the samples were archived for future study. Channel samples were later collected in the laboratory, each representing integrated ribbons of sediment weighing about 50 g and spanning ~3.5 m of core (~7,500 yrs of sedimentation). These samples were collected by pushing a U-shaped spatula along the face of the split core. The spacing between point samples and the length of channel samples, were chosen to ensure good resolution of climatic fluctuations operating on time scales of 10 k.y. or greater. Nearly 100 point and 100 channel samples were analyzed.

Methods

Grain-size analysis of point samples. About 10 g of each point sample was treated with Morgan's solution (weak acetic acid buffered to pH 5 with sodium acetate) and peroxide to remove carbonate and organic material (no attempt was made to remove biogenic silica, such as diatom tests). Periodically stirred samples sat in this solution for two days and then were boiled to remove excess peroxide. The remaining sediment was wet sieved to separate the gravel, sand, and silt-plus-clay fractions. The clay and silt fraction of each sample was collected in a 1,000-ml graduated cylinder. Sands and gravels were poured into evaporating dishes, dried, and weighed.

Dried gravels and granules (>-2 Φ; >2 mm, following Wentworth, 1922, size conventions) were sieved at 0.5-Φ intervals (Φ units are defined as minus log base 2 of the grain size in millimeters). Sands (−1 to +4 Φ; 0.0625–2 mm) were introduced into settling tubes and their grain sizes determined at 0.5-Φ intervals. To prevent flocculation, 5 ml of 5% sodium hexa-metaphosphate ("calgon") dispersant solution was added to each clay and silt solution (>+4 Φ; <0.0625 mm), and the graduated cylinders were filled with deionized water to 1,000 ml. Each solution was agitated, after which 20 ml was removed with a pipette submerged to 20-cm depth (Folk, 1968). The 20-ml aliquot was dried in an oven, and the weight of silt and clay in each sample was determined by multiplying the dried mass value by 50. Each solution was re-agitated and another 20-ml sample drawn off. These aliquots were placed in a hydrophotometer and grain sizes determined at 0.5-Φ intervals. For a more complete description of methods and instrumentation used, see Menking et al. (1993b).

A few replicate measurements on sample splits were carried out to determine precision, and these analyses indicate an average precision of about ±0.25 Φ. Torresan (1987) determined a precision of about ±0.5 Φ for the hydrophotometer used in these analyses, which is probably more indicative for the whole data set.

Sand, silt, and clay contents of channel samples. Sand-plus-gravel, silt, and clay contents were determined on 91 of the 3.5-m-long channel samples. A 10-gram split of each sample was subjected to the same chemical treatments used on the point samples. Sand and gravel were separated from silt and clay by wet sieving. The sands and gravels were collected together in evaporating dishes and weighed after drying. Concentrations of coarse silt, fine silt, and clay in each sample were determined by a scaled-down pipette analysis (Galehouse, 1971). Part of the clay-size fraction of each channel sample was set aside for X-ray diffraction (XRD) analysis.

Results of grain-size analyses

Grain-size analysis of point samples defines two distinct depositional regimes (Fig. 1). With the exception of a coarse-grained oolite layer at the top of the core, mean grain size between 7 m and 200 m in depth fluctuates between 2 µm and 22 µm (clay- to silt-sized material). In contrast, mean grain size between 200 m and the base of the core at 323 m fluctuates between 10 µm (medium fine silt) and 30–130 µm (coarse silt to fine sand). Sand-plus-gravel, silt, and clay contents of the 3.5-m-long channel samples broadly mimic the point-sample trends; fine silts and clays predominate from the base of the oolite bed to 190 m depth, and coarser grained material predominates between 190 m and 323 m depth (Fig. 1). Some mismatch exists between the percentage of sand in the channel samples and the mean grain size in microns of the point samples between 20 m and 50 m in depth and again from 95 m to 100 m. This mismatch is caused by inadvertent oversight of a few coarse-grained layers in the subsampling of the point-sample sediments. This oversight, however, does not invalidate the conclusion that the top 200 m of the core are distinctly different from the lower 123 m. A closer examination of the top 200 m of the core reveals a crude periodicity in mean grain size with depth in the point-sample record (Fig. 2).

Table 1 lists grain-size statistics (derived from point samples) for the top ~200 m of the core (excluding the very coarse oolite layer near the core top) and the bottom 123 m of the core. The difference in depositional style between these two core sections is evident in the mean grain size and skewness parameters, but the sorting and kurtosis are much the same in the two sections. All sediments are poorly sorted, whether they be the fine-grained sediments from the top of the core or the coarser grained sediments of the lower core. However, the poor sorting may be an artifact of the sampling technique because each point sample represents 2–3 cm of core length, and, therefore, may

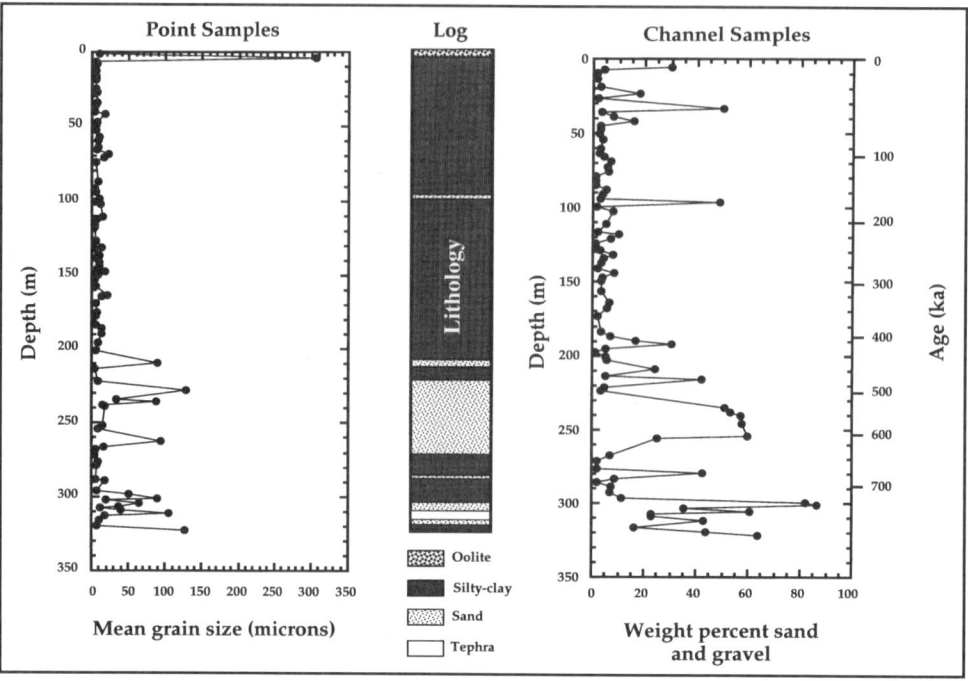

Figure 1. Mean grain size in point samples, and percentage of sand-, and gravel-sized grains in channel samples, core OL-92. Generalized lithologic log is based on Smith (1993). Note change in depositional style from the mean-grain-size record from predominantly silt and clay deposition between ~7 m and 200 m to silt and sand deposition below 200 m. Channel samples and lithologic log show similar changes in depositional style at ~200 m depth. Mean grain size of point samples was determined by sieving, hydrophotometer, and settling tube; percentage of sand and gravel was determined by sieving.

comprise more than one sedimentation unit, each with its own degree of sorting. Those sediments in the top 200 m of the core are positively skewed, implying a weighting toward fine grains. Sediments in the bottom 123 m show a nearly normal grain-size distribution. Both sections of the core are leptokurtic, meaning that the central part of the grain-size distribution is better sorted than the tails of the distribution (Folk and Ward, 1957).

CLAY MINERALOGICAL ANALYSES

Methods

X-ray diffraction techniques. Channel-sample clays were first mounted on glass microscope slides, using the filter-peel technique (Moore and Reynolds, 1989), and were then step-scanned by X-ray diffractometer (XRD). All minerals were identified by reflection peak locations (Table 2), and some were further identified by various thermal and chemical treatments (Moore and Reynolds, 1989). To test for the presence of smectite minerals, samples were glycolated and then rescanned; changes in peak location and intensity were recorded. To determine whether the 7-Å (Angstrom) and 3.5-Å reflections were the result of the presence of chlorite or kaolinite, clays were mounted on X-ray amorphous-tile slides and heated to 550 °C for 1 hour. At this temperature, kaolinite becomes amorphous, causing a reduction in peak intensity in those samples containing it. The chlorite (001) peak also intensifies and shifts to 6.3–6.4 Å caused by dehydroxylation of the inner brucite layer (Moore and Reynolds, 1989). Furthermore, a few samples were boiled in 1N HCl for 2 hours. This treatment dissolves chlorite but leaves kaolinite unaffected. An examination of the 7 Å peak before and after boiling in acid determines the presence of chlorite and/or kaolinite. For further description of methods and instrumentation used, see Menking et al. (1993a).

Artificial mixtures. To determine the relative abundances of true clays and nonclays in the clay-size fraction of the core and thereby assess the relative proportions of glacial rock flour versus chemically derived clays in the core sediment, a series of artificial mixtures of various minerals was prepared and analyzed by XRD. The <4-μm fractions of standard illite, smectite, plagioclase, and K-feldspar were separated by centrifugation of wet suspensions of commercially available standard mineral powders. Concentration of each mineral in suspension was determined by placing a known volume into a previously weighed aluminum vessel, drying the suspension in an oven heated to 60 °C, and then reweighing the vessel. Once mineral concentrations were known, it was possible to mix suspensions to give known proportions of clay and nonclay minerals. The resulting mixtures were mounted on glass slides and scanned with the XRD, as described earlier. Areas of selected peaks were measured, and peak-area ratios of the artificial mixtures were compared to peak-area ratios of the channel sample sediments.

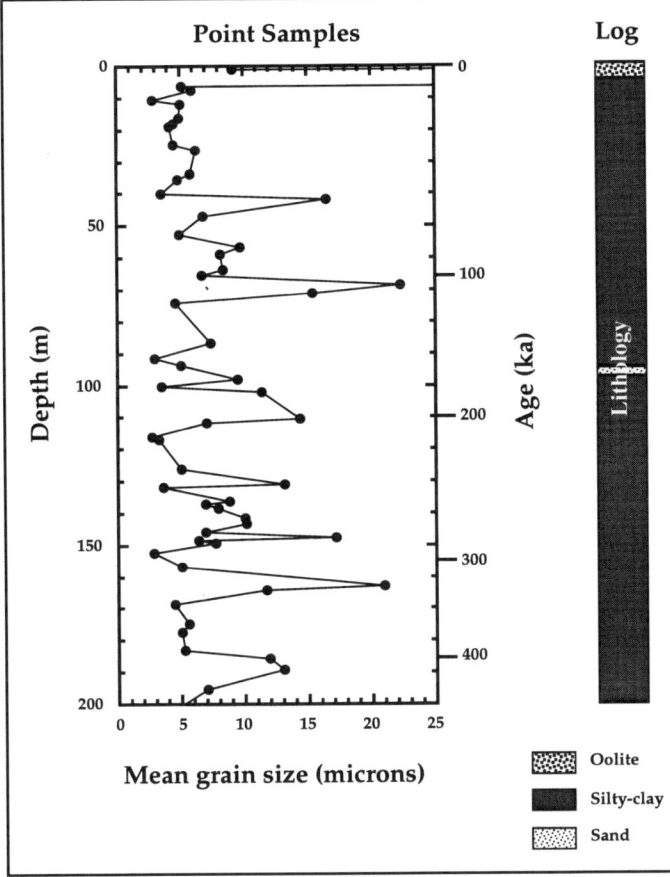

Figure 2. Enlarged point-sample grain-size record and lithologic log for top 200 m of core OL-92. Mean grain size of point samples was determined by sieving, hydrophotometer, and settling tube. Note apparent periodicity in mean-grain-size variation with depth.

TABLE 1. GRAIN-SIZE PARAMETERS FOR CORE OL-92

Statistical Parameter*	Core Depth	
	7–200 m	200–323 m
Mean grain size (Φ)†	7.40 ± 0.79	5.73 ± 1.56
Mean sorting (Φ)†	1.80 ± 0.43	1.90 ± 0.52
Mean skewness	0.16 ± 0.14	0.04 ± 0.25
Mean kurtosis	1.53 ± 0.39	1.41 ± 0.60

*Statistical parameters from Folk and Ward, 1957.
†Phi grain-size scale in which Φ = minus log base 2 of the grain size in millimeters.

Results of the X-ray diffraction analysis

The XRD determinations of the clay mineralogy of channel samples show that illite, smectite, chlorite, and kaolinite are the primary clay minerals. Most samples also contain some clay-size quartz, plagioclase, and K-feldspar. The X-ray diffraction patterns are typically weak, and only fair to poor in quality (i.e., peaks often are only 2 to 3 times larger than background), which may indicate poorly crystallized materials. Attempts to determine whether clay minerals are dioctahedral or trioctahedral, on the basis of (060) reflections, were unsuccessful owing to weak reflections and to the overlapping ranges of (060) reflections in this mineralogically complex suite of samples. Shrinkage of some illite peaks upon glycolation indicates some illite/smectite mixed layering.

The clay-size fractions are characterized by mixtures of two end members. These consist of an end member displaying small smectite but large illite, quartz, and feldspar peaks (Figure 3A), and an end member displaying large smectite but small illite, quartz, and feldspar peaks (Figure 3B). After noting these two end-member compositions, I measured the peak areas of several clay and nonclay minerals in each sample to see if any regular pattern or periodicity existed in the deposition of clays and nonclays. Chlorite and kaolinite are both present, but because of overlapping reflection peaks they are not easily decoupled from one another. Therefore, I measured the combined chlorite-kaolinite reflection at 7.11 Å (12.4 °2Θ). In the Owens Lake core sediments, the ratios illite/quartz and chlorite-kaolinite/quartz correlate positively to each other and to the ratios K-feldspar/quartz and plagioclase/quartz (Fig. 4). The ratio smectite/quartz (Fig. 4), however, correlates positively sometimes to the other ratios, negatively sometimes, and not at all sometimes. Furthermore, the smectite/quartz and the plagioclase/K-feldspar ratios (Fig. 5) behave inversely; the smectite/quartz ratio shows a low value during the last glacial maximum (ca. 20 ka, and 13 m depth in the core), whereas the plagioclase/K-feldspar ratio is high.

In addition, the multiplicative factors of Hallberg et al. (1978), modified by Hay et al. (1991), were used to determine the relative abundances (±~30 wt %) of clay minerals in the clay-size fraction of the core samples. The results of this analysis (Fig. 6) show that smectite and illite are the dominant phases and that chlorite and kaolinite account for only ~10–20% of the clay mineralogy. Because of this, smectite and illite abundances vary inversely; smectite concentrations are very low in samples from the last glacial maximum (~13 m depth, ca. 21 ka), whereas illite concentrations are high. Furthermore, a strong periodicity in the smectite and illite curves is evident with

TABLE 2. REFLECTION PEAK LOCATIONS USED TO IDENTIFY MINERALS IN CORE OL-92 CHANNEL SAMPLES

Mineral	(hkl)*	°2Θ†	Angstroms
Illite	(001)	4.4	10.1
Chlorite	(001)	6.2	14.2
	(002)	12.4	7.11
	(003)	18.8	4.72
	(004)	25.1	3.54
Smectite	(001)	5.2	17.0
Kaolinite	(001)	12.4	7.11
	(002)	25.1	3.54
Quartz	(100)	20.8	4.26
K-feldspar	(002)	27.5	3.24
Plagioclase	(002)	27.9	3.18

*Mineralogical index used to describe crystal-face location.
†The angle that the incident X-ray makes with the sample.

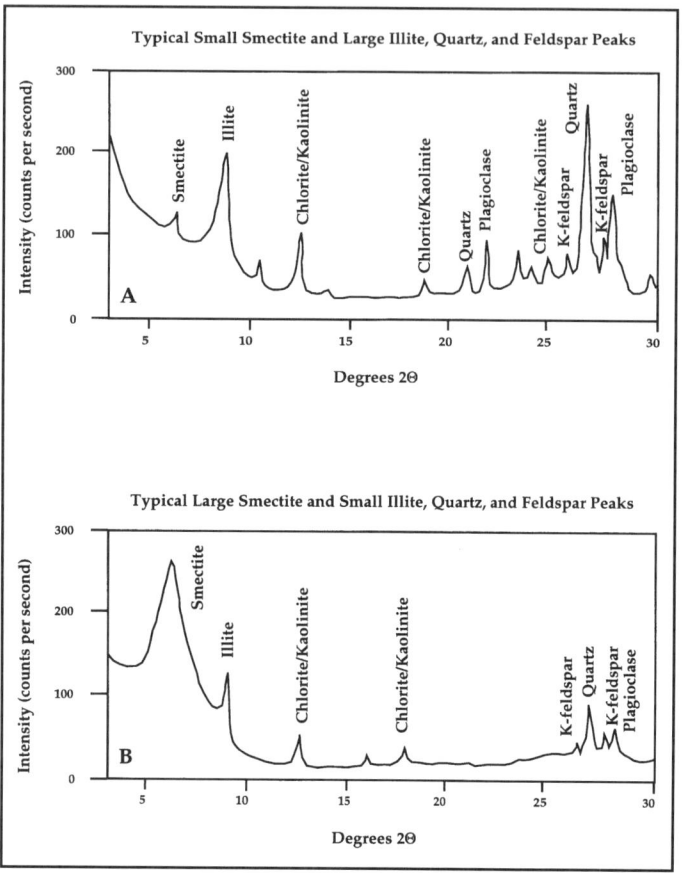

Figure 3. Typical X-ray diffraction scans from core OL-92. Tracings from air-dry mounts. All samples were scanned with a Norelco Step-Scanning X-Ray Diffractometer employing Cu Kα radiation, a 0.02 °2θ step size, and a 5-second dwell time. A: Small smectite but large illite, quartz, and feldspar peak scan; sample from 78 m depth. Peak at ~10.5 °2θ remains unidentified. Mineralogic assemblage is consistent with a glacial rock-flour source for the sediments. B: Large smectite but small illite, quartz, and feldspar peak pattern; sample from 192.13 m depth. Mineralogic assemblage is consistent with either a soil-clay origin or a combination of sediment eroded from the Sierra Nevada plus authigenic precipitation of smectite within Owens Lake.

depth. Cation exchange capacity (CEC) (Bischoff et al., 1993a; this volume, chapter 4) follows the smectite curve (Fig. 6).

Results of the artificial clay-mixture experiments

Results of the experiments using artificial mixtures of illite or smectite, and albite or K-feldspar, to determine the mean areas of their peaks on XRD charts are shown in Figure 7. The lines are derived from the experiments using known ratios of mineral pairs; the gray boxes represent the mean peak-area ratios of the different mineral mixtures on XRD charts of Owens Lake channel-sample sediments. The concentrations, in weight-percentages of each mineral in the channel-sample sediments, are then determined from the graphs. For example, in those samples, the mean peak area of albite, divided by the peak areas of albite + illite, equals 0.27; this is represented by the solid box plotted on the albite-illite line. This value corresponds to a weight-percent albite, in the albite + clay mixture, of ~0.05, or to an albite abundance relative to illite of ~5%. Thus, the average Owens Lake sample contains only ~5% as much albite as illite; it also contains only ~30% as much albite as smectite (Fig. 7).

Comparison of peak-area ratios for artificial mixtures with those for samples of unknown mineral abundances leads to the conclusion that no more than ~30% of the clay-size fraction of any sample consists of nonclay minerals (K-feldspar, plagioclase, quartz). Thus, variations in smectite and illite concentrations represent most of the total variation in mineralogy of the <4-µ fraction. These results confirm the findings of Droste (1961b), and Hay and Moiola (1963).

DISCUSSION

Grain-size analyses of the top 200 m of the core indicate that the mean sizes of sediments deposited during glacial periods vary from 2–5 µm, whereas those of interglacial periods vary from 11–23 µm (Fig. 2). For example, the clay- and silt-sized material in a sample from 13-m depth in core OL-92 (ca. 21 ka, during the last glacial maximum), has an average size of ~5 µm. Other core sections, identified on the basis of other paleoclimatic proxies as representing glacial periods (Smith et al., this volume), are similar. Interglacial periods, in contrast, are marked by both coarser deposition (mean grain size of ~15 µm) and more size variation. Below 200 m, no correlation between grain size and climatic regime is apparent.

As suggested by Picard and High (1972, 1981), Smith (this volume), and Boggs (1987), lacustrine grain size is a function of wave-base depth, fluvial inflow rates that allow some plumes of fine sediment to penetrate far into a lake, long-shore currents, ice-rafting of pebbles and coarse sand lenses, and turbid underflows. In general, these processes, with the exception of turbid underflows and ice-rafting, result in a pattern of deposition of concentric rings of sediment grains that increase in size toward shore. The difference in grain size variability in the Owens Lake core, as well as the predominance of finer deposition during glacial periods and of coarser deposition during interglacial periods, might best be explained by these processes and by consideration of the hypsometry of the Owens Lake basin.

The slopes of the Owens Lake basin resemble those of a bathtub; there is little change in altitude for several kilometers in all directions from the basin center and then a rapid increase in altitude over a small distance. A north-south transect between the core site and the north shore of the lake shows 4 m of elevation gain in ~17 km followed by 60 m of elevation gain over the next 8 km—most of that 60-m elevation gain occurring within ~1 km. High lake levels, such as are thought to have characterized glacial periods, should have resulted in deposition of fine sediments at the core site, as the distance to the shoreline was large and the wave base normally was far above the lake floor. During interglacial periods, if lake levels were lower, mean grain size would have increased. If the lake fell to

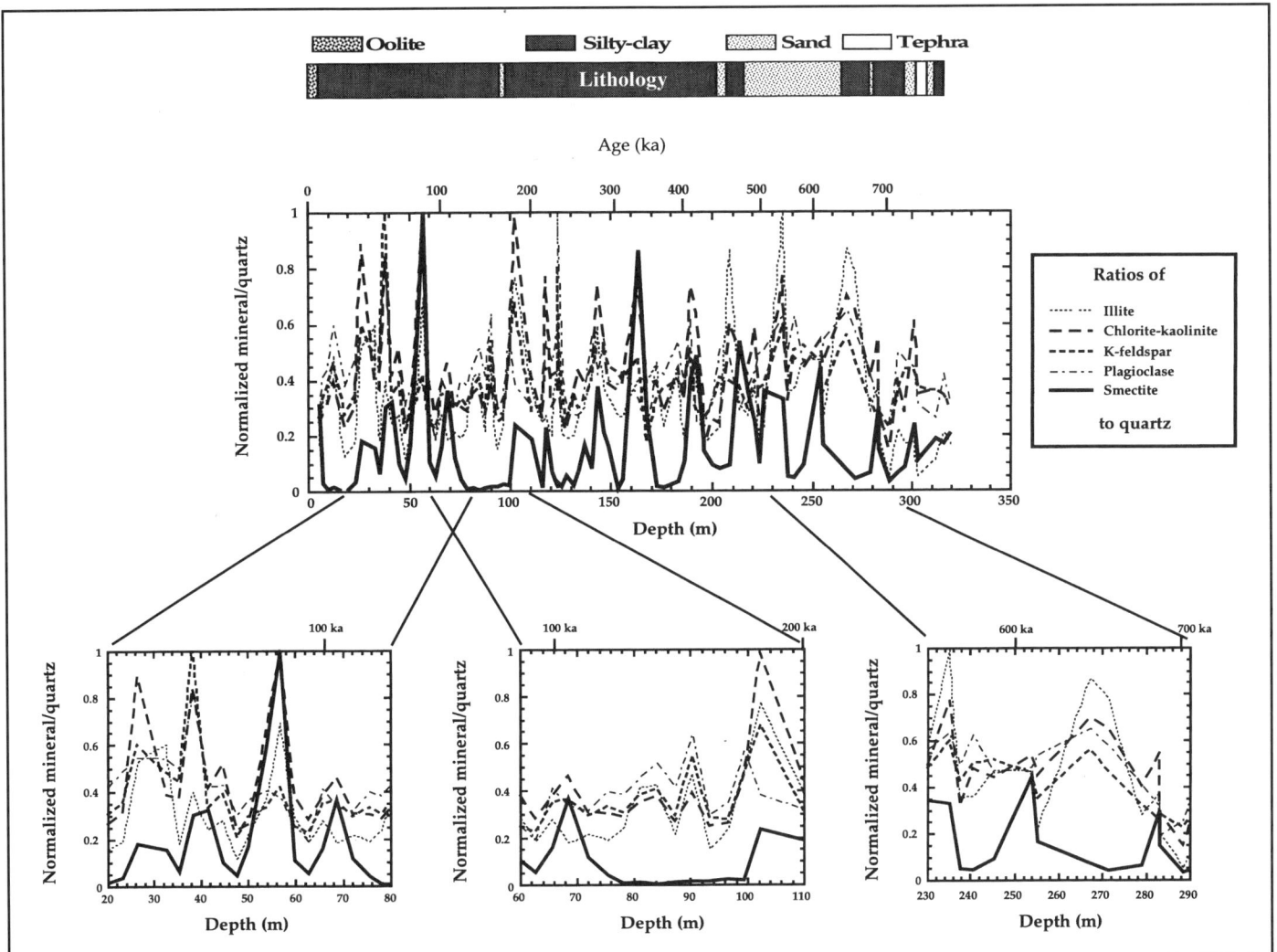

Figure 4. Plot of normalized clay peak-area to quartz peak-area ratios from core OL-92; plot shows that the ratios of illite, chlorite-kaolinite, K-feldspar, and plagioclase to quartz exhibit similar behavior with depth. Lithology is also shown. Similarity of illite and chlorite-kaolinite curves to the plagioclase and K-feldspar curves may indicate a mechanical origin for much of the illite and chlorite. Extended plots show that the smectite/quartz ratio correlates positively sometimes (20–80 m), negatively sometimes (230–290 m), and not at all sometimes (60–110 m) to other clay minerals or feldspars.

the point where the basin floor was nearly flat, further lowering would have caused a large contraction of the lake, and the shorelines would have rapidly approached the core site. As a result, coarser grains could be transported to the core site by waves and near-surface currents.

The bathtub morphology could also explain why the mean grain size in the glacial sediments varies less than that in the interglacial sediments. During glacial periods, when the lake was high, small changes in lake level would lead to little change in shoreline distance, and sediments having nearly uniform grain sizes would be deposited at the core site. During interglacials, if lake levels were lower, fluctuations could have produced large variations in shoreline distance relative to the core site, and mean grain sizes of particles deposited at the core site might have var-

ied more. Derivation of absolute lake level from mean grain size may be problematic, however, because other processes such as turbid underflows, ice-rafting, and aeolian transport might also have influenced grain sizes. Nevertheless, mean grain size appears to be a useful first approximation of lake level.

In core OL-92, subtle variations in grain size between depths of ~7 m and 200 m, although mostly in the clay- and silt-size ranges, show a definite climatic signal; deposits were finer grained during glacial periods and coarser during interglacial periods. However, the abrupt transition from fine sediments above ~200 m, to silt- and sand-rich sediments below 200 m remains unexplained. The transition may be the result of climatic or tectonic factors, although Smith (this volume) attributes this coarser section to a near filling of the lake basin by waning-stage

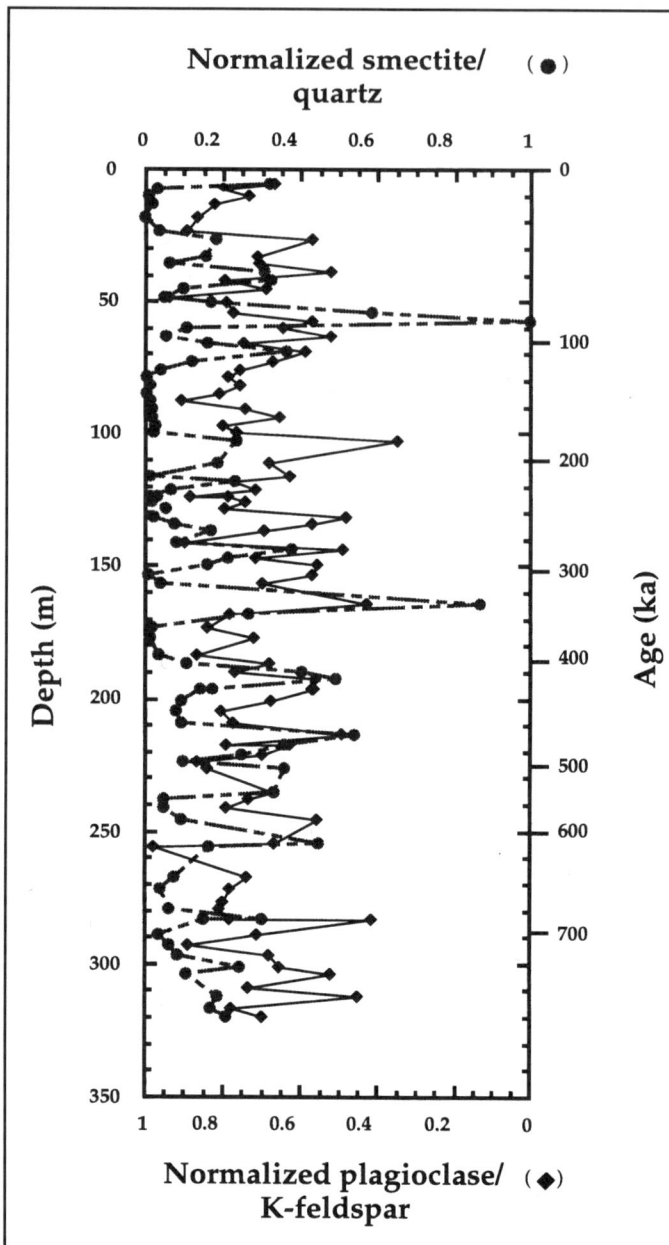

Figure 5. Plot of normalized smectite/quartz and normalized plagioclase/K-feldspar ratios in core OL-92. Note inverse behavior of the two curves with depth (normalized plagioclase/K-feldspar axis is plotted inversely). High smectite/quartz values should correspond to low plagioclase/K-feldspar values if the smectite was produced by the chemical weathering of plagioclase (see discussion). Likewise, if smectite and K-feldspar are primarily authigenic precipitates formed when Owens Lake became saline, high smectite/quartz values should still correspond to low plagioclase/K-feldspar values (see discussion).

outwash from the earlier Sherwin glaciation. However, aside from the size difference between the top and bottom sections of the core, both sections appear to represent poorly sorted lacustrine materials, although, as mentioned earlier, the poor sorting may be an artifact of the sampling technique.

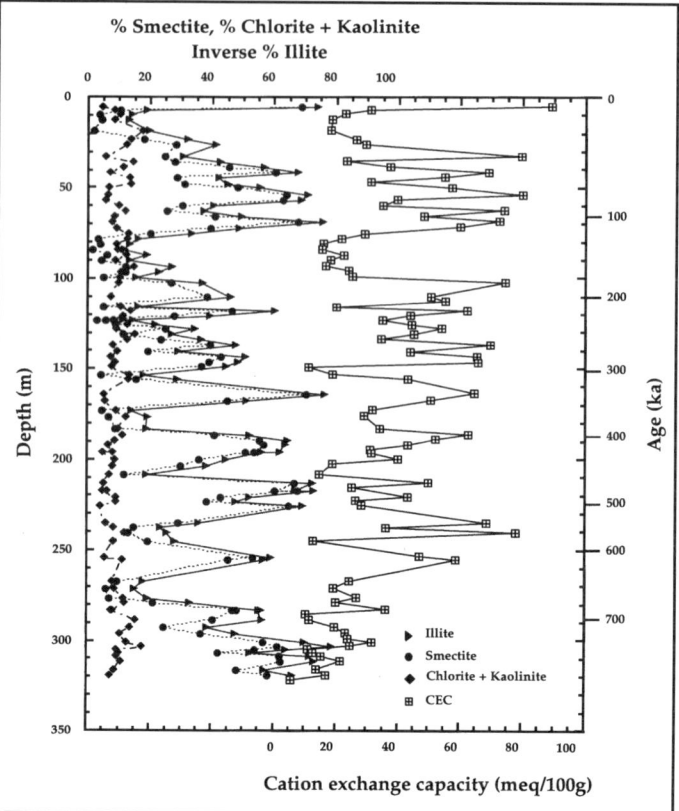

Figure 6. Weight percentages of smectite and chlorite-kaolinite versus depth in core OL-92 as calculated by the methods of Hallberg et al. (1978) and Hay et al. (1991). Inverse weight percent of illite is also plotted. Smectite and illite are the dominant clay minerals and are, therefore, inversely abundant. Cation exchange capacity (CEC) of channel-sample sediments as determined by Bischoff et al. (1993a; this volume, chapter 4) is also shown. CEC is high during periods in which mineralogy is dominated by smectite and low during periods dominated by illite, plagioclase and K-feldspar.

Like Newton (1991), I interpret the presence in the clay-size fraction of nonclay minerals (quartz, plagioclase, and K-feldspar) to be the result of glacial abrasion in the Sierra Nevada that produced large volumes of glacial flour. Indeed, it is rare to find quartz and feldspars in the clay-size fraction of any sediment unless that sediment had a glacial origin (Moore and Reynolds, 1989). Of the true clay minerals, illite and chlorite are typically found as products of intense physical weathering of micas by glaciers in the absence of strong chemical weathering (Chamley, 1989; Biscaye, 1965). Smectite and kaolinite, on the other hand, are more likely produced from the weathering of silicate minerals during soil formation, and they may, therefore, be more indicative of climates conducive to chemical weathering (Chamley, 1989). Smectite is also a common weathering product of volcanic ash, and care must be taken to distinguish between true climatic variations and volcanic events (Chamley, 1989).

Note that the illite/quartz and chlorite-kaolinite/quartz peak-area ratios correlate positively with the K-feldspar/quartz and plagioclase/quartz ratios in the Owens Lake samples (Fig. 4).

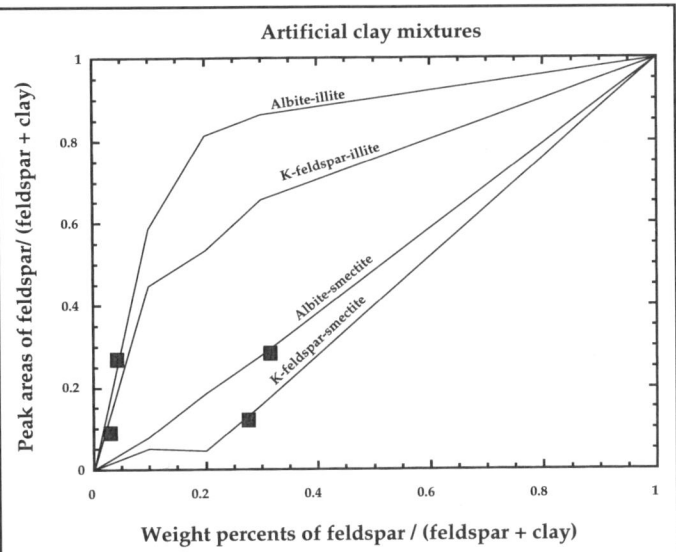

Figure 7. Plot of ratios of XRD peak areas versus ratio of weight percents of feldspars to clay minerals for known mixtures of illite and smectite with albite and K-feldspar. Note nonlinearity of relations. Solid squares represent the mean peak-area ratios for channel-sample sediments from core OL-92, whose mineral weight abundances we want to determine. Boxes are plotted so they fall on their corresponding peak-area ratio versus weight-percent ratio lines (derived from artificial mixtures), and at their measured peak-area ratio values. Weight percent of feldspar to clay can be read off the graph. These peak area ratios suggest that no more than ~30% of any channel-sample sediment consists of nonclay minerals in the <4-μm fraction.

Because feldspars may indicate immature, mechanically produced sediments, their correlation with illite and chlorite (although probably not kaolinite) may indicate a mechanical origin for these clays. In addition, samples having well-defined feldspar peaks always exhibit large illite peaks, while samples displaying weaker and fewer feldspar peaks show large smectite peaks. This peak distribution suggests that the illite is a low-grade alteration product of igneous muscovite or biotite and that the illite-plus-feldspar XRD end member represents glacial flour. The poor correlation of smectite/quartz to illite/quartz, K-feldspar/quartz, or plagioclase/quartz ratios (Fig. 4) suggests that smectite formed by a different process. Furthermore, the smectite/quartz ratio correlates negatively with the plagioclase/K-feldspar ratio. Because plagioclase is less resistant to chemical weathering than K-feldspar is (Loughnan, 1969), plagioclase/K-feldspar ratios should be lower in chemically weathered sediments than in those sediments unaltered by chemical processes. Because smectite is a chemical-weathering product, ratios of smectite/quartz should be high in those sediments that were chemically altered and low in those sediments that were unaltered. Therefore, smectite/quartz and plagioclase/K-feldspar ratios should, and do, behave inversely (Fig. 5).

An alternative explanation for the variation in sedimentation that is predominantly smectite or illite is that smectite may have formed authigenically in Owens Lake during periods of high salinity during low lake levels (interglacial periods). The curve of percent smectite versus depth (Fig. 8; derived by using the multiplicative factors of Hallberg et al., 1978) closely follows the abundance of $CaCO_3$ (Bischoff et al., 1993a; this volume, chapter 4), suggesting an authigenic origin for much of the smectite in the core. Bischoff et al. (1993a; this volume, chapter 4) also report the existence of a potentially authigenic acid-soluble Mg silicate whose abundance correlates with that of $CaCO_3$. The mineral that I describe as smectite may include or be this acid-soluble Mg silicate.

In addition to authigenic clay minerals, authigenic K-feldspar and zeolites are sometimes formed in saline lakes. Hay et al. (1991) found authigenic K-feldspar in Searles Lake sediments, and Banfield et al. (1991) found it in Abert Lake, Oregon. Even if Owens Lake smectite and K-feldspar are largely authigenic, the smectite/quartz and plagioclase/K-feldspar ratio curves would still vary inversely. In this instance, the varying smectite/quartz and plagioclase/K-feldspar values would result not from the differential chemical weathering of plagioclase and K-feldspar but from the periodic precipitation of K-feldspar and smectite in an authigenic setting. Presently, I cannot rule out either a detrital or authigenic origin for clay minerals and K-feldspars in the core, and my attempts at differentiating between detrital and authigenic K-feldspar by XRD have proven unsuccessful. However, evidence for the high salinities necessary for authigenic K-feldspar precipitation is scant. The only evidence for high salinities is the evaporite minerals and oolites found between 1-m and 5-m depth. No other evaporite minerals exist in the core, and evidence of dissolution of evaporites is lacking.

No zeolites were found in the clay-size fraction. These minerals are frequently found in saline lakes (Surdam, 1977) and in association with tephra layers that have been altered by saline brines. The absence of zeolites in the clay-size fraction of the Owens Lake core may indicate that authigenic mineral formation is not a significant process in the lake. However, Hay (1966) found four different zeolites in the silt-size fraction of sediments from a core of Owens Lake drilled in the 1950s, indicating that I may not have been looking for these minerals in the proper size fraction.

The initial hypothesis that the mineralogy of the <4-μm (clay) fraction would reflect climate appears confirmed. Clay-size particles are composed of varying proportions of rock flour and pedogenic or authigenic clays; most of the mineralogic variation occurs in the smectite and illite components of the sediments. CEC measurements confirm this mineralogic signal with values consistent with a variation in illite and smectite (Fig. 6).

COMPARISON OF GRAIN SIZE AND CLAY-MINERAL VARIATIONS TO OTHER PALEOCLIMATE PROXIES

Bulk grain size and clay mineralogy correlate to, and reflect, other paleoclimatic proxies in the Owens Lake core (Fig. 8). Glacial periods, as represented by a vigorously overflowing

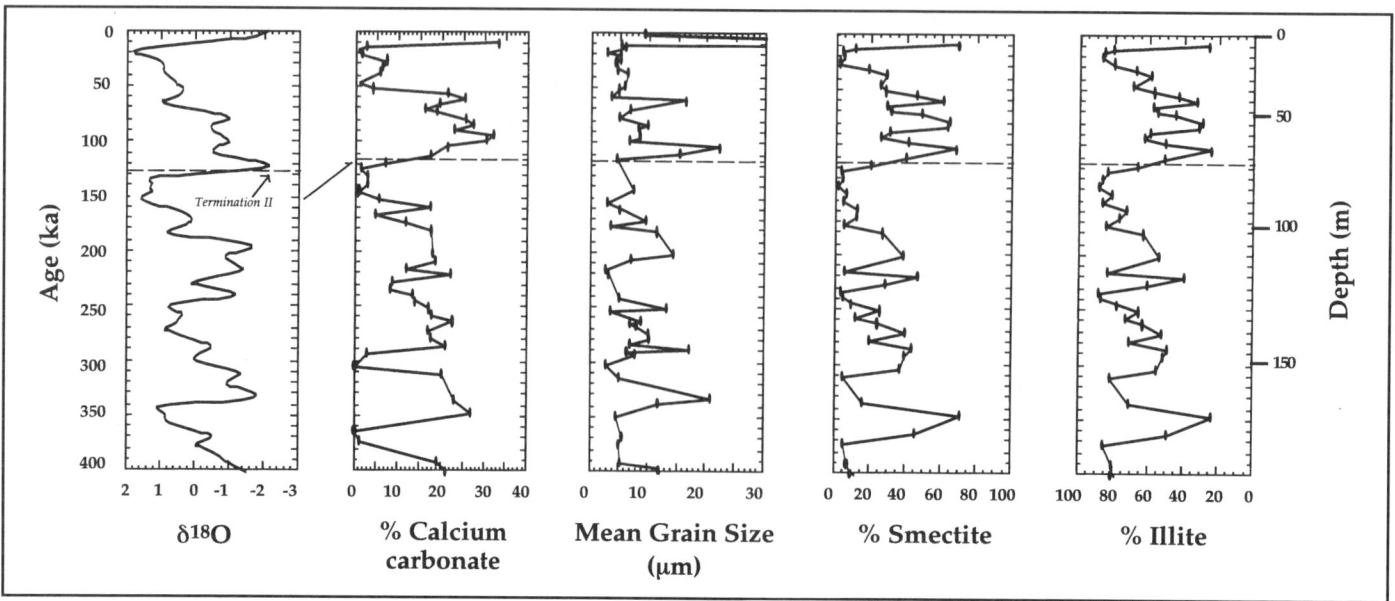

Figure 8. Comparison of deep-sea sediment $\delta^{18}O$ record and four different Owens Lake paleoclimatic proxies from core OL-92, plotted against age (derived using the age-depth model of Bischoff et al., this volume). The $\delta^{18}O$ record of deep sea sediments is from Imbrie et al. (1984), calcium carbonate content is from Bischoff et al. (1993a; this volume, chapter 4), curves of mean grain size, smectite, and illite (plotted inversely) are from this report. Note the strong similarity of Owens Lake proxies to oxygen isotopic record of the deep sea. Also note Termination II offset between the marine and Owens Lake records. This offset may result from an error in the age-depth curve determined for the OL-92 core, or it may represent an actual time lag between global and local glaciations.

Owens Lake, can be identified by combinations of the following criteria: fine mean grain size (~5 μm); a dominance of illite, feldspar, and quartz over smectite in the clay-size fraction; very low CEC values that probably indicate rock flour; as well as low carbonate and organic carbon contents that indicate a very fresh and low-productivity lake (Bischoff et al., 1993a; this volume, chapter 4); abundant fresh-water planktonic diatoms (Bradbury, 1993; this volume), and by a decrease in pine pollen (Litwin et al., 1993; this volume). Interglacial periods as represented by a less vigorously overflowing or nonoverflowing lake are represented by coarser mean grain sizes (~15 μm) and clays dominated by smectite, by high CEC values plus high carbonate and organic carbon contents (Bischoff et al., 1993a; this volume, chapter 4), by a greater frequency of saline diatoms (Bradbury, 1993; this volume) and ostracodes (Carter, 1993; this volume), and by increases in pine pollen (Litwin et al., 1993; this volume).

Both carbonate content and mean grain size appear to reflect variations in lake level. For example, the last glacial maximum (~13 m, ca. 21 ka) is marked in the core by very low carbonate contents, which indicate a very fresh overflowing Owens Lake (Bischoff et al., 1993a; this volume, chapter 4). The dark muds deposited during the onset of the Holocene interglacial period, on the other hand, show an increase in carbonate content which indicates a shallower lake in which the water eventually reached saturation, depositing carbonate oolites. Mean grain size also appears to reflect variations in lake level; coarse-grained materials were deposited during times of lake lowstands, when the shores of Owens Lake were closer to our core site, and fine-grained sediments were deposited during highstands, when the shoreline was farther from the core site.

Clay mineralogical variations do not indicate changes in lake level but reflect changes in the weathering environment in the Sierra Nevada or changes in salinity in Owens Lake. Comparison of the OL-92 record with the marine oxygen isotope record of Imbrie et al. (1984), using the age-versus-depth model (Bischoff, 1993; this volume, chapter 8) and the carbonate content (Bischoff et al., this volume, chapter 4), the mean grain size, and the smectite and illite contents of core OL-92 sediments suggests that the climatic conditions that led to glaciation in the Sierra Nevada and to lake-level changes in Owens Lake are the same as those that led to global climate cycles (Fig. 8). However, the age control on the Owens Lake core implies an average age difference of about 15 k.y. between the marine and terrestrial records (Smith et al., this volume). For example, the deglaciation from isotope stage 6 to stage 5, known as "Termination II," occurs in the marine record at ca. 128 ka; the same feature is found in the Owens Lake core at ca. 120 ka. This time lag may be the result of an error in dating or may represent an actual age difference between high-latitude ice-cap growth and low-latitude montane glacial response to global climatic change.

CONCLUSIONS

1. Variations in grain size of the sediments between ~7–200 m in core OL-92 reflect variations in depth of Owens Lake. Fine sediments (average grain size ~5 µm) were deposited at the core site during periods of high lake level, when the shores and wave base of Owens Lake were far from the site; coarse sediments (average grain size ~15 µm) were deposited during lower lake stands, when the shores of Owens Lake may have approached the core site. However, variations in grain size indicate only relative variations in lake level rather than absolute lake depths.

2. Alternations in clay mineralogy reflect cyclic glaciation in the Sierra Nevada. The clay-size mineral assemblage dominated by illite, K-feldspar, and plagioclase most likely indicates rock flour produced during glacial periods. The assemblage dominated by smectite may represent either chemical weathering processes on hillslopes active during interglacial periods or authigenic phyllosilicate formation within a shallow, saline Owens Lake.

ACKNOWLEDGMENTS

This work was supported by a National Defense Science and Engineering Graduate Fellowship. Hannah Musler and Jeff Fitts provided many hours of laboratory help, and Mark Johnsson and Mike Torresan provided much technical expertise. Discussions with Robert Anderson, James Bischoff, Jonathan Glen, and George Smith were very helpful. Many thanks to James Bischoff, Richard Hay, George Smith, and Joseph Smoot whose reviews improved the manuscript.

REFERENCES CITED

Banfield, J. F., Jones, B. F., and Veblen, D. R., 1991, An AEM–TEM study of weathering and diagenesis, Abert Lake, Oregon: II. Diagenetic modification of the sedimentary assemblage: Geochimica et Cosmochimica Acta, v. 55, p. 2795–2810.

Birman, J. H., 1964, Glacial geology across the crest of the Sierra Nevada, California: Geological Society of America Special Paper 75, 80 p.

Biscaye, P., 1965, Mineralogy and sedimentation of recent deep-sea clay in the Atlantic Ocean and adjacent seas and oceans: Geological Society of America Bulletin, v. 76, p. 803–832.

Bischoff, J. L., 1993, Age-depth relations for the sediment column at Owens Lake, California: OL-92 drill hole: U.S. Geological Survey Open-File Report 93-683, p. 251–260.

Bischoff, J. L., Fitts, J. P., Fitzpatrick, J. A., Menking, K. M., and King, B. W., 1993a, Geochemistry of sediments in Owens Lake drill hole OL-92: U.S. Geological Survey Open-File Report 93-683, p. 83–99.

Bischoff, J. L., Stafford, T. W., Jr., and Rubin, M., 1993b, AMS radiocarbon dates on sediments from Owens Lake Drill Hole OL-92: U.S. Geological Survey Open-File Report 93-683, p. 246–250.

Blackwelder, E., 1931, Pleistocene glaciation in the Sierra Nevada and basin ranges: Geological Society of America Bulletin, v. 42, p. 865–922.

Boggs, S., Jr., 1987, Principles of sedimentology and stratigraphy: New York, Macmillan Publishing Company, 784 p.

Bradbury, J. P., 1993, Diatoms in sediments: U.S. Geological Survey Open-File Report 93-683, p. 261–302.

Carter, C., 1993, Ostracodes present in sediments: U.S. Geological Survey Open-File Report 93-683, p. 303–306.

Chamley, H., 1989, Clay sedimentology: New York, Springer-Verlag, p. 7, 13, 58, 70.

Droste, J. B., 1961a, Clay minerals in the playa sediments of the Mojave Desert, California: California Division of Mines and Geology Special Report 69, 21 p.

Droste, J. B., 1961b, Clay minerals in sediments of Owens, China, Searles, Panamint, Bristol, Cadiz, and Danby Lake basins, California: Geological Society of America Bulletin, v. 72, p. 1713–1722.

Feth, J. H., Roberson, C. E., and Polzer, W. L., 1964, Sources of mineral constituents in water from granitic rocks, Sierra Nevada, California and Nevada: U.S. Geological Survey Water-Supply Paper 1535-I, 70 p.

Friedman, I., Johnson, C. A., and Fitts, J. P., 1993, Deuterium-hydrogen ratios of interstitial fluids from Owens Lake core OL-92: U.S. Geological Survey Open-File Report 93-683, p. 110–118.

Folk, R. L., 1968, Petrology of sedimentary rocks: Austin, Texas, Hemphill's, 170 p.

Folk, R. L., and Ward, W. C., 1957, Brazos River bar (Texas)—A study in the significance of grain size parameters: Journal of Sedimentary Petrology, v. 27, n. 1, p. 3–26.

Galehouse, J. S., 1971, Sedimentation analysis, in Carver, R. E., ed., Procedures in sedimentary petrology: New York, Wiley-Interscience, p. 69–94.

Gillespie, A. R., 1982, Quaternary glaciation and tectonism in the southeastern Sierra Nevada, Inyo County, California [Ph.D. thesis]: California Institute of Technology, 695 p.

Glen, J. M., Coe, R. S., Menking, K., Boughn, S. S., and Altschul, I., 1993, Rock- and paleo-magnetic results from core OL-92, Owens Lake, CA: U.S. Geological Survey Open-File Report 93-683, p. 127–183.

Hallberg, G. R., Lucas, J. R., and Goodmen, C. M., 1978, Semi-quantitative analysis of clay mineralogy, in Hallberg, G. R., ed., Standard procedures for evaluation of Quaternary materials in Iowa: Iowa Geological Survey Technical Information Series No. 8, p. 5–21.

Hay, R. L., 1966, Zeolites and zeolitic reactions in sedimentary rocks: Geological Society of America Special Paper 85, 130 p.

Hay, R. L., and Moiola, R. J., 1963, Authigenic silicate minerals in Searles Lake, California: Sedimentology, v. 2, p. 312–332.

Hay, R. L., Guldman, S. G., Matthews, J. C., Lander, R. H., Duffin, M. E., and Kyser, T. K., 1991, Clay mineral diagenesis in core KM-3 of Searles Lake, California: Clays and Clay Minerals, v. 39, p. 84–96.

Imbrie, J., Hays, J. D., Martinson, D. G., McIntyre, A., Mix, A. C., Morley, J. J., Pisias, N. G., Prell, W. L., and Shackleton, N. J., 1984, The orbital theory of Pleistocene climate: Support from a revised chronology of the marine $\delta^{18}O$ record, in Berger, A. L., et al., eds., Milankovitch and Climate, Part 1: Hingham, Massachusetts, D. Riedel, p. 269–305.

Litwin, R. J., Frederiksen, N. O., Adam, D. P., Andrle, V. A. S., and Sheehan, T. P., 1993, Continental-marine correlation of late Pleistocene climate change: Census of palynomorphs from core OL-92, Owens Lake, California: U.S. Geological Survey Open-File Report 93-683, p. 333–391.

Loughnan, F. C., 1969, Chemical weathering of the silicate minerals: New York, American Elsevier Publishing Company, Inc., 154 p.

Menking, K. M., Musler, H. M., Fitts, J. P., Bischoff, J. L., and Anderson, R. S., 1993a, Clay mineralogical analyses of the Owens Lake core: U.S. Geological Survey Open-File Report 93-683, p. 75–82.

Menking, K. M., Musler, H. M., Fitts, J. P., Bischoff, J. L., and Anderson, R. S., 1993b, Sediment size analyses of the Owens Lake core: U.S. Geological Survey Open-File Report 93-683, p. 58–74.

Moore, D. M., and Reynolds, R. C., Jr., 1989, X-ray diffraction and the identification and analysis of clay minerals: New York, Oxford University Press, 332 p.

Newton, M., 1991, Holocene stratigraphy and magnetostratigraphy of Owens and Mono Lakes, eastern California [Ph.D. thesis]: Los Angeles, University of Southern California, 330 p.

Phillips, F. M., Zreda, M. G., Smith, S. S., Elmore, D., Kubik, P. W., and Sharma, P., 1990, Cosmogenic chlorine-36 chronology for glacial deposits at Bloody Canyon, eastern Sierra Nevada: Science, v. 248, p. 1529–32.

Picard, M. D., and High, L. R., Jr., 1972, Criteria for recognizing lacustrine rocks, in Rigby, J. K., and Hamblin, W. K., eds., Recognition of ancient sedimentary environments: Society of Economic Paleontologists and Mineralogists Special Publication 16, p. 108–145.

Picard, M. D., and High, L. R., Jr., 1981, Physical stratigraphy of ancient lacustrine deposits, in Ethridge, F. G., and Flores, R. M., eds., Recent and nonmarine depositional environments: Models for exploration: Society of Economic Paleontologists and Mineralogists Special Publication 31, p. 233–259.

Sarmiento, R., and Kirby, R. A., 1962, Recent sediments of Lake Maracaibo: Journal of Sedimentary Petrology, v. 32, no. 4, p. 698–724.

Sarna-Wojcicki, A. M., Meyer, C. E., Wan, E., and Soles, S., 1993, Age and correlation of tephra layers in Owens Lake drill core OL-92-1 and -2: U.S. Geological Survey Open-File Report 93-683, p. 184–245.

Sharp, R. P., 1972, Pleistocene glaciation, Bridgeport Basin, California: Geological Society of America Bulletin, v. 83, p. 2233–2260.

Smith, G. I., 1984, Paleohydrologic regimes in the southwestern Great Basin, 0–3.2 my ago, compared with other long records of "global" climate: Quaternary Research, v. 22, p. 1–17.

Smith, G. I., 1993, Field log of Core OL-92: U.S. Geological Survey Open-File Report 93-683, p. 4–57.

Surdam, R. C., 1977, Zeolites in closed hydrologic systems, in Mumpton, F. A., ed., Mineralogy and geology of natural zeolites: Mineralogical Society of America Short Course Notes, v. 4, p. 65–91.

Torresan, M. E., 1987, A review and comparison of the hydrophotometer and pipette methods in the analysis of fine-grained sediment: U.S. Geological Survey Open-File Report 87-514, 38 p.

Waitt, R. B., Jr., 1980, About forty last-glacial Lake Missoula jökulhlaups through southern Washington: Journal of Geology, v. 88, p. 653–679.

Waitt, R. B., Jr., 1985, Case for periodic, colossal jökulhlaups from Pleistocene glacial Lake Missoula: Geological Society of America Bulletin, v. 96, p. 1271–1286.

Wentworth, C. K., 1922, A scale of grade and class terms for clastic sediments: Journal of Geology, v. 30, no. 5, p. 377–392.

MANUSCRIPT ACCEPTED BY THE SOCIETY JUNE 17, 1996

Responses of sediment geochemistry to climate change in Owens Lake sediment: An 800-k.y. record of saline/fresh cycles in core OL-92

James L. Bischoff, Jeffrey P. Fitts,* and John A. Fitzpatrick
U.S. Geological Survey, 345 Middlefield Road, MS 910, Menlo Park, California 94025

ABSTRACT

Geochemical parameters of sediments from drill hole OL-92 indicate that Owens Lake was saline, alkaline, and highly productive during interglacial periods, and was hydrologically open and relatively unproductive during glacial periods. Abundance of $CaCO_3$, organic carbon, and cation-exchange capacity of the clay fraction show cyclic variation down the core. Six minima in these components during the past 500 k.y. are interpreted as caused by intensive overflow that occurred during Sierran glacial advances. Maxima in these components indicate closed-lake conditions, reflecting warmer and more arid interglacial climates. The pattern of $CaCO_3$ abundance suggests that closed lake conditions predominated over the past 500 k.y. The absence of gaylussite and gypsum in the sediments, however, indicates lake salinity never exceeded about 15 wt %, a limit which requires flushing of accumulated salts every 10 k.y.

Oscillations of $CaCO_3$ generally indicate a 100-k.y. dominant cycle, a characteristic of the marine $\delta^{18}O$ record. Four of the last five marine isotope terminations are clearly shown in the Owens Lake record. The last interglacial at Owens Lake appears to have occurred between 120 ka and 50 ka. The roughly 10-k.y. offset between this interval and marine oxygen-isotope stage 5 reflects either error in the age-depth model, or alternatively, a time lag between changes in Northern Hemisphere ice volumes and the manifestation of local climate change in lake geochemistry and sedimentology.

INTRODUCTION

Prior to its desiccation in the decade following 1913, a result of source-water diversion by the City of Los Angeles, Owens Lake was historically below its spill point and was the terminus of the Owens River. As a consequence, the lake was saline and alkaline (Gale, 1914). Saline conditions likely characterized the second half of the arid Holocene and, presumably, the earlier interglacials as well. During earlier wet periods that are thought to be coeval with the periodic Sierran glacial advances, however, down-stream evidence at Searles Lake (Bischoff et al., 1985; Smith et al., 1983; Smith, 1991) shows that Owens Lake had been intensely overflowing, and therefore, must have been flushed with fresh water.

Owens Lake core OL-92 was located near the depocenter of the Owens Lake basin in order to obtain a complete record of the alternating closed and overflowing cycles in a stratigraphic and chronological framework. The postulate of the present study is that the sediment composition should reflect the cycling between relatively wet and dry periods.

---

*Present address: Department of Geological and Environmental Sciences, Stanford University, Stanford, California 94305.

Bischoff, J. L., Fitts, J. P., and Fitzpatrick, J. A., 1997, Responses of sediment geochemistry to climate change in Owens Lake sediment: An 800-k.y. record of saline/fresh cycles in core OL-92, *in* Smith, G. I., and Bischoff, J. L., eds., An 800,000-Year Paleoclimatic Record from Core OL-92, Owens Lake, Southeast California: Boulder, Colorado, Geological Society of America Special Paper 317.

Sediments from Owens Lake drill hole OL-92 were analyzed for major oxides and minor elements, acid-leachable cations, total organic carbon (TOC), carbonate (CO_3), and cation-exchange capacity (CEC). Results show cyclic variation of sediment composition down the core, reflecting alternating global glacial and interglacial conditions that correspond, in general, to the marine isotope cycles. Details of sampling, analytical methods, along with complete data in depth format are given in the data depository of the Owens Lake Drilling Project (Bischoff et al., 1993a). Sediment data are given chronological format in this presentation using the time-depth relations for core OL-92 presented elsewhere in this volume (Bischoff et al., this volume, Chapter 8).

SAMPLING

Both channel and point samples were taken from the drill core. Channel samples, which are composite strip samples, were taken in a continuous series to represent the entire recovered sedimentary column without gaps (91 samples). This methodology avoids the bias of single point samples which may not represent the entire sedimentary unit from which they were taken (for point-sample studies of core OL-92 see Bischoff et al., 1993b, and Menking et al., 1993b). The advantage of the channel samples is that each represents a smoothed or running mean of conditions represented by the time span of the sample. The advantages are that no important events are missed (except for gaps in core recovery) and that geochemical budget calculations can be carried out. The time resolution of such samples, however, is inverse to the thickness of the section sampled. In the present study, we took strip samples approximately 3 m long, each deemed to represent about 7,000 years of deposition. Samples were taken from the working half of the core with a U-shaped spatula, resulting in a continuous semi-cylindrical strip about 1.5 cm wide, 1 cm deep, and about 3 m long. The point samples were taken during the drilling operations at every 2 or 3 m (120 samples). These samples were specially preserved and used for determination of water content and pore water chemistry as reported in the accompanying report (Bischoff et al., 1993c; Friedman et al., this volume). These samples were also analyzed for organic carbon and for carbonate content (but not the other components). These results are reported here to supplement similar analyses of the channel samples.

LABORATORY PROCEDURES

Initial processing and splits

Samples were rinsed in distilled water, dried at 60 °C and mechanically homogenized. One aliquot was split for cesium (Cs) treatment and major oxide analysis, and the remaining sample was used for determination of total organic carbon (TOC) and carbonate (CO_3), x-ray diffraction (XRD) determination of carbonate minerals, and acid-leachable Mg and Ca.

Cesium-ion treatment and major elements

An aqueous CsCl solution was used to displace all exchangeable cations in the sample with Cs ions (following Beetem et al., 1962). Thus, the amount of Cs taken up by the sample is a measure of its cation-exchange capacity (CEC). The CEC, in turn, should be a measure of the relative abundance of weathering-zone clay minerals such as smectite and a measure of warm weathering conditions. A sample split was suspended and stirred in 0.09 molal CsCl solution for 24 hours, after which it was collected on filter paper and rinsed with distilled water. The sample was then dried at 60 °C and analyzed for Cs and the major rock-forming oxides SiO_2, Al_2O_3, Fe_2O_3, MgO, CaO, Na_2O, K_2O, TiO_2, P_2O_5, and MnO by X-ray fluorescence (XRF) spectrometry. A split of each channel sample was also analyzed for minor elements by semiquantitative optical emission spectroscopy.

Other properties

Organic carbon and carbonate were analyzed from the bulk sample by standard coulometry which successively measures the carbonate as CO_2 released by strong acid attack and then the total carbon in the sample (Engleman et al., 1985). Splits of bulk samples were leached in 3 molar HCl overnight for analysis of acid-leachable Ca and Mg by standard atomic-absorption spectroscopy. Standard XRD scans of powder mounts were performed on a selection of 36 carbonate-rich samples.

RESULTS

Bulk composition

Analytical results for individual samples are given in Bischoff et al. (1993a). Contents of CO_3 (7.5% average) and TOC (0.92% average) vary widely as summarized in Table 1. The composition of the acid-insoluble residue, however, is remarkably homogeneous. The average bulk composition, normalized after removing carbonate, organic carbon, and acid-soluble Ca and Mg (Table 2), is similar in all major oxide components to granodiorite, the predominant rock of the Sierra Nevada batholith (Bateman et al., 1963). The minor compo-

TABLE 1. AVERAGED CHARACTERISTICS OF SEDIMENTS FROM OWENS LAKE OL-92 DRILL HOLE*

Carbonate (CO_3)	7.5 ± 6.3%
Organic carbon (TOC)	0.92 ± 0.87%
CEC (cfb)†	32.7 ± 6.8 meq/100g
Grain density	2.63 ± 0.05 g/cc

*Values are reported as mean ± 1s. Data averaged from tabulated results presented in Bischoff et al., 1993a.
†Cation exchange capacity for clay-size fraction on a carbonate-free basis.

TABLE 2. AVERAGED COMPOSITION OF SEDIMENT FROM OWENS LAKE OL-92 DRILL HOLE, ON AN ACID-INSOLUBLE BASIS (CARBONATE-FREE), COMPARED TO GRANODIORITES AND AVERAGE SHALE*

	Owens Lake Sediment	GSP-1[†]	Lamarck Granodiorite[§]	Average Shale[**]
(wt %)				
SiO_2	68.60	67.38	66.92	58.1
Al_2O_3	14.77	15.25	15.19	15.4
Fe_2O_3t	5.05	4.32	5.05	7.52
CaO	2.92	2.02	3.79	3.11
MgO	2.41	0.96	1.74	2.44
Na_2O	2.67	2.80	3.16	1.3
K_2O	3.53	5.53	3.82	3.24
TiO_2	0.62	0.66	0.47	0.65
P_2O_5	0.26	0.28	0.18	0.17
MnO	0.13	0.04	0.08	0.06
Total	99.96	99.24	100.40	100.00
(ppm)				
B	61	<3		310
Ba	980	1,300		460
Be	1.3	1.5		<4
Co	14	6		8
Cr	31	13		500
Cu	26	33		192
Ga	31	22		50
Mo	4	1		
Ni	21	13		24
Pb	22	51		20
Sc	7	7		7
V	75	53		120
Y	14	30		28
Zr	56	500		120

*Data averaged from tabulated results reported by Bischoff et al., 1993a. Ignition loss and acid-leachable CaO and MgO were subtracted from bulk sediment composition and the difference normalized to 100%.
[†]GSP-1 is a granodiorite collected near Silver Plume, Colorado (Flanagan, 1976).
[§]Lamarck Granodiorite was collected from east central Sierra Nevada, California (Bateman et al., 1963).
[**]Oxides from Clark, 1924, minor elements from Rankama and Sahama, 1950.

nents of Owens Lake sediment (Table 2) show the same strong granodiorite affinity and a contrast to average shale with the single exception of Zr (56 ppm versus 500 in granodiorite and 120 ppm in shale). Granodioritic Zr is contained primarily in zircon which is transported with and concentrated in the sand fraction, and therefore is relatively depleted in the dominantly silt and clay sediment found in the lake basin. Triangular diagrams of all the samples show the general affinity to granodiorite, but with small scale variability attributable to grain-size variations. The Al_2O_3-Na_2O-K_2O triangular plot (Fig. 1A) shows a grouping of points toward the Al_2O_3-rich side but close to granodiorite. The Fe_2O_3-Na_2O-K_2O triangular plot (Fig. 1B) shows an elongate pattern with respect to the Fe_2O_3 apex with the Lamarck Granodiorite at the center of the trend. Samples relatively enriched in Fe_2O_3, tending toward average shale, are clay-rich samples, whereas those in the opposite direction are rich in arkosic sand. The Fe_2O_3-SiO_2-Al_2O_3 (truncated) triangular plot (Fig. 1C) shows an elongate trend with respect to the SiO_2 apex with granodiorite at the midpoint. The SiO_2-depleted side trends toward average shale and represents more clay-rich samples, whereas the SiO_2-enriched side represent sandy units. The bifurcation of the trend on the SiO_2-enriched side distinguishes between quartz sands and arkosic sands.

Variation of carbonate and organic carbon fractions

The CO_3 and TOC co-vary down core (Fig. 2). It is remarkable that both point samples (representing 40 years of sediment accumulation) and channel samples (representing 7,000 years of sediment accumulation) show almost coincident patterns, indicating that even on the small scale of the point samples the sediments are representative of larger thicknesses of sediment. This observation points to slowly changing (millennial scale) homogeneous depositional conditions. The CO_3 and TOC display sharp minima, very close to zero values during the most recent glacial maximum that occurred between 25 ka and 17 ka. These results suggest that the residence time of water in the lake was short, preventing the buildup of dissolved salts sufficient for $CaCO_3$ precipitation and high biologic productivity. Five similar minima in these parameters occur between 25 ka and 500 ka are interpreted as successively older glacial maxima. Conversely, the recurring and broader maxima in CO_3, and TOC are interpreted to represent full interglacial conditions during which the lake was the terminus and was saline and biologically highly productive. Before 500 ka (below 230 m), there is a striking change in depositional conditions from silty clays to thick sandy units as described in the accompanying reports (Smith, this volume; Smith et al., this volume; Menking, this volume), which may signal irregular fluctuations from lacustrine to nonlacustrine conditions for this lower part of the core. For the section representing the most recent 500 k.y., however, lacustrine conditions apparently prevailed.

Composition of the acid-soluble fraction

Results of XRD indicate that calcite is the dominant carbonate mineral, with detectable dolomite occurring in about a third of the samples, and aragonite in only two (Bischoff et al., 1993a). Minerals characteristic of higher salinity and/or playa conditions, such as gypsum and gaylussite, were sought but not detected. Gaylussite, in particular, would be the first mineral to form (after calcite) at increasing salinity. Even if later leached by low-salinity waters, gaylussite would leave distinctive pseudomorphs composed of $CaCO_3$ (see below, and Bischoff et al., 1991). The lack of gaylussite or its pseudomorphs suggest the lake had not attained the required salinity throughout the past 800 k.y. The rel-

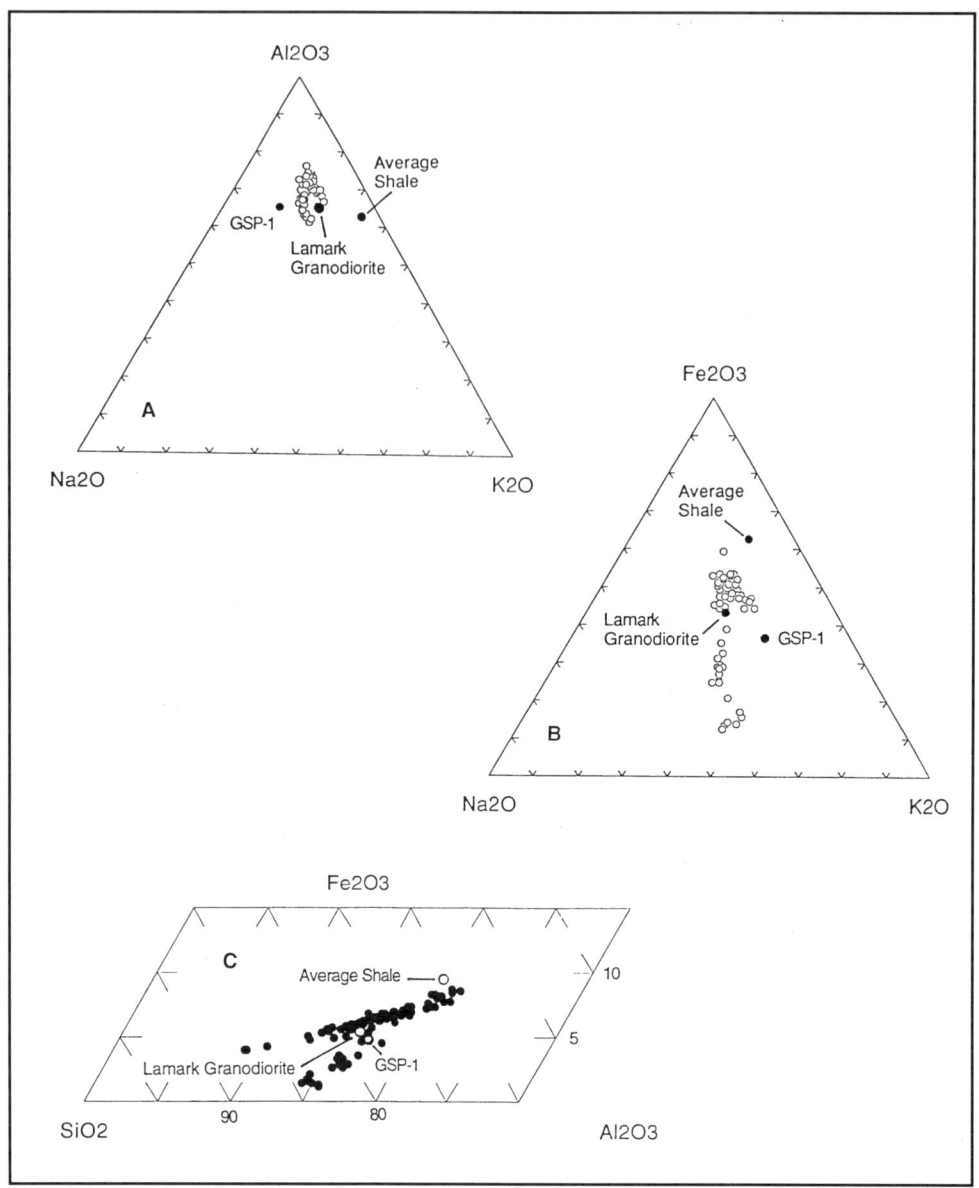

Figure 1. Triangular diagrams showing major oxide composition of sediments from Owens Lake core OL-92 compared to average shale and granodiorite (GSP-1 and Lamarck Granodiorite, see Table 2). (A) Al_2O_3-Na_2O-K_2O. (B) Fe_2O_3-Na_2O-K_2O. (C) Truncated triangle (lower left corner) of Fe_2O_3-SiO_2-Al_2O_3. Diagrams show the general similarity of Owens Lake sediment to granodiorite, but with small-scale variability attributable to sediment-size fractionation.

ative amounts of Ca and Mg carbonates, calculated by balancing the analyzed acid-leachable Ca and Mg against analyzed CO_3 (Bischoff et al., 1993a), indicate an average of 95 mole % $CaCO_3$ and 5 mole % $MgCO_3$, for the total carbonate fraction. This suggests that 86% of the total acid-leachable Mg is actually noncarbonate (Fig. 3), which we postulate to be authigenic Mg hydroxysilicates, varyingly crystalline and amorphous, and including such phases as sepiolite, kerolite, and stevensite (Jones, 1986). Such acid-soluble authigenic phases form in saline lakes by reaction of dissolved Mg and silica in alkaline solution, both reacting directly with each other and/or reacting with preexisting clastic phyllosilicates that were either suspended in the water column or at the sediment-water interface. In general, it is not possible to distinguish detrital and authigenic phyllosilicates in mixtures by XRD. Thus, calculating averages from data given in Bischoff et al. (1993a) on a weight basis indicates that 9% of the total bulk Mg is carbonate, 57% is acid-soluble authigenic silicate, and 34% is in the nonleachable clastic component. Figure 3

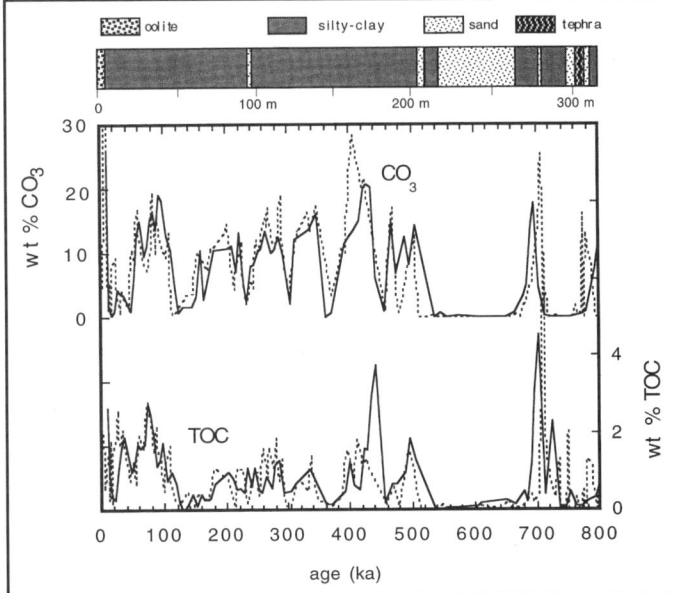

Figure 2. Variation of carbonate (CO$_3$) and total organic carbon (TOC) with depth and age of sediments from Owens Lake drill hole OL-92. Solid lines are from channel samples, dotted lines from point samples. Lithology generalized from core log presented in Smith (this volume). Lacustrine conditions prevailed back to 500 ka (220 m) below which fluvial and lacustrine conditions may have alternated. Both point and channel samples show almost coincident patterns. Values of CO$_3$ and TOC are very close to zero during the most recent glacial maximum at 17–25 ka suggesting the lake was intensely overflowing during glacial advances.

shows that abundance of both carbonate Mg and authigenic Mg silicate follows that for total carbonate, and that both Mg phases are likely indicators of saline and alkaline conditions.

Cation-exchange capacity (CEC)

The Cs content of a Cs-exchanged sample is a function of the amount of clay in the sample, and of the CEC of the clay. The CEC of the clay fraction was obtained by normalizing the Cs content (carbonate-free basis) to the weight fraction of the <2µ component of the sample (reported in Menking et al., 1993a). The clay-normalized CEC is seen to co-vary with the relative abundance of smectite in the clay fraction (Fig. 4), an independent confirmation of its validity. The CEC is in phase with the carbonate cycles (Fig. 2) back to 500 ka. As does CO$_3$, CEC indicates a minimum coinciding with the last glacial maximum (25–17 ka), and other minima occur successively deeper in the core coinciding with carbonate minima. This correlation suggests that during glacial maxima the clay-size material has a low exchange capacity, perhaps representing a glacial rock-flour component. Beyond 500 ka (~230 m), the normalized CEC pattern is de-coupled from the carbonate pattern. If CEC is a reflection of drainage-basin conditions rather than basin-deposition conditions, then CEC cycles might reflect climatic cycles even though the depositional basin is alternating between lacustrine and nonlacustrine conditions. The average CEC of clay material in the core is 32.7 meq/100 g (Table 1), which compares to a

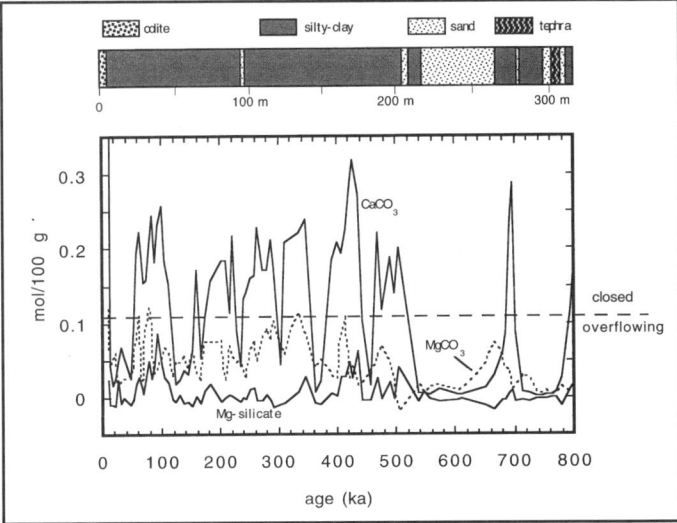

Figure 3. Variation in relative amounts of CaCO$_3$, MgCO$_3$, and acid-soluble Mg-silicate component with depth and age of sediments from Owens Lake drill hole OL-92. Lithology generalized from core log presented in Smith (this volume). Abundance of both carbonate Mg and authigenic Mg silicate follows that for total carbonate. High contents of these phases are likely indicators of elevated salinity caused by terminal lake conditions. Peaks in the abundance of these phases mark interglacial conditions. Based on Ca budget calculations (see text) sediments with less than about 0.11 mol/100 g CaCO$_3$ represent overflowing conditions, while those with more CaCO$_3$ represent closed lake conditions.

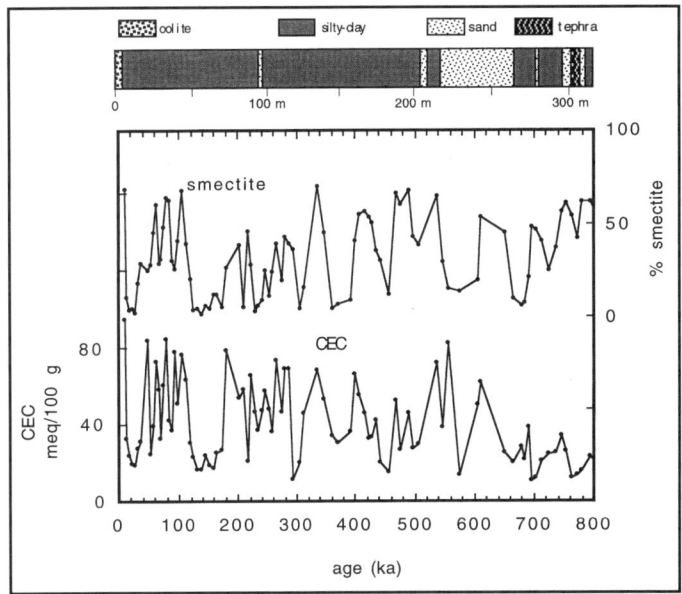

Figure 4. Co-variation of cation exchange capacity (CEC in meq/100 g) of clay fraction (measured by bulk Cs uptake divided by wt % clay size) and smectite abundance in clay fraction with age and depth in sediments from Owens Lake drill hole OL-92. Sediment-size data and smectite abundance taken from Menking et al. (1993b) and Menking et al. (1993a). Lithology generalized from core log presented in Smith (this volume). The CEC co-varies with smectite abundance, and both reflect interglacial conditions.

range of 80–150 meq/100 g for pure smectites and to about 10–40 meq/100 g for pure illite (Grim, 1968). Menking et al. (this volume) report that illite and smectite are the dominant clay minerals in the Owens sediment. During the interglacials, CEC reaches values within the pure smectite range, whereas during the glacials CEC is within the range of pure illite.

DISCUSSION

Evolution of closed lake salinity

Closed lake conditions give rise to sediments rich in Ca-Mg carbonates, TOC, authigenic Mg silicates, and high CEC smectites, in contrast to overflowing lake conditions where these components approach zero values. Because minerals characteristic of high salinity and/or playa conditions, such as gypsum and gaylussite, were not detected suggests the lake had not attained this salinity at any time during the past 800 ka. If closed conditions prevailed for sufficient time, even with the lake level immediately below the sill, such minerals would eventually form. It is, therefore, possible to establish the maximum time between spilling events using a simple evaporation model for the Owens River to determine the order of formation of minerals more soluble than $CaCO_3$. We consider salinity evolution for two water levels of Owens Lake: at its spill level and at its historic, pre-diversion level. In both cases, we begin with the lake composed of fresh Owens River water, and we assume that the flux of dissolved salts from the Owens River into Owens Lake has been about the same as that of provided by the modern Owens River. Records of river discharge and the dissolved load composition are essentially continuous from the present back to 1906 and include unpublished records of the Los Angeles Department of Water and Power (LADWP) from 1934–1992 and published records of Gale (1914) for 1906–1912, of Wilcox (1946) for 1929–1944, and of Hollett et al. (1991) for 1974–1985. We take the composition of Owens River water as that averaged by Hollett et al. (1991) corrected by us for water imports from the Mono Basin as shown in Table 3. The salinity (total dissolved salts, TDS) is 241 mg/L and is characterized by Na and Ca as major cations with HCO_3 dominating the anions followed by SO_4 and Cl. Average discharge, corrected for imports from Mono Basin, is 3.9×10^{11} L/yr (Table 3), which results in a salt flux of 9.40×10^{13} mg/yr. The volume of the lake in 1872, according to Gale (1914) was 2.92×10^{12} L, and we estimate the volume of the lake at spill point to be approximately 2.21×10^{13} L (Table 3). Thus the filling time for the lake to its spill point, assuming no evaporation, is only 57 years. According to Gale (1914) the area of the 1872 lake was 2.91×10^8 m². At steady state, evaporation rate times lake area equals discharge, indicating the historic evaporation rate to be on the order of 1.3 m/yr decrease in depth. This rate compares well with measurements made during 1939–1940 and 1969–1970, which averaged 1.27 m/yr decrease in depth (Smith and Street-Perrott, 1983). At this evaporation rate, the increased discharge of the Owens River needed to maintain the

TABLE 3. CHEMICAL COMPOSITION OF LOWER OWENS RIVER, HISTORIC OWENS LAKE, AND THEORETICALLY EVAPORATED OWENS RIVER WATER OF SIMILAR CHLORINITY*

	Modern Owens River†	Historic Owens Lake§	Evaporated Owens River**
pH	8.1	9.9	9.8
	mg/L	mg/L	mg/L
Ca	24	...	3
Mg	4	...	...
Na	35	21,400	32,908
K	4	2,700	4,053
HCO_3	133	633	460
CO_3	<1	13,000	12,160
Cl	15	13,300	13,300
SO_4	25	9,300	23,053
TDS‡	241	60,336	85,934

Owens River discharge§§ = 3.9×10^{11} L/yr

	Historic (1872)	At spillway
Lake area***	2.91×10^8 m²	6.22×10^8 m²
Lake volume***	2.92×10^{12} L	2.21×10^{13} L

*Also shown are average discharge of the Owens River and area and volume estimates for Owens Lake at historic and spill levels.
†Average for 1974-1985 from Hollett et al., 1991, and corrected for water imports from Mono Basin.
§Based on analysis of anhydrous salts reported for 1876 in Gale, 1914, and recalculated for 6 wt % salinity. The pH and CO_3/HCO_3 speciation calculated from PHRQPITZ (Plummer et al., 1988) assuming equilibrium with atmospheric CO_2.
**Theoretical evaporation of Owens River water to the Cl concentration of Owens Lake in 1876, (13,300 mg/L), allowing calcite to precipitate at 10-fold supersaturation, and equilibrating with atmospheric CO_2.
‡Total dissolved solids, mg/L.
§§Mean discharge for period 1934-1992 from unpublished data of the Los Angeles Department of Water and Power corrected for water imports from Mono Basin. The same value is calculated from data given by Hollett et al., 1991, for the period 1974-1985.
***Area and volume of historic lake and area of lake at spill taken from Gale, 1914. Volume of lake at spill was calculated by estimating the incremental increase in volume (as a frustum of a cone) and adding it to the historic volume, using areas of the two lake surface areas as bounding planes of the frustum, and the increase in lake level as its height.

lake at its spill point is simply the ratio of surface areas between spill point and historic level, or 2.13 times (see Benson and Paillet, 1989). During cooler glacial periods, the evaporation rate was probably less, and the river discharge required to maintain overflow, therefore, was somewhat less than 2.13 times historic.

The initial salt buildup in the lake is simply the flux of salts divided by the lake volume which calculates to be 4.25 mg/L/yr. Thus, we allow an annual increase of 4.25 mg/L/yr dissolved salt of the composition given in Table 3 and calculate the progression of the water composition (Fig. 5) using the computer program PHRQPITZ (Plummer et al., 1988). Because of inhibition of nucleation, most $CaCO_3$-producing alkaline lakes seem to be at about ten-fold supersaturation with respect to calcite

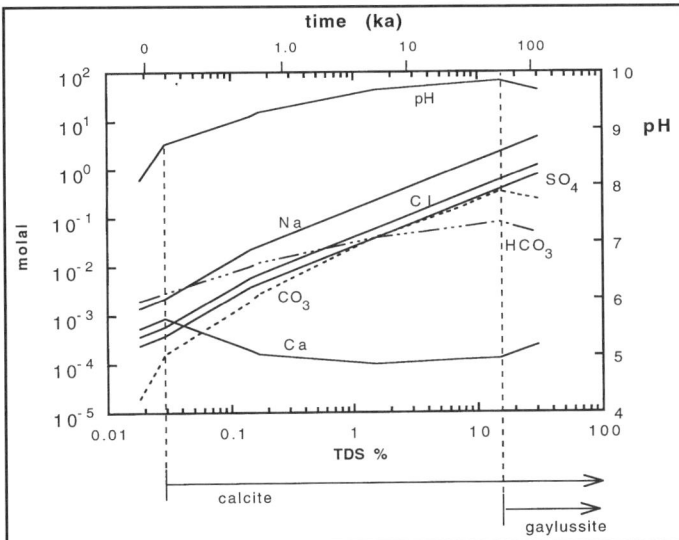

Figure 5. Chemical evolution of Owens Lake water as a consequence of salt buildup in a terminal lake. Calculations are made using PHRQPITZ (Plummer et al., 1988), assuming constant annual input of dissolved load from modern Owens River (Hollett et al., 1991). Amount of CO_2 is fixed at atmospheric level, calcite precipitates at ten-fold supersaturation, and Mg phases are not considered. Lake volume is taken as that at the spill level, 2.54×10^{13} L, approximately ten times the volume of the historic (1872) lake. Calcite begins precipitating after only 35 years at a concentration factor of 1.4, whereas gaylussite does not become saturated until a concentration factor of 1,600 (15 wt % salinity) is achieved, which requires about 55 k.y. to attain. For the volume of the historic lake, gaylussite saturation is attained in 8 k.y. The lack of gaylussite in sediments from OL-92 suggests that the lake had never remained at its historic level for longer than 10 k.y. during the past 800 k.y. TDS = total dissolved solids.

(Galat and Jacobsen, 1985; Bischoff et al., 1991). A closed Owens Lake will reach such supersaturation with respect to calcite in only 38 years, at a concentration factor of 1.4 of the original Owens River water (TDS = 344 mg/L). This result shows that $CaCO_3$ is an extremely sensitive indicator of closed lake conditions. Beyond this point, each parcel of river water entering the lake will quantitatively precipitate all its Ca and equivalent HCO_3 (an amount equal to about 48% of the dissolved solids) according to the reaction

$$Ca^{+2} + 2HCO_3^- = CaCO_3 + H_2O + CO_2 \qquad (1)$$

As we have outlined, Mg seems to be removed to the sediment in closed lake conditions via reaction with dissolved silica and detrital clays as well as with reaction with dissolved carbonate. Therefore, dissolved Mg and silica are also removed during the salt buildup. According to Jones and Van Denburgh (1966), such Mg removal is complete and occurs at the same time as Ca removal. Thus, salinity increases at 3.18 mg/L/yr beyond this point and the concentration of Ca progressively declines (Fig. 5). Dissolved CO_2 is allowed to continuously equilibrate with the atmosphere and calcite is allowed to precipitate at ten-fold supersaturation. Thus, from this point the lake water becomes increasingly saline and alkaline (pH 9–10), characterized by linear increases in Na-CO_3-Cl-SO_4, and with Ca and Mg becoming vanishingly small. Salinity increases continuously for the next 55 k.y. (or 8 k.y. for the 1872 lake level) until it reaches 150,000 mg/L at which point gaylussite ($Na_2Ca(CO_3)_2 \cdot 5H_2O$) begins to precipitate. No other new minerals become saturated during the next two-fold concentration of the brine to 300,000 mg/L, at which point the simulation was terminated (Fig. 5).

Mono Lake, 187 km to the northwest of Owens, a lake with a similar water chemistry and a similar history to Owens (Phillips, 1877), is presently at saturation with gaylussite, which is forming abundantly in the lake (Bischoff et al., 1991). Gaylussite began forming after the salinity exceeded about 100,000 mg/L in the 1970s as a consequence of the diversion of source waters by LADWP begun in 1941. Thus, gaylussite would be expected to form in Owens Lake not far beyond its point of saturation, and its (or its psuedomorph's) absence in the Owens sediments, in particular, or other soluble minerals in general, implies that Owens Lake could not have been at its historic level for any time during the past 800 k.y. for longer than about 8 k.y., or closed at its spill level for longer than 55 ka.

The salinity of the pre-diversion Owens Lake was about 6–8 wt % (Gale, 1914), only about halfway to gaylussite saturation. Table 3 compares the reconstructed historic-lake composition (from an 1876 analysis of the anhydrous salts given in Gale, 1914) to that calculated from the evaporation model above for the same Cl concentration (13,300 mg/L). The model predicts pH, CO_3, and HCO_3 of the natural lake rather well, but it overestimates Na by a factor of 1.5 and SO_4 by a factor of 2.5. The abundance of reduced sulfur in the sediments (Tuttle, 1993) corroborated by SO_4 deficiencies in the pore waters (Bischoff et al., 1993b) suggests that significant SO_4 was removed by reduction in the bottom waters of the lake. The Na deficiency is explained, in part, by ion-exchange reactions in which river-suspended Ca smectites convert to Na smectites upon flocculation into the Na-dominated lake water (Jones and Van Denburgh, 1966). Clays suspended in Owens River water will have 99% of their exchange sites occupied by Ca, and conversely, those suspended in lake water of 1876 will be 99% occupied by Na, according to criteria of the U.S. Salinity Laboratory Staff (1954). The liberated Ca ions react to precipitate additional $CaCO_3$.

$CaCO_3$ budget and criteria of overflow

The time periods subtended by carbonate minima are short relative to the time periods of the carbonate maxima down the core, suggesting that closed lake conditions were periodically interrupted by brief periods of overflow (Fig. 2). The $CaCO_3$ content in the sediment can provide some information about when the lake was overflowing. The maxima of sediment $CaCO_3$ content in the core signal times of minimal overflow or closed lake conditions. It is also possible to approximate the limiting content of $CaCO_3$ that marks the transition from overflow to closed conditions, given knowledge of lake area at spill, Ca flux from the Owens River, and sediment mass-accumulation rate. Owens Lake precipitates $CaCO_3$ within a minimum of 38 years after termination of intense spilling. Because the lake water is carbonate-dominated and Ca-limited, the rate of production of $CaCO_3$ in the closed lake is equal to the Ca flux provided by the Owens River. The flux of Ca is given by the product of the average river Ca concentration and average discharge. We calculate average Ca concentration for the lower Owens River for the years 1934–1992 to be 0.6 mmolal from unpublished LADWP data for 1934–1992. This concentration is almost identical to an average of 0.6 mmolal given by Hollett et al. (1991) from U.S. Geological Survey bimonthly analyses for 1974–1985. An average of 0.44 mmolal is calculated from data given in Gale (1914) for the years 1906–1912, and 0.72 mmolal from the data given by Wilcox (1946) for the years 1929–1946. Thus, average Ca concentration is between 0.44 and 0.70 mmolal. An additional source of Ca derives from Na-Ca exchange on suspended clays as they enter Owens Lake. Based on the ratio of CEC to acid-leachable Ca observed for the sediment, on the order of 5% of the river-supplied Ca could have been supplied by ion-exchange, or an additional 0.035 mmolal of Ca. Thus, including this ion-exchange of Ca for Na in the upper limit, the modern Ca flux is between 2 and 3.4×10^8 mol/yr.

Because the Owens Lake basin is relatively flat, we consider the sediment area to be approximately equal to the surface area of the lake, or 6.22×10^{12} cm^2 for the lake just at spill level. Therefore, the mass-accumulation rate (MAR) of $CaCO_3$ per unit area will be 0.03–0.055 mol/cm^2/k.y. For the past 800 k.y. the bulk-sediment MAR has been approximately constant at 51.4 g/cm^2/k.y. (Bischoff et al., this volume chapter 8). Therefore, assuming that $CaCO_3$ is evenly distributed over the floor of the lake, the limiting $CaCO_3$ content was between 0.06–0.11 mol/100 g sediment, or 6–11 wt % $CaCO_3$. Thus, sediments with less than about 0.11 mol/100 g represent overflowing conditions, whereas those with more represent closed lake conditions (Fig. 3). The pattern shown in Figure 3 for the past 500 k.y. indicates that the lake was overflowing about 34% of the time, and closed for about 66% of the time. The lake appeared to be continuously closed for periods of as much as 50- to 70-k.y. duration punctuated by briefer, 20- to 30-k.y. periods of intense overflow. In particular, the entire period from 120–50 ka seems to represent closed conditions. Data from downstream in Searles Lake, however, indicates that it was receiving spill water, at least periodically, during this same interval (Smith et al., 1983, Smith, 1991; Bischoff et al., 1985). Also, the patterns observed in the freshwater diatom remains suggest that there were several short-lived spilling events between 80 ka and 90 ka (Bradbury, this volume). Thus, it appears that these events were too short to be discernible within the 8-ka resolution of the channel samples.

In addition, the limiting $CaCO_3$ content is probably somewhat underestimated because of the assumptions that the Owens River is the only source of dissolved Ca, that the present Ca flux has been approximately constant through time, and that $CaCO_3$ is evenly distributed over the floor of the lake. For example, we do not consider eolian input of pedogenic $CaCO_3$. In addition, much of the dissolved load of the Owens River is supplied by the geothermal system of Long Valley (Smith, 1976). Thus, Ca flux of the Owens River may vary through time with the fluctuations in magmatic activity in the Long Valley caldera.

Paleoclimate indicators and correlation to SPECMAP

Cycling between closed and open lake conditions, best shown by $CaCO_3$ and CEC contents, likely reflects the cycles between glacials and interglacials, and these correlate reasonably well with the pattern of global glaciations. The oscillations of the oxygen-isotope record of marine foraminifers through time are now widely accepted as recording the mass of water stored in polar ice. In order to understand how the Owens Lake record reflects the global cycles, we compare the CO_3 and CEC patterns of OL-92 with the marine $\delta^{18}O$ standard of SPECMAP (Imbrie et al., 1984) back to the major break in Owens sedimentary conditions at 500 ka (Fig. 6). The marine record back to about 700 ka is characterized by distinct 100-k.y. major cycles which terminate abruptly the progressively developing glacials (Termination), and with minor cycles of 41, 23, and 19 k.y. The minor cycles apparently reflect variations in insolation caused by astronomical rhythms, whereas the cause of the 100-k.y.

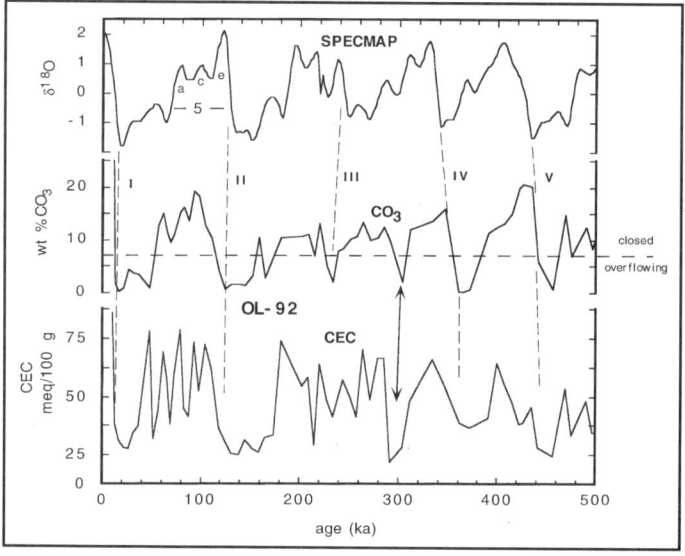

Figure 6. Carbonate and CEC (in meq/100 g) variations in sediments from OL-92 compared to SPECMAP (Imbrie et al., 1984) for the past 500 k.y. of lacustrine conditions. Cycles in carbonate and CEC seem to reflect the dominant 100-k.y. cycles of SPECMAP. Terminations I, II, IV, and V, seem to be represented by abrupt increases in CEC and carbonate in the Owens record. Moreover, both carbonate and CEC indicate a marked glacial-interglacial transition at about 290 ka, which seems to have no counterpart in SPECMAP. The overflow model based on the $CaCO_3$ budget suggests that Owens Lake was overflowing about 34% of the time during the past 500 k.y.

cycle remains in dispute (see, for example, Ruddiman and Wright, 1987; Winograd et al., 1992; Imbrie et al., 1993). The 100-k.y. cycles are reflected in the Owens Lake record (Fig. 6) where abrupt increases in CO_3 and CEC seem to correlate with at least four of the last five Terminations of SPECMAP. The minor cycles, however, are not revealed in spectral analysis of the Owens data, but the results are inconclusive because of the coarse resolution of the channel samples (8 k.y.) and the less-than-complete core recovery. Moreover, because the marine record reflects polar ice volume, and the Owens Lake record presumably reflects precipitation in the eastern Sierra Nevada, exact correlation should not be expected. Terminations I, II, IV, and V seem to be strongly reflected in the Owens record (Fig. 6), whereas Termination III is less obvious. Plots of CO_3, TOC, and CEC (Figs. 2 and 6) show a marked glacial/interglacial transition at about 290 ka in the Owens record, which seems to have no counterpart in SPECMAP. Chronological uncertainties in both SPECMAP, which assumes astronomical forcing for the 100-k.y. cycles, and the Owens core, which is based on inferences of constant mass accumulation rate between widely spaced controls (Bischoff et al., this volume, chapter 8) may explain the approximate ±5% differences in the positions of the Terminations. Changes in Owens sediment signaling the end of the last glacial are coincident with SPECMAP Termination I at about 13 ka. Changes that appear to relate to Termination II (oxygen-isotope stages 5/6 boundary) appear at 120 ka at Owens compared to 128 ka in SPECMAP. This Termination is shown very markedly in OL-92 for several other parameters as well, including smectite abundance (Menking, this volume), pollen, in which Pinus replaces cedars and junipers (Litwin et al., this volume), diatoms, in which saline forms replace freshwater forms (Bradbury, this volume), and ostracodes, in which the saline forms replace freshwater forms (Carter et al., this volume). In contrast, the fresh-water $\delta^{18}O$ record from nearby Devils Hole, Nevada, dated by high-precision U/Th (Winograd et al.,1992; Ludwig et al., 1992) show changes corresponding to Termination II at $140 \pm 3(2s)$ ka. The 120-ka timing of this event at Owens is within 6% and 7%, respectively, of SPECMAP and Devils Hole timing, a discrepancy perhaps within the uncertainty of the Owens chronology.

The warm period of oxygen-isotope stage 5 in SPECMAP extends from 128–70 ka (58 k.y. duration) where it is bounded by an abrupt cooling event, the stages 4/5 boundary. As reviewed by Muhs (1992), there is debate about the meaning of the "last interglacial" in various marine and continental settings, whether it was short (10 k.y.) and corresponds to stage 5e, or whether it was long (60 k.y.) and corresponds to all of stage 5. The corresponding conditions at Owens Lake appear to span the entire time from 120 ka to about 50 ka, a similar time span to all of stage 5. Moreover, the Owens record seems to lack the prominent 5e of SPECMAP (Fig. 6). Thus, the Owens record seems to suggest that the "last interglacial" was long in the Western Great Basin. The approximately 10-k.y. offset between this span and marine oxygen-isotope stage 5 reflects either the error in the age-depth model (i.e., 10% error), or alternatively, a real time lag between changes in Northern Hemisphere ice volumes and the manifestation of local climate change in lake geochemistry and sedimentology.

Discrepancies notwithstanding, the general pattern of the global glacial cycles as recorded in the marine record appear to be reflected in the sediments of Owens Lake.

SUMMARY

The chemical composition of sediments from OL-92 and their variation through time allow the following conclusions:

1. Owens Lake sediments are composed predominantly of silts and clays of bulk composition close to granodiorites of the eastern Sierra Nevada. The sediment also contains significant but variable amounts of authigenic Ca and Mg carbonates and Mg silicates.

2. Records of CO_3 and TOC show cyclic co-variation back to about 500 ka. Maximum glacial conditions are clearly seen at 25–17 ka, where CO_3 and TOC display conspicuous and sharp minima, very close to zero values. These results suggest that at glacial maxima the lake was overflowing with cold fresh water and was relatively nonproductive. Five similar minima in CO_3 and TOC, going back to 500 ka are interpreted as successively older glacial maxima. Conversely, the recurring and broader CO_3 and TOC maxima are interpreted to represent full interglacial conditions during which the lake was closed, saline, and biologically productive. Closed lake conditions give rise to sediments rich in CO_3 and TOC, authigenic Mg-silicates, and clays of high CEC, in contrast to overflowing lake conditions where these components decrease to close to zero.

3. The CEC of the clay fraction reflects smectite abundance and shows a minimum coinciding with the glacial maximum at 25–17 ka. Other CEC minima occur in the core at the same points of CO_3 and TOC minima. This correlation suggests that during glacial advances the clay-size material has a low exchange capacity, perhaps representing a glacial rock-flour component.

4. Minerals characteristic of extreme salinity and/or playa conditions, such as gypsum and gaylussite, were not detected, suggesting that Owens Lake had not attained the required salinity at any time during the past 800 k.y. Calculations of progressive salinization of Owens Lake indicates that the first mineral to form beyond $CaCO_3$ is gaylussite, but not until the salinity reaches 15 wt %. For the lake to acquire this salinity at its spill level with present Owens River flux of dissolved salts 55 k.y. are required, but only 8 k.y. are required to reach the historic (1872) level.

5. The limiting sediment-content of $CaCO_3$ that separates closed and overflowing is about 11 wt %, calculated from a mass balance between the Ca flux from the modern Owens River, the mass accumulation rate of the lake sediments, and the area of the Owens Lake basin at spill. Applying this limit to the pattern of $CaCO_3$ abundance suggests that the lake was

closed about 66% of the time over the past 500 k.y., periodically interrupted by briefer periods of overflow.

6. The $CaCO_3$ oscillations generally track and correlate with those of the marine $\delta^{18}O$ cycles back to about 500 ka. Four of the last five marine isotope terminations are clearly shown in the Owens core. Terminations I, II, IV, and V are clearly shown at the appropriate times by abrupt increases in the carbonate content of the sediments, reflecting change from overflowing to closed lake conditions. The last interglacial at Owens Lake appears to span from 120 ka to about 50 ka. The roughly 10-k.y. offset between this span and marine oxygen-isotope stage 5 reflects either the error in the age-depth model (i.e., 10% error), or alternatively, a real time lag between changes in Northern Hemisphere ice volumes and the manifestation of local climate change in lake geochemistry and sedimentology.

ACKNOWLEDGMENTS

We thank Rod Kurimoto and Marty Adams of LADWP for generously providing us with unpublished data on discharge and chemical composition of the Owens River. We are particularly indebted to Larry Benson and Blair Jones for sharing with us their extensive knowledge about Great Basin lakes and for in-depth review of the manuscript. We thank Kirsten Menking, George Smith, and Isaac Winograd for many stimulating discussions and conversations about the meaning of our data. Special thanks are also due Robert Rosenbauer for assisting in the drilling operations and for lending his computing wizardry to the evaporation modeling.

REFERENCES CITED

Bateman, P. C., Clark, L. D., Huber, N. K., and Moore, J. G., 1963, The Sierra Nevada Batholith: U.S. Geological Survey Professional Paper 414-D, 46 p.

Beetem, W. A., Janzer, V. J., and Wahlberg, J. S., 1962, Use of Cesium-137 in the determination of cation exchange capacity: U.S. Geological Survey Bulletin 1140-B, 7 p.

Benson, L. V., and Paillet, F. L., 1989, The use of total lake-surface area as an indicator of climatic change: Examples from the Lahonton Basin: Quaternary Research, v. 32, p. 262–275.

Bischoff, J. L., Rosenbauer, R. J., and Smith, G. I., 1985, Uranium-series dating of sediments from Searles Lake: Differences between continental and marine records: Science, v. 227, p. 1222–1224.

Bischoff, J. L., Herbst, D. B., and Rosenbauer, R. J., 1991, Gaylussite formation at Mono Lake, California: Geochimica et Cosmochimica Acta, v. 55, p. 1743–1747.

Bischoff, J. L., Fitts, J. P., Fitzpatrick, J. A., and Menking, K., 1993a, Sediment Geochemistry of Owens Lake Drill Hole OL-92: U.S. Geological Survey Open-File report 93-683, p. 83–99.

Bischoff, J. L., Fitts, J. P., and Menking, K., 1993b, Sediment pore-waters of Owens Lake drill hole OL-92: U.S. Geological Survey Open-File report 93-683, p. 100–105.

Bischoff, J. L., Stafford, T. W., and Rubin, M., 1993c, AMS radiocarbon dates on sediments from Owens Lake drill hole OL-92: U.S. Geological Survey Open-File report 93-683, p. 246–250.

Clark, F. W., 1924, The data of geochemistry: U.S. Geological Survey Bulletin 616.

Engleman, E. E., Jackson, L. L., and Norton, D. R., 1985, Determination of carbonate carbon in geological materials by coulometric titration: Chemical Geology. v. 53, p. 125–128.

Flanagan, F. J., 1976, Descriptions and analyses of eight new USGS rock standards: U.S. Geological Survey Professional Paper 840, 192 p.

Galat, D. L., and Jacobsen, R. L., 1985, Recurrent aragonite precipitation in saline-alkaline Pyramid Lake, Nevada: Archiv für Hydrobiologia, v. 105, p. 137–159.

Gale, H. S., 1914, Salines in the Owens, Searles, and Panamint Basins, southeastern California: U.S. Geological Survey Bulletin 580-L, p. 251–323.

Grim, R. E., 1968, Clay mineralogy: New York, McGraw-Hill, 596 p.

Hollett, K. J., Danskin, W. R., McCaffrey, W. F., and Walti, C. L., 1991, Geology and water resources of Owens Valley, California: U.S. Geological Survey Water-Supply Paper 2370, p. B1–B77.

Imbrie, J., and eight others, 1984, The orbital theory of Pleistocene climate: Support from a revised chronology of the marine ^{18}O record, in Berger, A. L., Imbrie, J., Hays, J. D., Kukla, G., and Saltzman, B., eds., Milankovitch and climate, Part 1: Dordrecht, The Netherlands, D. Riedel, p. 269–305.

Imbrie, J., and 18 others, 1993, On the structure and origin of major glaciation cycles 2. The 100,000-year cycle: Paleoceanography, v. 8, p. 699–735.

Jones, B., 1986, Clay mineral diagenesis in lacustrine sediments, in Mumpton, F. A., ed., Studies in diagenesis: U.S. Geological Survey Bulletin 1578, p. 291–300.

Jones, B. F., and Van Denburgh, A. S., 1966, Geochemical influences on the chemical character of closed lakes: International Association for Scientific Hydrology, Publication no. 70, p. 435–446.

Ludwig, K. R., Simmons, K. R., Szabo, B. J., Winograd, I. J., Landwehr, J. M., Riggs, A. C., and Hoffman, R. J., 1992, Mass-spectrometric ^{230}Th-^{234}U-^{238}U dating of the Devils Hole Calcite Vein: Science v. 258, p. 284–287.

Menking, K., Musler, H. M., Fitts, J. P., Bischoff, J. L., and Anderson, R. S., 1993a, Clay mineralogical analyses of the Owens Lake core: U.S. Geological Survey Open-File report 93-683, p. 75–82.

Menking, K., Musler, H. M., Fitts, J. P., Bischoff, J. L., and Anderson, R. S., 1993b, Sediment size analyses of the Owens Lake core: U.S. Geological Survey Open-File report 93-683, p. 58–74.

Muhs, D. R., 1992, The last interglacial-glacial transition in North America: Evidence from uranium-series dating of coastal deposits, in Clark, P. U., and Lea, P. D., eds., The last interglacial-glacial transition in North America: Geological Society of America Special Paper 270, p. 31–35.

Phillips, F. A., 1877, The alkaline and boracic lakes of California: The Popular Science Review, v. 16, p. 153–164.

Plummer, L. N., Parkhurst, D. L., Fleming, G. W., and Dunkle, S. A., 1988, PHRQPITZ—A computer program incorporating Pitzer's equations for calculation of geochemical reactions in brines: U.S. Geological Survey Water Resources Investigations Report 88-4153, 305 p.

Rankama, K., and Sahama, T., 1950, Geochemistry: Chicago, Illinois, University of Chicago Press, 911 p.

Ruddiman, W. F., and Wright, H. E., Jr., 1987, Introduction, in Ruddiman, W. F., and Wright, H. E., Jr, eds., North America and adjacent oceans during the last deglaciation: Boulder, Colorado, Geological Society of America, The Geology of North America, v. K-3, p. 1–12.

Smith, G. I., 1976, Origin of lithium and other components in the Searles Lake evaporates, California, in Vine, J. D., ed., Lithium resources and requirements by the year 2000: U.S. Geological Survey Professional Paper 1005, p. 92–103.

Smith, G. I., 1991, Continental paleoclimate records and their significance, in Morrison, R. B., ed., Quaternary nonglacial geology: Conterminous U.S.: Boulder, Colorado, Geological Society of America, The Geology of America, v. K-2, p. 35–41.

Smith, G. I., and Street-Perrott, F. A., 1983, Pluvial lakes of the Western United

States, *in* Porter, S., ed., Volume 1, The Late Pleistocene, *in* Wright, J. E., Jr., ed., Late-Quaternary environments of the United States: Minneapolis, University of Minnesota Press, p. 190–212.

Smith, G. I., Barczak, V. J., Moulton, G. F., and Liddicoat, J. C., 1983, Core KM-3, a surface-to-bedrock record of late Cenozoic sedimentation in Searles Valley, California: U.S. Geological Survey Professional Paper 1256, p. 1–24.

Tuttle, M. L., 1993, The distribution and isotopic composition of sulfur in Owens Lake core OL-92: U.S. Geological Survey Open-File report 93-683, p. 119–126.

U.S. Salinity Laboratory Staff, eds., 1954, Diagnosis and improvement of saline and alkali soils: U.S. Department of Agriculture, Agricultural Handbook No. 60, 160 p.

Wilcox, L. V., 1946, Boron in the Los Angeles (Owens River) aqueduct: U.S. Department of Agriculture, Bureau of Plant Industry Research Report 78, 66 p.

Winograd, I. J., Coplen, T. B., Landwehr, J. M., Riggs, A. C., Ludwig, K. R., Szabo, B. J., Kolesar, P. T., and Revesz, K. M., 1992, Continuous 500,000-year climate record from vein calcite in Devils Hole, Nevada: Science v. 258, p. 255–260.

MANUSCRIPT ACCEPTED BY THE SOCIETY JUNE 17, 1996

Movement and diffusion of pore fluids in Owens Lake sediments from core OL-92 as shown by salinity and deuterium-hydrogen ratios

Irving Friedman
U.S. Geological Survey, Federal Center, Box 25046, MS 963, Denver, Colorado 80225
James L. Bischoff
U.S. Geological Survey, 345 Middlefield Road, MS 910, Menlo Park, California 94025
Craig A. Johnson
U.S. Geological Survey, Federal Center, Box 25046, MS 963, Denver, Colorado 80225
Scott W. Tyler
Desert Research Institute and Department of Environmental and Resource Sciences, University of Nevada, Reno, Nevada 89506
Jeffrey P. Fitts
Department of Geological and Environmental Sciences, Stanford University, Stanford, California 94305

ABSTRACT

Chemical and δD data on the pore fluids extracted from the Owens Lake core were used to derive snapshots of the hydrology of the ancestral lake. Models of evaporation processes that are relevant to lakes in arid environments were developed and applied to paleo–Owens Lake. Based on pore fluid data and the models we conclude the following: (1) At least some of the interstitial fluids extracted from the sediments sampled in the Owens Lake core OL-92 were not deposited contemporaneously with the sediments. (2) Both vertical and horizontal movement of the fluids within the sediments have been demonstrated. (3) This fluid movement might result in fluid transport out of the basin. (4) During the interval in which fluids now found between 240 m and 320 m were deposited, one half of the inflow to Owens Lake escaped as outflow. The chlorinity of the inflow to the lake during this time period was much higher than the present inflow, probably the result of chloride emitted by the eruption of the Bishop ash. (5) The δD values in most of the core, although enriched as compared to that of Owens River inflow, did not reach high values, indicating that the lake either discharged, or was in a steady state in regard to volume such that evaporation equaled inflow. (6) The desiccation of the lake shortly after 1913 resulted in a brine now found in the upper few meters of core with a δD of –50‰. This relatively light δD is the result of dilution with precipitation of the heavy δD brine formed by Rayleigh evaporation.

INTRODUCTION

In this paper we use chemical and isotopic data on the pore fluids extracted from the Owens Lake core (OL-92) to derive snapshots of the hydrology of the ancestral lake. We first discuss the experimental data, present models of evaporation that are relevant to lakes in arid environments, and then apply these models to paleo–Owens Lake.

In order to reconstruct the paleoclimatic history of Owens Lake from the chemistry of the pore fluids now present in the sediments, it is necessary to examine the post depositional processes that would cause changes in their chemistry and their posi-

Friedman, I., Bischoff, J. L., Johnson, C. A., Tyler, S. W., and Fitts, J. P., 1997, Movement and diffusion of pore fluids in Owens Lake sediments from core OL-92 as shown by salinity and deuterium-hydrogen ratios, *in* Smith, G. I., and Bischoff, J. L., eds., An 800,000-Year Paleoclimatic Record from Core OL-92, Owens Lake, Southeast California: Boulder, Colorado, Geological Society of America Special Paper 317.

tion in the sediment column. These processes include (1) upward movement of fluid expelled during compaction of the sediment—in their discussion of compaction of deep sea cores, Friedman and Hardcastle (1988) observed that most of the compaction occurred in the upper 10–20 cm and that any fluid that was expelled beyond this point did not mix with the overlying pore water but migrated upward through cracks or along paths of high permeability; (2) interaction with the host sediments; (3) transport of fluid by advection and diffusion from the position in the sediment column where the fluids were originally deposited. Because interaction with the enclosing sediments will not alter the chloride and deuterium concentrations, these constituents can be used to model transport processes that modify the chemistry of the pore fluids.

Diffusion of dissolved constituents will occur at any time a chemical-activity gradient is established and will act to eliminate the gradient. Therefore the persistence over long time periods of steep concentration gradients in the core indicates that fluid is moving by advection. Using several arguments, we show that the pore fluids present in at least sections of the core are younger than the sediments in which they are found. We also show that movement of pore fluid has occurred both by advective movement downward and horizontally, as well as by diffusion both upward and downward.

EXPERIMENTAL

Sampling procedures

Samples of fresh wet sediment (60 cc) for the determination of water content and pore water chemistry were taken during the drilling operations at every 2–3 m from clay-rich horizons for the entire length of the core (120 samples). The samples were taken in the field from the freshly split core within minutes of exposure of the fresh sediment. Each sample was trimmed of disturbed sediment adjacent to the core liner, immediately sealed within a 75 mL air-tight glass bottle, and kept refrigerated at ~5 °C until laboratory processing some three months later.

Chemistry

For the determination of water content, 1–2 cc splits of the fresh sediment were weighed in ceramic crucibles and their weight loss recorded after heating for two days at 100 °C. Water content as weight percent is simply the percentage of weight loss of the sediment. In the upper 10 m of the core, where the salinities of the pore fluids are high, the salt that remained after drying added significantly to the weight of the sediment, and the calculated water content values for these samples will be as much as 10% too low.

An aliquot of the sample was transferred into a stainless steel cylindrical squeezer (modified after that of Manheim, 1966) that was then pressurized with a simple laboratory hydraulic press (12-ton capacity). The pore water squeezed from the sediment passed through three layers of filter paper and into a poly-ethylene syringe. Squeezing for 10–30 minutes yielded from 3–25 mL of pore water, depending on the water content of the sample. The sample was then passed through a swinney-mounted membrane filter (0.45 μm pore size). Two drops were used for immediate measurement of refractive index for salinity, a 1 mL aliquot was used for immediate pH determination by micro electrode, 1–2 mL were injected into a septum-capped evacuated blood tube for isotopic analyses, and the remainder stored in a tightly capped polyethylene bottle for further chemical analyses. Total dissolved inorganic carbon (Ct) was determined by an infrared CO_2 analyzer. Cl was determined by potentiometric titration with an auto-titrator; and sulfate concentration was determined by ion chromatography. The chemical data are shown in Table 1, and plotted in Figures 1, 2, and 3.

Deuterium-hydrogen analysis

Two μL aliquots of the pore waters, extracted as described by Bischoff et al. (this volume, Chapter 4), were first distilled under vacuum to separate salts and then converted to hydrogen gas by reaction with zinc metal at 500 °C (Kendall and Coplen, 1985). The deuterium content of the hydrogen gas was then measured using a mass spectrometer that separated the HD^+ from H_2^+, determined the ratio of the two ions, and compared the ratio to that in a sample of standard hydrogen gas. Corrections were made for H_3^+ as well as for other mass spectrometric errors. The results are reported in per mil (‰) units relative to Vienna Standard Mean Ocean Water (V-SMOW):

$$\delta D, \text{ per mil} = \frac{R_{\text{sample}} - R_{\text{standard}}}{R_{\text{standard}}} \times 1,000,$$

where R_{sample} is the ratio of deuterium to hydrogen in the sample, and R_{standard} is the ratio in V-SMOW. The δD values are precise to ±2‰ (2σ). It should be noted that deuterium is present almost entirely as HDO, and not as D_2O.

Determination of diffusion coefficient for chloride

Estimation of effective diffusion coefficients in the Owens Lake sediments requires a knowledge of both the tortuosity of the media (τ) and the free water coefficient (D_0) of the solute of interest. Boron diffusion measurements on core samples from depths of 36 m, 148 m, and 295 m from core OL-92-2 were used to calculate tortuosities. Boron was chosen as the tracer for diffusion experiments because of its ease of analysis and its relatively high concentration in the pore fluids. Subcores (1.8 cm diameter by 2.8 cm long) were taken from the original cores and placed in clear acrylic plastic sleeves for diffusion cell experiments. An additional subsample from each core was squeezed to extract pore water for analysis of ionic strength and boron concentration by inductively coupled plasma emission spectroscopy. Pore water boron concentrations (as boron) ranged from 22 mg/L to 236 mg/L. At the high pH and high concentrations of the pore fluids, boron should behave conser-

TABLE 1. WATER CONTENT AND PORE-WATER COMPOSITION OF SEDIMENTS FROM OL-92*

Sample	Depth (m)	H_2O (wt %)	pH	Salinity (wt %)	Cl (mmolal)	Ct† (mmolal)	SO_4 (mmolal)	δD (‰) V-SMOW
463	1.00	62.7	9.5	14	1,450	511.8	118	-52
464	2.00	58.2						
465	3.00	27.1						
466	4.00	29.0	9.3			450.8		-50
200	6.09	56.4	9.5	12	1,162	506.4	96.3	-48
201	7.48	58.2	9.4	10	929.0	438.2	74.4	-54
202	8.11	46.8	9.4	8.7	829.0	489.1	94.2	-58
203	9.51	58.0	9.5	9.8	898.0	428.4	77.6	-58
204	10.49	44.0	9.1	6.9	733.3	333.1	25.5	-57
205	11.92	63.1	9.5	7.7	695.5	332.9	57.6	-69
208	16.06	59.0	9.5	9.9	894.3	405.5	76.0	-62
211	17.94	52.9	7.9	2.4	304.2	109.1		-70
213	18.70	58.6	7.8	2.0	260.6	109.4		-70
216	24.61	51.3	7.7	1.4	153.0	85.88	0.885	-76
218	26.46	60.0	7.7	1.2	134.0	87.89	0.883	-83
221	33.59	53.5	8.1	1.0	92.63	95.77	0.122	
223	35.56	60.7	8.2	1.0	87.47	106.1	0.143	-79
226	39.85	53.5	8.7	1.0	84.03	85.22	0.338	
228	41.72	55.1	8.8	1.1	85.75	89.53	0.658	-75
231	46.99	53.3	8.9	1.2	92.63	131.9	0.188	-78
233	52.37	63.4	8.9	1.5	103.0	129.7	1.52	-74
236	56.65	52.9	8.9	1.5	106.4	161.2	1.07	
238	58.80	47.2	8.7	1.7		147.0	1.08	-68
240	61.76	54.9	8.9	1.9	116.7	164.5	0.576	
243	63.48	53.5	8.9	1.8	121.9	163.3	0.883	
245	65.23	51.6	8.8	1.8	123.6	163.8	0.752	-70
248	68.35	33.1	9.0	2.0	128.8	172.9	0.775	-78
250	70.78	44.2	9.4	2.0	206.7	192.5	7.71	-84
253	73.79	42.7	9.2	2.0	139.2	163.5	0.515	
255	75.76	39.0	9.2	2.1	151.3	182.1	2.56	
258	79.72	37.9	9.3	2.4	175.5	186.3	4.80	-80
260	86.61	41.3	9.3	2.4	170.3	208.3	0.874	-77
263	91.26	30.4	9.4	2.6	187.6	218.7	3.97	
265	93.78	48.8	9.4	2.7	180.7	230.7	1.90	
268	98.18	42.6	9.5	3.0	194.5	232.3	8.47	
270	100.18	33.2	9.2	3.0	194.5	240.7	1.08	-76
273	102.18	34.6	9.4	3.0	201.5	241.6	7.30	
275	103.36	29.9	9.2	3.5	201.5	264.9	4.64	
278	110.40	40.1						
280	111.65	45.6	9.5	3.4	204.9	282.7	9.22	
283	115.78	27.3	9.7	3.5	204.9	291.8	2.15	-77
285	117.00	31.3	9.6	3.6	206.7	293.3	7.25	
288	120.09	29.9	9.7	3.9	210.1	297.8	5.26	
290	122.86	42.4	9.7	3.8	213.6	287.5	1.24	
293	126.22	46.5	9.6	3.9	215.4	309.1	5.47	-82
295	128.21	48.1	9.7	3.9	215.4	308.7	1.94	
298	130.78	46.8	9.7	4.0	215.4	306.6	4.08	
300	131.77	25.8	9.8	4.0	213.6	299.9	3.51	
303	134.49	33.1		4.1				-77
305	136.24	44.2	9.7	4.4	225.8	303.9	4.96	
308	137.18	45.6	9.8	4.2	222.3	326.2	2.44	
310	138.31	41.3	9.8	4.2	218.8	337.0	1.97	-79
313A	141.27	29.4						

TABLE 1. WATER CONTENT AND PORE-WATER COMPOSITION OF SEDIMENTS FROM OL-92* (continued - page 2)

Sample	Depth (m)	H_2O (wt %)	pH	Salinity (wt %)	Cl (mmolal)	Ct† (mmolal)	SO_4 (mmolal)	δD (‰) V-SMOW
315A	143.01	38.6	9.6	4.4	227.5	347.5	2.28	
318A	145.55	34.7	9.9	3.7	225.8	343.1	4.21	-83
320A	147.27	40.7	9.7	4.4	234.5	314.2	9.06	
322B	148.47	27.8	9.5	4.8	234.5	347.3	1.97	-82
323A	149.23	24.1	10	5.0	231.0	351.3	3.35	-79
325A	152.32	28.4	9.8	4.5	220.6	339.8	2.85	
328A	156.74	37.3	9.8	4.5	222.3	339.9	4.75	-88
330A	162.92	26.5	9.9	4.5	217.1	334.0	5.00	
333A and B	164.54	29.2	9.8	4.4	211.9	314.9	7.31	-92
335A	168.28	19.5		4.5	232.7	333.7		
338A	174.63	36.2	9.8	4.4	199.7	324.6	4.46	-90
340A	177.08	28.9	9.8	3.9	192.8	315.7	1.05	
343A	182.95	38.7	9.7	3.8	187.6	311.6	3.61	-97
345A	185.36	37.4	9.8	3.6	184.1	290.5	5.49	
348A	189.25	33.7						
350A	191.54	26.0		4.0				
353A	195.09	33.2	9.7	2.8	158.2	239.4	2.66	-101
355A	196.28	40.6						
358A	200.92	32.6	9.7	2.4	142.6	205.9	4.02	-99
360A	203.62	41.1	9.5	1.8	134.0	180.2	8.39	
363A	209.04	32.2	9.8	0.70				-100
365A	213.07	33.9						
368A	217.03	16.6	8.9	1.3	172.0	125.8	32.9	
370A	221.41	44.4	9.3	0.90	132.3	125.5	6.92	-101
373A	225.29	29.7	9.0	1.5	111.6	115.0	4.99	-100
375	227.30	14.7		2.1	144.4	122.6	10.6	
378A	233.51	22.5	9.2	1.7	121.9	107.2	18.7	
380	234.89	16.6	8.6	1.4	111.6	39.08	51.2	
383A	237.53	24.2	9.8	1.5	111.6	108.6	2.24	
385A	238.46	13.4		2.2	115.0	113.8	3.23	-106
388A	244.77	16.8	8.9	1.5	115.0	79.32	2.57	
390A	245.67	19.5	9.8	1.6	111.6	107.8	0.629	
392A	250.90	21.7	9.8	1.7	115.0	101.6	0.914	-104
395B	253.17	28.8	9.6	1.6	113.3	102.1	6.79	
398A	262.07	23.3	7.9	1.5	106.4	4.814		
400A	266.12	27.7	9.2	1.3	108.1	96.24	17.3	
402A	266.97	29.8	10.0	1.1	103.0	70.27	3.38	
405	270.86	31.0	9.4	1.1	99.52	71.32	1.86	
406A	271.85	33.5	9.8	1.1	96.07	69.68	1.64	
407A	276.00	35.4	8.6	1.1	94.35	71.19	21.0	
409A	278.36	33.5	8.8	1.2	89.19	64.33	5.67	
410A	279.15	16.7		1.7	88.85	2.290		-104
412A	283.24	34.7	8.5	0.80	87.43	51.69	11.3	
415A	286.53	45.2		0.70	84.02	47.38	7.16	
416A	287.07	52.1	8.4	0.70	82.32	49.00	4.22	
418A	288.14	54.8	8.1	0.60	81.19	30.09	17.2	
419A	288.89	40.4	8.4	0.60	82.89	30.99	11.3	
421A	291.85	40.6	7.8	0.90	84.31	39.77	40.7	
423A	295.46	34.8	8.8	0.60	88.85	39.13	5.68	
426A	297.99	29.0	8.3	0.60	93.39	29.04	11.4	
428A	298.86	30.2	7.9	0.60	91.97	11.29	20.5	
430A	300.75	28.2	8.6	0.60	88.85	29.61	8.41	-106

TABLE 1. WATER CONTENT AND PORE-WATER COMPOSITION OF SEDIMENTS FROM OL-92* (continued - page 3)

Sample	Depth (m)	H$_2$O (wt %)	pH	Salinity (wt %)	Cl (mmolal)	Ct† (mmolal)	SO$_4$ (mmolal)	δD (‰) V-SMOW
431A	301.83	34.7	8.3	0.60	91.97	28.32	6.41	
433A	302.64	51.5	8.2	0.60	90.27	24.50	10.8	-106
436A	304.01	24.4	8.6	0.70	93.39	16.37	12.9	
438A	305.98	37.9	8.6	0.50	87.43	40.31	1.22	-107
441A	307.11	31.6	8.4	0.90	96.23	23.69	2.38	
443A	308.36	32.4	8.9	0.80	88.85	36.47	2.14	-107
446A	310.39	24.0	8.1	1.0	91.97	4.144	24.5	
447A	310.82	23.5	8.2	0.60	93.39	18.96	18.4	
448A	311.21	24.3	8.3	0.60	84.31	7.935	20.4	
449A	311.74	26.6	8.6	0.80	88.85	18.84	6.79	-105
451A	312.83	37.0	8.2	0.60	90.27	16.02	19.1	
453A	315.58	41.4	8.2	0.50	90.27	24.65	7.57	-107
456A	318.46	18.2	8.2	0.70	93.39	3.520	28.5	-95
458A	319.66	29.9	8.4	0.40	93.39	17.30	2.04	
460A	321.24	11.8	8.4	1.2	84.31	2.200		-90
462A	322.18	19.7	7.8	0.80	75.23	14.30	40.0	-93

*Absence of data entry indicates component was not analyzed.
†Ct = total dissolved CO$_2$.

vatively as B(OH)$_4^-$ and was used for analysis of the medium's tortuosity.

Effective diffusion coefficients (τD_0) were measured using a diffusion cell apparatus modified from Stoessell and Hanor (1975). One end of each subcore was immersed in individual 500 mL reservoirs filled initially with a boron-free solution adjusted to the ionic strength of the pore fluids by the addition of sodium chloride. The reservoirs were agitated daily to maintain well-mixed conditions. Ten-mL samples from each reservoir were periodically withdrawn and analyzed for boron concentration. A fourth diffusion cell, without a subcore sample, was used as a control.

Effective diffusivities were calculated from the slope of the reservoir concentration versus time$^{1/2}$ during the period when the reservoir concentration remained less than 1% of the initial pore water concentration (approximately 30 days). The control cell showed no increase in dissolved boron in the reservoir. Effective diffusivities ranged from 3.03×10^{-6} cm^2/sec (148 m sample) to 4.56×10^{-6} cm^2/sec (295 m sample). Using a free water diffusivity for boron of 9.8×10^{-6} cm^2/sec, the resulting tortuosities calculated were 0.37, 0.31, and 0.47 for the depths of 36 m, 148 m, and 295 m, respectively. These tortuosities (and the Li and Gregory, 1974, value of 14.5×10^{-6} cm^2/sec for chloride D_0; see later section on chloride diffusion) yield effective diffusion coefficients for chloride of 5.4×10^{-6} cm^2/sec, 4.5×10^{-6} cm^2/sec, and 6.8×10^{-6} cm^2/sec for these three sections of core.

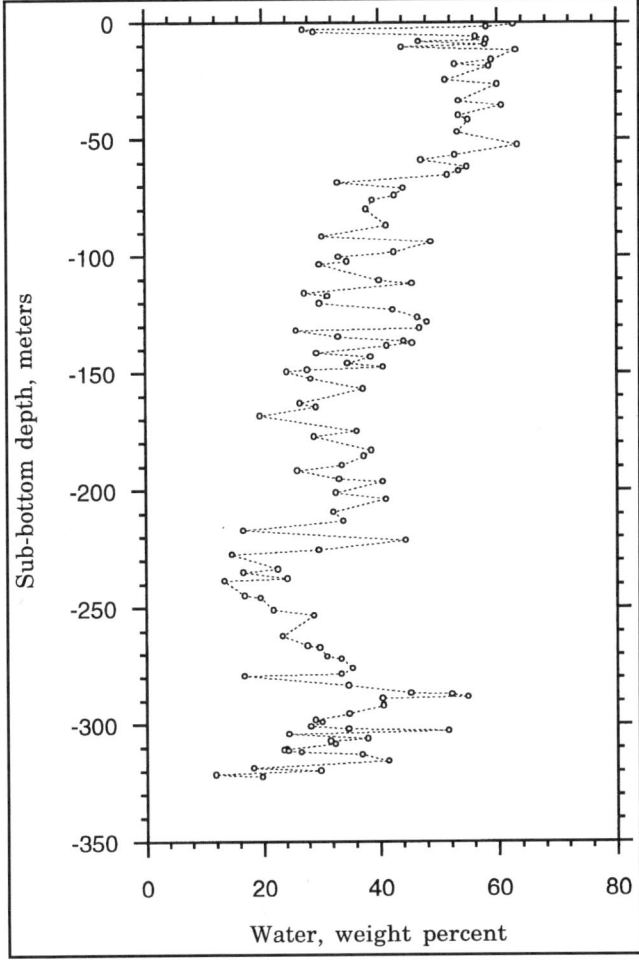

Figure 1. Graph showing the water content, in weight percent, of core samples as a function of depth below lake bottom.

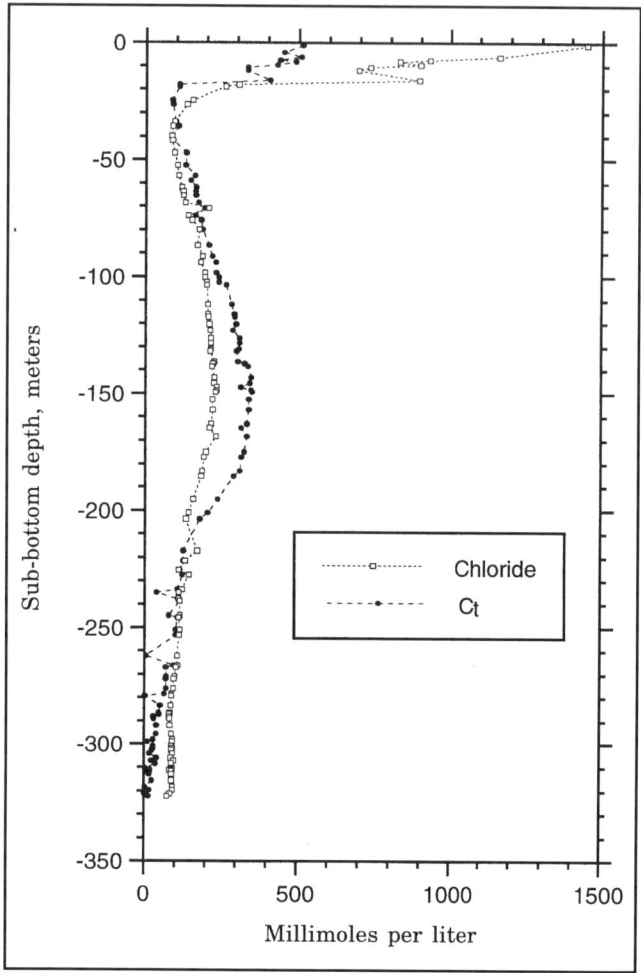

Figure 2. Graph of both chloride and Ct (total dissolved carbon), both in millimoles per liter, plotted against depth below lake bottom.

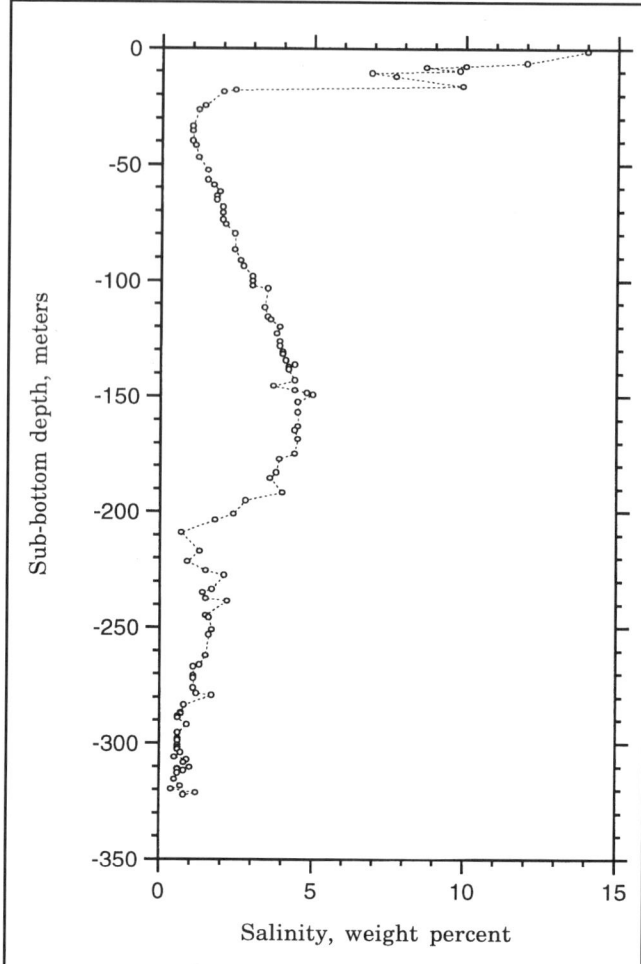

Figure 3. Graph showing salinity, in weight percent, plotted against depth below lake bottom.

DATA AND DISCUSSION

Water content

Water content, which for uniform sediments can be used as a measure of compaction, varies erratically down the core, generally decreasing from about 60 wt % at the top to about 20 wt % at 240 m (Table 1, Fig. 1). Below 240 m to the bottom of the core the water content sharply increases to between 40–60 wt %. This zone is characterized by an abundance of sandy units (Smith, this volume).

Chemistry of pore water

The pre-1872 water of Owens Lake was characterized by an anionic composition of Ct:Cl:SO_4 about 47:47:6 (mole basis), and Na was the only significant cation (Gale, 1914). With a salinity of about 9%, the lake was alkaline and must have had a pH on the order of 10. The average pore water of the sediment, on the other hand, has a pH between 8 and 10, a salinity of only 2.7%, and the anionic proportions are Ct:Cl:SO_4 of 52:45:3. The pore water, therefore, is similar in Ct:Cl to the modern lake, but has a lower salinity and a reduced proportion of SO_4, a probable consequence of the activity of sulfate-reducing bacteria that produce the abundant iron monosulfides that blacken the fine grained sediments. A depth plot of Ct and Cl (Fig. 2) shows that the relative proportions of the two change with depth. With the exception of the salinity peak at ~150 m, there is neither sedimentological nor mineralogical evidence in the entire length of the core that such concentrations of brine and precipitation of saline minerals had been attained before. Below about 50-m depth, both Ct and Cl increase in a smooth pattern. At 40 m, Ct and Cl are about equal as they were in the pre-1872 lake. Below this depth the Cl pattern is more spread out than Ct, and in the region of the salinity maximum at 150 m, Ct exceeds Cl, whereas in the low-salinity region below 240 m, Cl exceeds Ct. The pattern is most readily explained by relative diffusivity of Cl and HCO_3. With a decreasing salinity gradient in both directions from the maximum at 150 m, Cl and HCO_3 are diffusing from this maximum both downward and upward. The coefficient of ionic diffusion (at infinite dilution) for Cl is nearly twice that for HCO$_3$ (2.03 × 10^{-5} versus 1.18 × 10^{-5} cm^2/sec, Li and Gregory, 1974), so it is to be expected that Cl transport away from the salinity maximum will be about twice the rate of HCO$_3$ transport.

Age of pore water

It is usually assumed that the age of pore water is that of the enclosing sediment. Since the age of the base of core OL-92 is approximately 800 ka (Bischoff et al., this volume, Chapter 8), under this assumption the pore fluids in the lower part of the core would also be of approximately the same age. However, upon examining the chemistry of the pore fluids from this core, it became apparent that some of the fluids were appreciably younger than the host sediments. The reasons for this conclusion will be discussed in the following sections.

In his log of core OL-92-3, Smith (this volume) observed that calcareous oolites, found between the upper meter of salts and 5.32 m, were coated with, and contained, gaylussite (Na$_2$CO$_3$ · CaCO$_3$ · 5H$_2$O). He interpreted this phase to have formed early in the twentieth century when diversion of inflow water to Owens Lake into the Los Angeles Aqueduct caused the lake to desiccate, forming a brine of sufficient density to displace the original interstitial fluid with brines that crystallized gaylussite. The brine may have penetrated to depths greater than the limit of gaylussite-coated oolites (5.32 m), but the near absence of CaCO$_3$ between 5.32 m and 41.72 m (Bischoff et al., this volume, Chapter 4) might have prevented the formation of gaylussite below 5.32 m. In any case, the presence of an 80-yr brine in the pores of a sediment radiocarbon dated as 5 ka (Bischoff et al., this volume, Chapter 8) is an indication that the pore fluids have moved downward in the core. The brine probably sank as discrete thin fingers rather than piston like in a broad area (Lyons et al., 1995), resulting in localized convection cells.

Another indication that vertical movement has occurred is to be found by examining the salinity-depth profile. Salinity varies continuously with depth (Table 1, Fig. 3) from a maximum at the top of the core, decreasing rapidly to a minimum at 30 m, gradually increasing to a single broad maximum at about 150 m in depth, and declining thereafter to steady low values at 210 m and below. The salinity of the historic lake (1872) prior to agricultural activity in the watershed, was about 9% (Gale, 1914). If elevated salinity characterizes the various interglacial times when Owens Lake was the terminus of the chain of lakes, and if fresh water characterized the glacial periods of active overflow, one might expect about 8 salinity oscillations during the 800-k.y. time span of the core. Such cycles are seen in the solid components of the sediments, particularly for calcium carbonate and organic carbon content (Bischoff et al., this volume, Chapter 4), indicating the lake did indeed experience such changes. The salinity-depth profile, therefore, has been drastically smoothed by post-depositional processes, including the diffusion of dissolved salts, as well as by possible downward movement of pore waters. The smooth and gradual increase of salinity in the older sediments below 40 m to a maximum at about 150 m, then decreasing to ~250 m, is likely the result of diffusional smoothing of an older cycle. Diffusion should have had more than sufficient time, therefore, to smooth salinity gradients even lower in the core. The role of diffusion will be discussed in detail in later sections.

In the following sections, we discuss the chloride and deuterium-versus-depth profiles in order to assign dates to the pore fluids and to characterize the hydrologic conditions in the lake where the fluids originated.

Chlorinity profile

Sub-bottom depth 10–20 m. The pore water chlorinity peak at 16 m (Fig. 4) is contained in sediments radiocarbon dated at approximately 20 ka (Bischoff et al., this volume, Chapter 8). Could this be the actual age of these pore fluids? To answer this question it is instructive to model the diffusion profile that would be expected if a few meters of saline brine

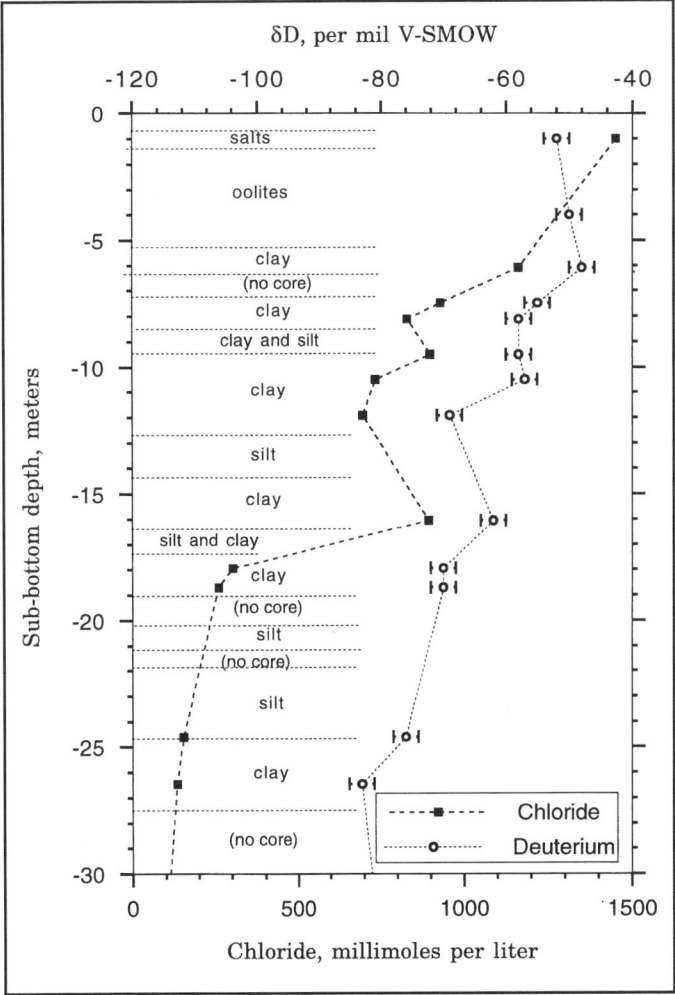

Figure 4. Graph of both chloride concentration and δD as a function of depth below lake bottom from 0–30 m. The core log description for the core sections are also shown. The data points for deuterium are plotted with error bars depicting a two sigma error of ±2‰.

had been deposited over a pore fluid of low salinity during a partial (or total) desiccation of Owens Lake at 20 ka, and then covered with water of a lower chlorinity when the lake was rejuvenated. To model the diffusive loss of chloride from a layer of chloride-rich brine sandwiched between two layers of low-salinity fluid, we used the following equation from Crank (1975, equation 2.15) to describe diffusion from a finite region where the brine with a chloride concentration of C_0 was initially confined in the region $-h < x < +h$.

$$\frac{C}{C_0} = .5\left[\text{erf}\left(-\frac{h-x}{2\sqrt{Dt}}\right) + \text{erf}\left(\text{L}\,\frac{h+x}{2\sqrt{Dt}}\right)\right], \quad (1)$$

where C_0 = original concentration; C = concentration at depth x; x = length of diffusion path in cm; h = thickness (in cm) of the saline layer, t = time in seconds; D = diffusion coefficient of chloride; erf = error function. The value of 3.6×10^{-6} cm^2/sec determined on the 36 m core section was used.

We found that a diffusion time of only ~50 years will yield a diffusion profile that matches the chlorinity change from 16 m to 19 m. This diffusion age agrees approximately with the the probable 80 yr age of the brine, but not with the radiocarbon age of the host sediment.

One explanation for this feature is that the steep chlorinity gradient between 16 m and 19 m is the result of horizontal movement of fluids within the lake sediments, and that the chlorinity peak at 16 m is the result of the the local convection cell caused by the sinking of the post-1913 brine as discussed previously.

The ages calculated from diffusion profiles will depend linearly on the value of the diffusion coefficient (D) that is used. Li and Gregory (1974) determined a value for the diffusion coefficient of chloride in sea water at 16 °C (present mean annual temperature for Owens Lake) as 14.5×10^{-6} cm^2/sec; in red clay it was reduced to approximately 7×10^{-6} cm^2/sec. As discussed previously, the effective diffusion constant is a function of the porosity and tortuosity of the sediments, and our determinations of the effective diffusion constants in the Owens Lake core range from 4.5×10^{-6} cm^2/sec for silty clay sections of the core to 6.8×10^{-6} cm^2/sec for sandy sediments.

It is possible to calculate diffusion ages in the Owens Lake sediments to within a factor of ~2, provided that experimentally derived values of D are determined using relevant sections of the core. Lacking such experimentally derived diffusion coefficients, we believe that the calculated ages are only precise to a factor of 5.

Sub-bottom depth 50–250 m. As mentioned previously, the gradual increase in chloride from approximately 50 m to a broad maximum at 150 m and gradual decrease to about 250 m probably represents a period in the history of the lake when partial or complete desiccation occurred, and the resulting chlorinity maximum was smoothed by diffusion. Although there is no evidence of desiccation in that part of the core, it is possible that a short interval during which desiccation occurred might not leave a record in the sediments.

We use Equation 1 to model the chloride profile to be expected when diffusion occurs in a layer of chloride-rich brine sandwiched between two layers of low salinity fluid. The diffusion model that we used assumes the existence of a sharp chloride concentration profile at time = 0. Because advective movement of fluids would tend to smooth the concentration profile, the initial conditions would have to be modified. The effect of a small amount of smoothing would be to lower the diffusion "age" slightly and would not seriously affect the conclusions regarding the age of this concentration profile.

After consideration of the particle size distribution in the cores which showed that the major portion of the core, particularly below 150 m, was composed of clay, silt, and sand (Smith, this volume), we used a value of 5×10^{-6} cm^2/sec for the diffusion coefficient, an average of the values determined on sections of the core taken at 36 m and 148 m. Plots were made for original thickness (h) of the brine layer in the core of 5 m to 20 m and for diffusion durations of 50 k.y. to 200 k.y. The best fit to the core data, shown in Figure 5, was found for an original thickness *in the core* of 13 m (6.5-m layer of brine in the lake) and a diffusion duration of 100 k.y. This match to the core data suggests that the "age" of the chloride peak at about 150 m is 100 ka ± 50 ka, which corresponds to the last interglacial. However, the age of the sediments at this depth in the core is far greater (ca. 300 ka) and leads to the conclusion that pore fluids must be advecting downward through the core. The downward moving fluids might be leaving the basin, or an alternate hypothesis is that these fluids move downward in the center of the lake (where core OL-92 was taken) and upward along the lake shore where the rising fluid would evaporate to dry salts.

It should be noted that in the Owens core the period 50 ka to 150 ka was characterized by sediments that indicate significant outflow, and in Searles this period was also characterized by a perennial (sometimes shallow, but not dry) lake during which the bottom mud was deposited. Both lines of evidence are in apparent conflict with the above model that would have Owens dry, and hence no inflow to Searles. A possible explanation of this conflict would be that the desiccation and refilling of the lake that we observe in the pore fluids now present at 150 m was very rapid, and as a result left no evidence of its occurrence in the Searles Lake sediments.

Although there is no evidence for this desiccation event in core OL-92, a well, drilled in the northern part of Owens Lake in the Owens River delta (Jacobson et al., 1992) shows evidence for low lake level at about 100 ka. This well, which for much of its length consists of layers of mixed clay and silty clay resembling core OL-92, contains a coarse sand layer from about 50 m to 90 m, dated at approximately 100 ka. This coarse sand indicates deposition under riverine conditions and therefore low lake level.

Deuterium-hydrogen profile

The results of the δD analyses are given in Table 1 and are plotted versus depth on Figure 6. In general, the δD versus depth

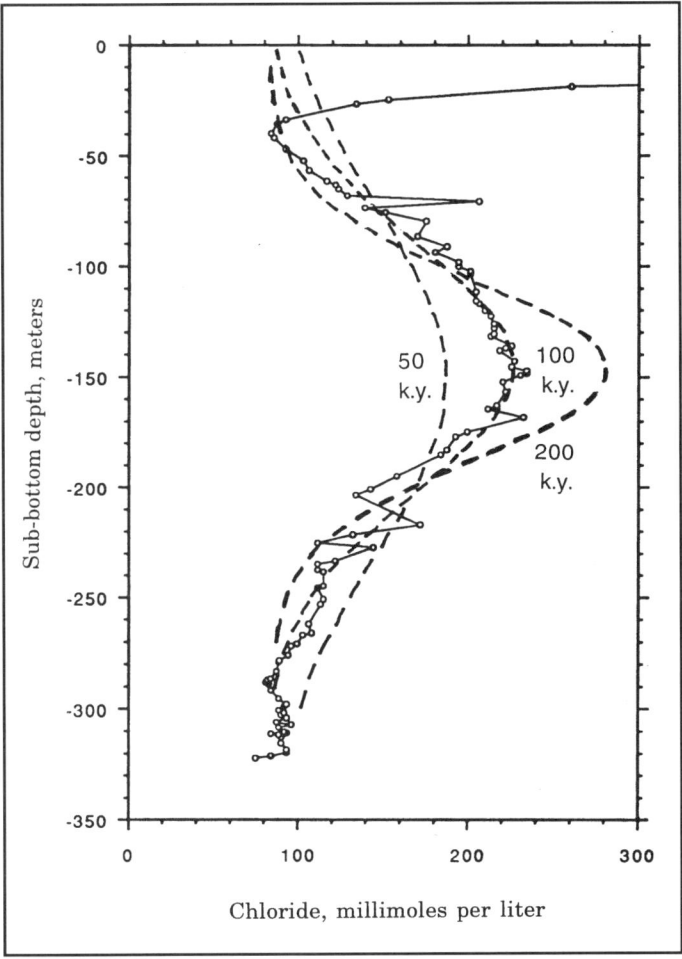

Figure 5. Plot showing chloride concentration profile against depth, with solutions of diffusion Equation 1 superimposed. The diffusion times for the diffusion plots are 50 k.y., 100 k.y., 200 k.y. Note the match of the 100 k.y. diffusion curve to the chloride peak at 150 m.

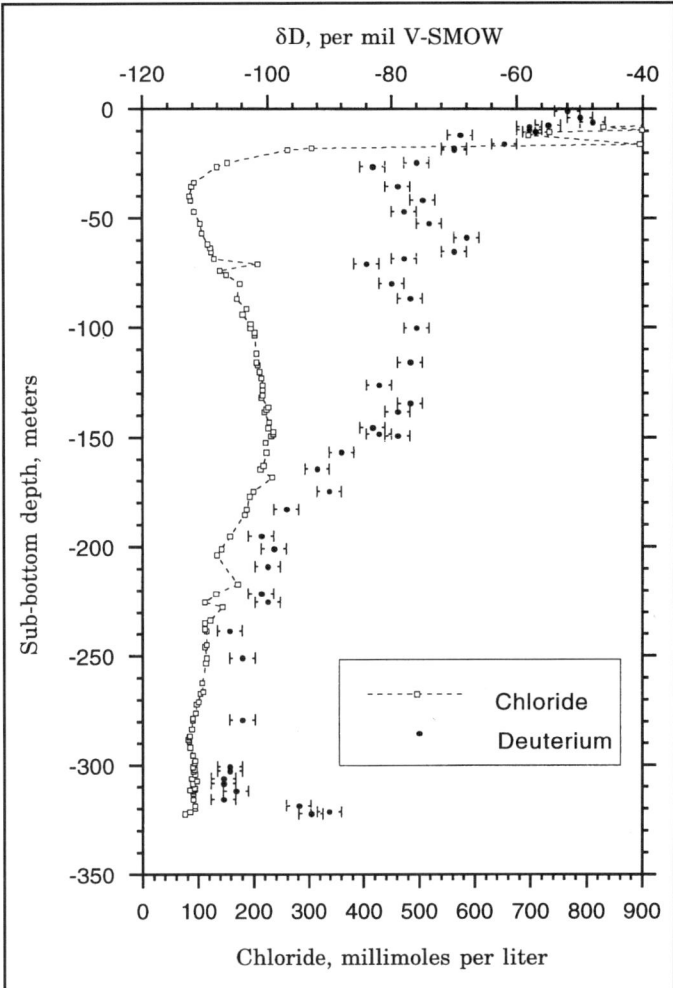

Figure 6. Plots of δD, in per mil V-SMOW, and chloride, in millimoles per liter, against depth below lake bottom. The δD points are plotted with error bars depicting a two sigma error of ±2‰.

profile appears different than the salinity profile. This is because there are processes (discussed later) that can act to greatly change the chloride concentration without a corresponding effect on the δD values. Several features in the δD-depth plot deserve comment.

Upper 11 meters of core. The relative enrichment in deuterium found in the upper 11 m of the core is the result of evaporative enrichment that occurred as the lake desiccated post-1870. The processes that enrich deuterium (and salt) will be discussed in a later section.

Sub-bottom depth 65–70 m. The rapid change in δD observed between ~65 m and 71 m was modeled using Equation 2 (Crank, 1975, equation 2.14) and an effective self-diffusion coefficient for HDO of 7×10^{-7} cm^2/sec (Friedman and Hardcastle, 1988, corrected for temperature). A series of diffusion curves were plotted and superposed on a plot of δD versus depth (Fig. 7).

$$\frac{C}{C_0} = .5\left[1 - \mathrm{erf}\left(\frac{A}{2\sqrt{Dt}}\right)\right], \qquad (2)$$

where C_0 = original concentration; C = concentration at depth A; A = length of diffusion path in cm; t = time in seconds; D = the self-diffusion coefficient for HDO. The initial conditions for this model are: at $t = 0$, a 5 m layer of of brine of a higher δD is deposited over pore fluids of low δD.

The close match between the diffusion curve for 1,000 years and the δD curve suggests that the δD profile cannot be older than a few thousand years, since this is the length of time that would be required for self-diffusion of water to produce the observed δD change with depth, assuming that initially a sharp δD interface existed at a depth of 65 m. This "age" does not agree with data developed in the previous sections on chloride; it cannot be explained by diffusion coefficient changes resulting

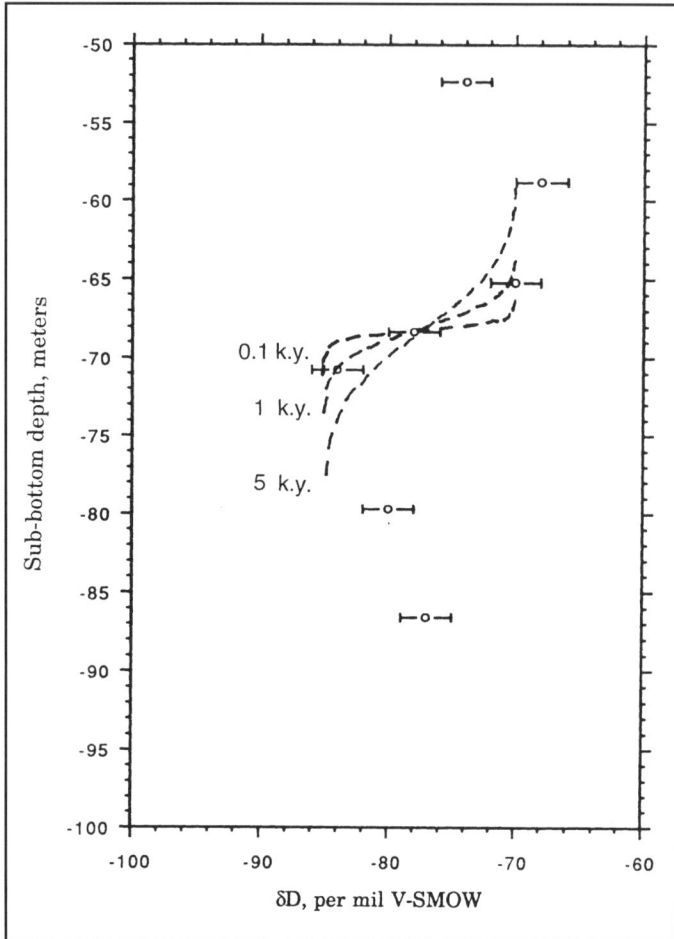
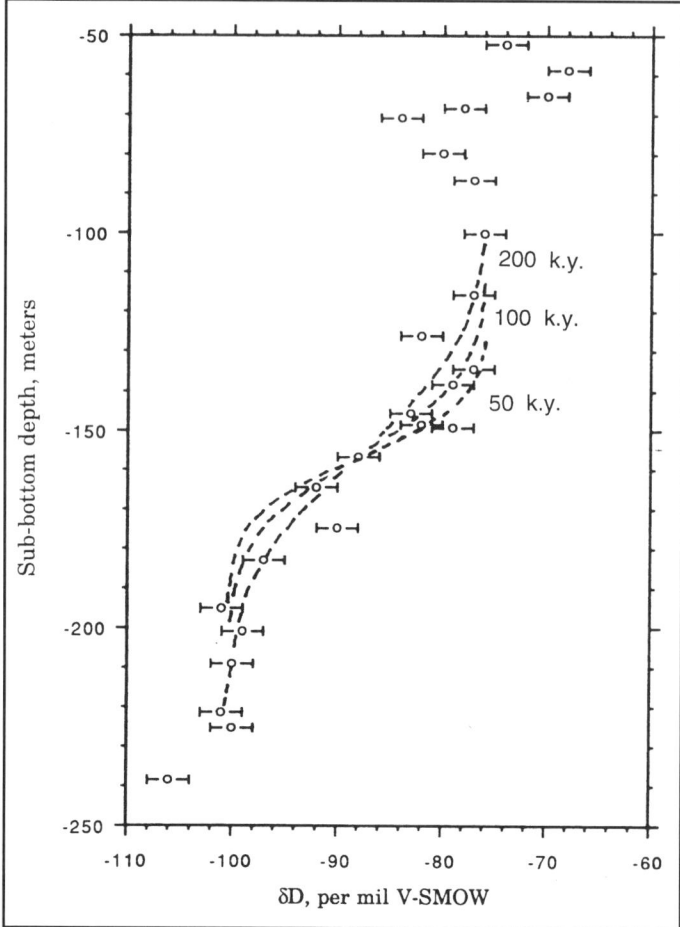

Figure 7. Plot of the δD values against depth, from 50–100 m, with plots of diffusion Equation 2 superimposed. The δD points are plotted with error bars depicting a two sigma error of ±2‰. The diffusion times for the diffusion plots are 0.1 k.y., 1 k.y., and 5 k.y.—see text for parameters used. Note the match of the 1 k.y. diffusion curve with the δD event at approximately 68 m.

Figure 8. Plots of the δD, in per mil V-SMOW and chloride, in millimoles per liter, against depth below lake bottom for depths between 50 m and 250 m. The δD points are plotted with error bars depicting a two sigma error of ±2‰, and for diffusion times of 50 k.y., 100 k.y., and 200 k.y.

from facies changes in the sediments, and it suggests that this sharp δD profile is the result of horizontal movement of fluid in the sediments, similar to the horizontal movement postulated to have occurred at 16 m. We will later show that horizontal movement of fluid also occurred at the bottom of the core.

Sub-bottom depth 150–250 m. Another feature that deserves comment is the depletion of deuterium in the fluids from 150 m sub-bottom depth to approximately 250 m. We have plotted diffusion profiles to these data using Equation 2, assuming that the deuterium changed abruptly at 150 m from −100‰ (its value below 200 m) to its value of −75‰ above 100 m. Because of the large error attached to the δD measurements relative to the δD change through this depth interval, it is difficult to make close comparisons between our data and diffusion curves (Fig. 8). Curves ranging from 50 ka to 300 ka are possible matches. This range of possible deuterium diffusion ages for this feature is in keeping with the diffusion age determined from the chloride profile.

Sub-bottom depth 240–315 m. The relatively low δD values of −104‰ to −107‰ for the 200–315 m interval in the core suggests only limited evaporative enrichment of deuterium in the lake during the time that these pore fluids were present in the lake. As will be discussed later, these δD values could have resulted from a lake where a large proportion of inflow escaped as outflow.

Sub-bottom depth 315–322 m. Due to the intermixing caused by the self-diffusion of water, the sharp δD gradient from 315–322 m could not persist for more than a few thousand years (see Fig. 9) and indicates that fluid is moving horizontally in the sediments adjacent to the bottom of the sediment column sampled by this core. This section of the core is fairly coarse, consisting of silty sand, clay, and silt from about 310 m to 316 m, with mainly sand below this depth.

Rate of vertical transport of pore fluids through the sediments

The rates of movement of pore fluids can be estimated from the diffusion ages calculated from δD and chlorinity profiles.

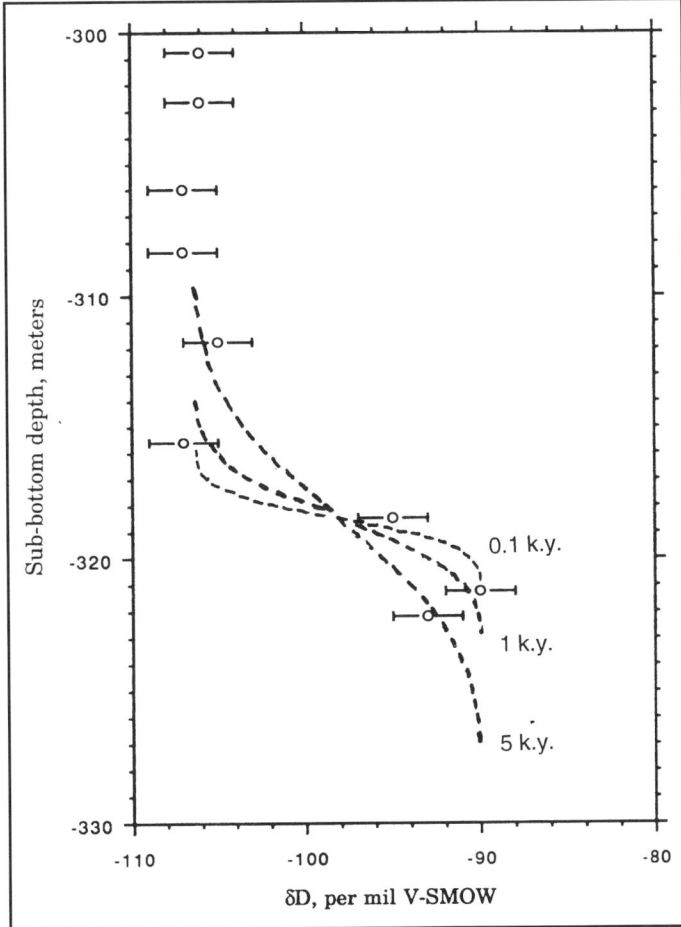

Figure 9. Plot of the δD, in per mil V-SMOW, against depth below lake bottom for depths between 300 m and the bottom of the core at 322 m, and for diffusion times of 0.1 k.y., 1 k.y., and 5 k.y. The δD points are plotted with error bars depicting a two sigma error of ±2‰. Note the correspondence between the δD event at 320 m and the 1 k.y. diffusion curve.

This assumes that these profiles have not been modified by horizontal transport of fluids. Since horizontal movement has been demonstrated for certain sections of the core, the calculated vertical convection rates are, at best, imprecise estimates.

Using the measured value for the chloride diffusion coefficient yields a calculated age of 100 ka for the event at 150 m resulting in a vertical rate of 1.5 m/k.y. and a maximum age of about 220 ka for the pore water in the lowest part of the core.

If fluids are leaving the Owens Lake basin, the annual loss of water based on the above rate of advection is equivalent to less than 0.1% of the (estimated) pre-1870 annual inflow, which is far too small to be measured using conventional water balance measurements.

MODELS OF EVAPORATING LAKES IN ARID ENVIRONMENTS

With the exception of the few, but important, observations of Gale (1914) on the chemistry of Owens Lake in 1870, no systematic studies of the lake were made before Dub (1947) described the evaporation of the lake following flooding of the dry lake during the winter of 1938–1939. We will use the data of an earlier chemical and isotopic study of the desiccation of the lake that followed its flooding in 1969 (Friedman et al., 1976) to model the paleo–Owens Lake.

A number of models have been developed to explain the hydrology of lakes in arid climates; these will be discussed in the following sections and applied to Owens Lake. The simplest of these is the Rayleigh distillation model.

Rayleigh distillation

Isotopic enrichment. Although this simple model has been used to explain many cases of natural evaporation, we believe that it is an example of evaporation that seldom applies directly to the natural environment. However, with some modification it is useful in describing evaporation in arid environments. It was derived by Lord Rayleigh to explain distillation of a binary liquid mixture, where the vapor pressure of the two constituents differed from each other; for example, a mixture of alcohol and water. It can also be applied to the distillation or evaporation of a mixture of isotopes, where the isotopic species have different vapor pressures—as do the isotopic molecules H_2O and HDO.

During evaporation of a binary mixture of two species, A and B, the composition $(A/B)_t$ of the remaining liquid at time t is given by:

$$\left(\frac{A}{B}\right)_t = \left(\frac{A}{B}\right)_0 f^{\alpha-1},$$

where $(A/B)_0$ is the initial composition; f is the fraction of the liquid remaining at time t; α is the ratio of the vapor pressures of species A and B.

In deriving this equation it is assumed that the mixture is a ideal solution, where the vapor pressures of A and B are unaffected by the composition of the mixture. This is true in the solutions under consideration, where the low abundance of the isotopic species HDO renders natural water a perfect solution in reference to deuterium. The assumption is also made that the vapor over the liquid is saturated and is at all times in equilibrium with the liquid. As evaporation proceeds, the residual liquid as it decreases in volume, first slowly, then rapidly, becomes enriched in the less volatile (lower vapor pressure) component HDO (see Fig. 10).

In a Rayleigh evaporation, the residual liquid can become highly enriched in the heavier (less volatile) isotopic species, whereas the vapor, which is being continuously removed, is enriched in the lighter (more volatile) species. The composition of the vapor in equilibrium with the liquid is at any time $(1 - \alpha)$ times that of the liquid. If the composition of the vapor in contact with the liquid, instead of being determined solely by that of the liquid, is in part vapor of unspecified composition (e.g., vapor present in ambient air passing over an evaporating lake), the evaporation will proceed to adjust the vapor composition so

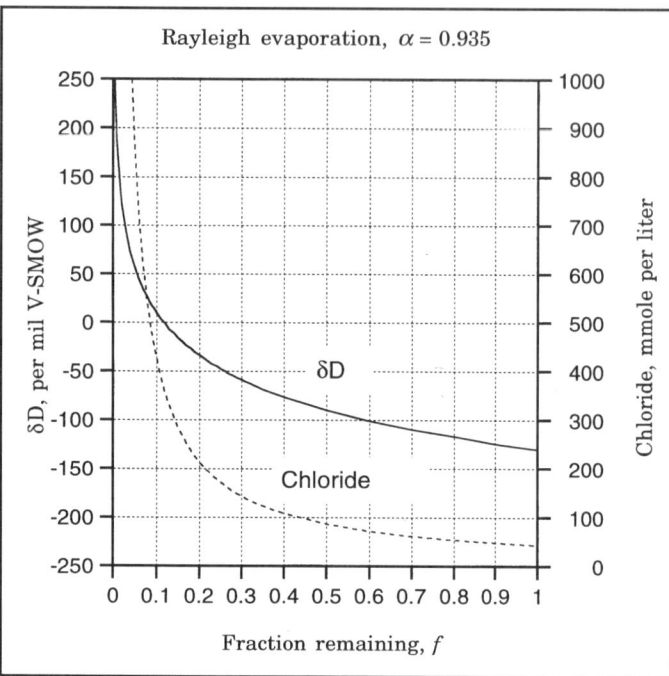

Figure 10. Rayleigh plots of δD, in per mil V-SMOW and chloride, in millimoles per liter, against f, the fraction of liquid remaining. The δD curve was calculated for a fractionation coefficient (α) of 0.935, and an original δD of $-130‰$. The chloride concentration of the inflow water was taken as 43 mmol/L.

that the saturated vapor, which now consists in part of ambient water vapor and in part of vapor that has left the liquid, differs in composition from the liquid by $1 - \alpha$.

Salt enrichment. During a Rayleigh evaporation, salt is not lost to the vapor, and salinity will increase in the remaining lake as $1/f$

$$S_t = S_0 \left(\frac{1}{f} \right),$$

where S_t = salinity of the lake at time = t; S_0 = salinity of original lake at time = 0; and f = fraction of the original lake volume remaining at time = t. Figure 10 is a plot illustrating the changes in δD and chloride during a Rayleigh evaporation, using a starting δD and a value of α that might reasonably apply to Owens Lake.

A modification of the Rayleigh evaporation process can be made by assuming that the α is not the true ratio of the vapor pressure of the two evaporating species, but is affected by two processes: (1) by kinetic effects—for example, differences in the diffusion rates of the two isotopic species (H_2O and HDO)—as diffusion occurs from a saturated vapor layer immediately above the liquid surface to the unsaturated air above, and (2) the isotopic exchange between the water vapor present in the ambient air and the water surface. These effects have been discussed by Craig et al. (1963); Craig and Gordon (1965); Ehhalt and Knott (1965); Eriksson (1965); Merlivat (1970); Gat (1970); Stewart (1975); Allison et al. (1979). For deuterium, the influence of kinetic factors on the isotopic fractionation during evaporation is small, and can usually be ignored.

The difficulty with using these models of evaporation is that they require knowledge of the relative humidity and the δD of atmospheric water vapor, these values integrated over the total time-period that the evaporation took place. These two parameters are difficult to measure, particularly since the values apply to the air immediately above the evaporating water surface. For examples of the difficulty in applying these "exact" models of evaporation, see Gat (1970), Zimmermann and Ehhalt (1970), and chapters in *Isotopes in Lake Studies* (1979) authored by Allison et al. (1979); Matsuo et al. (1979); Zimmermann (1979). It should be noted that evaporation models that correct for the assumed equilibration of the lake with ambient water vapor generally yield δD values in the lake that are not as enriched in deuterium as those calculated using a Rayleigh model, and the lake also tends to approach a limiting δD value as evaporation proceeds. This limiting value is determined by the δD difference between the lake and the ambient water vapor, and by the relative humidity. At relative humidities below about 25%, the influence of the ambient air on the δD of the lake is relatively small.

In applying deuterium abundance measurements to evaporation in arid environments, is possible to make use of the equilibrium Rayleigh process, with the modification that the α used is not the ratio of vapor pressures, but is the "effective alpha," α', that is calculated from parameters that are measured in the region under examination. This approach was used by Friedman and Redfield (1971) in their study of the hydrology of the lakes of the lower Grand Coulee, Washington, as well as by Friedman et al. (1976) in their study of the desiccation of Owens Lake in 1969–1971, and it is illustrated by unpublished data obtained on the Salton Sea, a large lake in the Imperial Valley of southeastern California (see discussion in the next section of the steady state).

Figure 11 shows a plot of the enrichment in deuterium that occurred when Owens Lake desiccated in 1969–1971 (data from Friedman et al. 1976), together with a plot of the enrichment to be expected if the evaporation proceeded by a Rayleigh evaporation. It can be seen from this figure that the evaporation of Owens Lake closely follows a Rayleigh evaporation. We have also plotted the δD data of Fontes and Gonfiantini (1967) on the wadi (Guelta) Gara Dibas in the northwestern Sahara, as well as the solution of the equation derived by the authors to correct for the effect of atmospheric water vapor on the evaporation. With the exception of the most enriched data point, their data also fit a Rayleigh evaporation with a constant α. As the authors observed, the α is not constant during evaporation, but decreases as the salinity of the lake reaches high values, and this change in α can explain the deviation of these data points from the constant-α Rayleigh curve.

The steady state

If the volume of an evaporating lake remains constant, the lake can be said to be in a steady state with regard to volume.

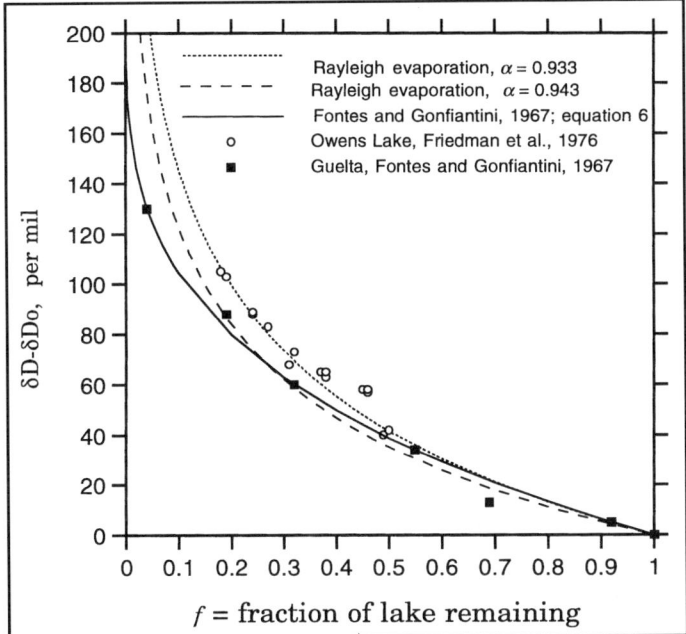

Figure 11. Plot showing (1) the enrichment in δD that occurred as Owens Lake desiccated following flooding of the dry lake in 1969 (open circles). Plots of the enrichment to be expected using a Rayleigh evaporation model with an α of 0.933 (dotted curve) and α of 0.943 (dashed curve) are also shown. Data from Friedman et al. (1976). (2) A similar plot for wadi (Guelta) Gara Diba (northwestern Sahara) from the data of Fontes and Gonfiantini (1967) (filled squares), together with a plot of their Equation 6 (solid curve).

Following the discussion presented by Friedman and Redfield (1971), two models will be developed. The first is where there is no outflow from the lake, and because inflow to the lake equals evaporative water loss, the volume of the evaporating water body remains relatively constant. The second model to be developed is where outflow occurs, but where evaporation plus outflow equals inflow, and the lake level again remains constant. These two models will be treated separately.

Inflow = evaporation. No outflow, lake level constant:

Q_i = rate of inflow;
Q_v = rate of evaporation;
R_i = deuterium/hydrogen ratio in inflow;
R_v = deuterium/hydrogen ratio in vapor;
R_L = deuterium/hydrogen ratio in lake;

Mass balance requires that

$$Q_i = Q_v.$$

The deuterium balance requires

$$R_i Q_i = R_v Q_v.$$

Since in a Rayleigh evaporation

$$\frac{R_v}{R_L} = \alpha,$$

and substituting

$$\frac{R_i}{R_L} = \alpha, \qquad (3)$$

then

$$R_i = \alpha R_L = R_v.$$

Under these conditions, the deuterium concentration in the vapor is equal to that in the inflow, and the deuterium concentration in the lake is equal to

$$R_L = \frac{R_i}{\alpha}. \qquad (4)$$

In this model, the deuterium value in the lake will reach a steady-state value equal to R_i/α, whereas the concentration of salts in the lake will continue to increase until the solubility products of solid phases are reached, and one or more salts precipitates.

The Salton Sea in the Imperial Valley of southeastern California affords an example of the use of this model. The present Salton Sea was created by inflow from the All American Canal (Colorado River water) in the early part of the 1900s, has no outflow, is in a steady state with a relatively constant area, and all inflow is removed through evaporation. More than 90% of the inflow is from the Alamo and New Rivers (almost all from the New River), the remainder coming from springs and precipitation on the lake surface. The estimated integrated inflow has a δD value of –96‰ (see Table 2). The lake is well mixed and has a δD value of –34‰. The α' calculated from the δD values of the inflow water and the lake water, using Equation 3, is 0.904/0.966 = .936. Assuming that the evaporation occurred at 30 °C to 33 °C (reasonable values for Salton Sea surface temperature), the theoretical α would be between 0.929 and 0.931 (Friedman and O'Neil, 1977). Considering the uncertainties in the inflow δD and the temperature, the effective α is in reasonable agreement with the theoretical value.

As mentioned previously, the effective α will be a function of the salinity and chemical composition of the salts dissolved in the lake. Its value will also be effected by kinetic effects that affect the evaporation process, as well as by the influence of water vapor present in the ambient air. In the case of lakes in arid climates, the effects of the ambient water vapor and kinetic effects during evaporation have minimal effect upon the value of the effective α, and values of this constant can be used that approximate the theoretical value. The change in the value of the effective α as evaporation raises the lake salinity was observed by Fontes and Gonfiantini (1967) in their investigation of the Sebkha el Melah, a saline basin in the Algerian Sahara, and by Friedman et al. (1976) in their study of the desiccation of Owens Lake. During most of the desiccation of Owens Lake, the α' for the evaporation process was constant at 0.935 (the authors used the inverse α, $1/\alpha$, in their discussions), and increased to 0.976 as the salinity increased to high values during the final desiccation of the lake.

TABLE 2. δD OF SALTON SEA AND INFLOW WATERS

Sample	Description	Collection Date	δD (‰)
3278-1	New River near Niland, California	February 2, 1967	-96
3278-10	New River near Niland, California	September 7, 1967	-97
3278-7	Alamo River, inflow to Salton Sea	February 2, 1967	-93
3278-11	Alamo River, inflow to Salton Sea	September 7, 1967	-85
3278-6	Whitewater River, inflow to Salton Sea	February 2, 1967	-80
3278-12	Whitewater River, inflow to Salton Sea	September 7, 1967	-85
3278-2	Salton Sea, site A	February 2, 1967	-37
3278-3	Salton Sea, site B	February 2, 1967	-34
3278-4	Salton Sea, site C	February 2, 1967	-33
3278-5	Salton Sea, site D	February 2, 1967	-33
3278-8	Salton Sea, site E	February 2, 1967	-36
3278-9	Salton Sea, site F	February 2, 1967	-35
3278-13	Salton Sea, site G	September 6, 1967	-33
3278-14	Salton Sea, site H	September 6, 1967	-34
3278-15	Water vapor collected near Niland*	September 6, 1967	-125

*This sample is probably a mixture of free air water vapor plus water vapor evaporated from the Salton Sea.

Inflow = evaporation + outflow. The previous treatment of the steady state can be expanded to include outflow.

S = concentration of salt
Q_o = rate of outflow;
Q_v = Rate of evaporation;
R_o = deuterium/hydrogen ratio in outflow;
R_L = deuterium/hydrogen ratio in lake

$$Q_i = Q_v + Q_o$$

$$R_i Q_i = R_v Q_v + R_o Q_o$$

and if the lake is well mixed

$$R_L = R_o$$

$$R_i Q_i = R_v Q_v + R_L Q_o.$$

If θ = fraction of inflow escaping as outflow,

$$Q_o = \theta Q_i$$

and $1 - \theta$ = fraction of inflow escaping as vapor,

$$Q_v = (1 - \theta) Q_i$$

$$\frac{R_i}{R_L} Q_i = \frac{R_v}{R_L} Q_v + Q_o$$

$$\frac{R_i}{R_L} = \alpha - \alpha\theta + \theta$$

$$R_L = \frac{R_i}{\alpha - \alpha\theta + \theta} \quad (5)$$

Under these conditions, the deuterium concentration in the lake will increase because R_L is less than in the previous case where $\theta = 0$.

Because no salt is lost to vapor, under steady-state conditions the amount of salt flowing into the lake equals that flowing out.

$$S_o Q_o = S_i Q_i$$

Since the salinity of the outflow is the same as the salinity of the lake

$$S_o = S_L$$

$$S_L = \frac{S_i Q_i}{Q_o}, \text{ and}$$

$$\frac{Q_i}{Q_o} = \frac{1}{\theta}, \text{ so that}$$

$$S_L = \frac{S_i}{\theta}. \quad (6)$$

Therefore, the steady-state salinity of the lake is equal to the salinity of the inflow divided by θ, the proportion of the inflow escaping as outflow.

Figure 12 is a plot illustrating the changes in δD and chloride during a steady-state Rayleigh evaporation with outflow, as a function of the proportion of inflow escaping as outflow, using the same starting δD and the value of α that were used in Figure 10. Note that in this model, the δD will undergo only a moderate increase as the proportion of outflow decreases, as compared to a Rayleigh evaporation with no outflow (compare Figs. 10 and 12).

The step state

Lakes in arid regions commonly overflow during pluvial periods. Later, as evaporation increases or inflow decreases, the outflow ceases, and the lake remains for some time in a steady state with no outflow. As inflow decreases further, and/or evaporation continues to increase, the lake may desiccate. During this drying-up period, two conditions may occur.

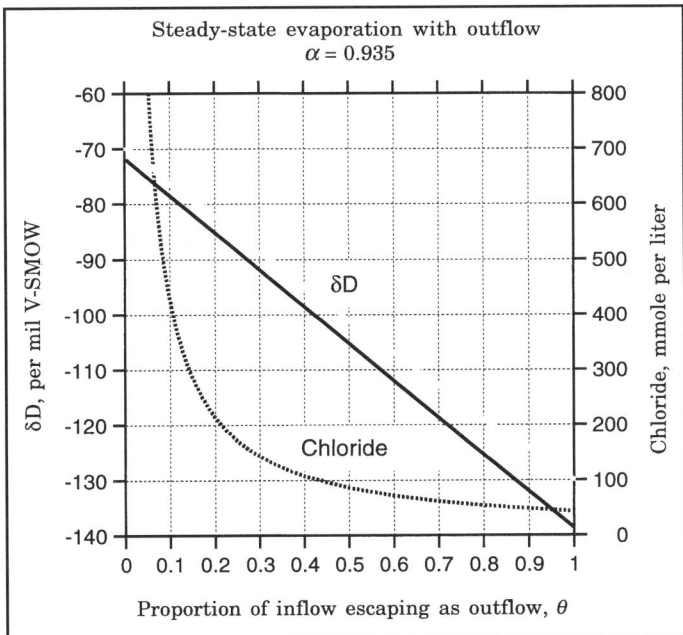

Figure 12. Plots of δD, in per mil V-SMOW, and chloride, in mmol/L, against θ, the proportion of inflow that exits as outflow during a steady-state evaporation with outflow. The δD curve was calculated for a fractionation coefficient (α) of 0.935, and an original δD of $-130‰$. The chloride concentration in the inflow was taken as 43 mmol/L.

1. The inflow is cut off completely, as for example, the level of a feeder lake upstream in a chain of lakes falls below the level of its outlet dam. If little or no inflow by local precipitation occurred in the lake drainage area, the desiccation of the lake would follow a Rayleigh-type evaporation, and high deuterium concentrations in the lake can be expected near the time of complete desiccation (Fig. 10).

2. If inflow does not cease, but is less than evaporation, and if lake level fluctuates from time-to-time because of variable rates of inflow and evaporation, a step-wise desiccation will occur. In order to clarify the deuterium relations in this model, let us first consider a case where the lake is at a steady state with no outflow (inflow = evaporation). The inflow δD will be taken as $-130‰$ ($R_i = 0.87$), and $\alpha = 0.935$. The lake will then stabilize according to Equation 4 at

$$R_L = \left(\frac{.870}{.935}\right) = .930 = -70‰.$$

If inflow now ceases, and the lake evaporates to one-half its original volume, its deuterium concentration will increase through a Rayleigh evaporation to $-27‰$. If some inflow is now reestablished, the lake will restabilize itself in a steady state (again, with no outflow) and will preferentially lose deuterium until it is again at $-70‰$. As long as significant inflow continues, there will be a tendency for the lake to maintain the steady-state δD value. However, the salt concentration will continue to increase. If we envision the final desiccation as taking place as a succession of steady states, we can arrive at a final near-desiccation without a large increase in deuterium over that in the steady-state condition. When inflow finally ceases, the δD will increase to high values.

We believe that this stepwise steady-state process commonly occurs in saline basins prior to desiccation, and that this process can explain the scarcity of highly enriched waters in these basins (compared to inflow stream and ground waters). This process, plus the reduction in the value of the effective α as the brines became very concentrated, can explain the relatively light δD value of $-50‰$ found in the upper few meters of the Owens Lake pore fluids when the lake had shrunk to a small volume. Another explanation for the lack of water that is highly enriched with deuterium in highly evaporated brines is the effect of atmospheric water vapor that will tend to buffer δD in the evaporating lake. This effect will be more pronounced the greater the difference in δD between the atmospheric water vapor and the lake, and the higher the relative humidity of the ambient air. However, it is interesting to note that during the 1969 desiccation of Owens Lake, a δD value of $+3‰$ was reached before rain diluted the δD and other constituents of the brine (Friedman et al., 1976). Buffering by atmospheric water vapor (whose δD was probably $<-125‰$) apparently was not an important process in this example.

MODELS OF THE ANCESTRAL OWENS LAKE

Models for the ancestral Owens Lake will be developed using the processes discussed in the previous sections.

Model of the pre-Holocene lake

The present Owens River inflow to Owens Lake has an estimated δD of $-120‰$ (Friedman et al., 1976). Smith et al. (1992) found that the δD of ground water that was postulated to have been recharged during the Pleistocene had δD values approximately $10‰$ or more depleted in deuterium as compared to modern recharge. Using this information, the Pleistocene Owens River would have had a δD value ranging from a low of at least $-130‰$ (glacial) to $-120‰$ (interglacial). The δD value of $-104‰$ to $107‰$ now present between 240 m and 320 m could be the result of evaporative enrichment in a steady-state lake with 50% of this inflow escaping as outflow ($\theta = 0.5$).

The relatively large outflow of Owens Lake for an extended period post-760 ka can explain the formation of both the bottom mud and the upper part of the mixed layer (Smith, 1979) in the Searles Lake sediments. The mixed layer consists of units with changes in evaporite mineral content indicating changing conditions in the lake during their deposition that, in turn, would require alternating periods of relatively high and low salinity in Owens Lake. However, because of the age of these deposits, diffusion in the pore fluids contained in the sediments beneath Owens Lake would have removed evidence of these salinity (as well as δD) variations. In addition, the horizontal movement of fluid in the bottom layers of sediment would also aid in the removal of these salinity and deuterium alternations.

The present chlorinity of the Owens River inflow to Owens Lake is 14 ppm (0.4 mmol/L; Friedman et al., 1976). The chloride concentration in the river was probably higher than the present concentration following volcanic eruptions in Long Valley that led to deposition of the 780 ka Bishop ash found at the base of the core. The steady-state model would suggest that the chloride concentration in the inflow for an extended time following the eruption of the Bishop ash was ~1,500 ppm (43 mmol/L) and that evaporation increased the chlorinity to that now found in the fluids in the 240–320 m interval. This conclusion is partially supported by a calculation of the amount of chloride deposited in downstream Searles Lake during a period of ~200 k.y. beginning 115 k.y. after the eruption in Long Valley; it finds that the chloride concentration in the Owens River during that period averaged ~2.3 times its modern concentration (Smith et al., this volume).

The chloride that is being expelled at present from the Yellowstone hydrothermal system illustrates the large quantities of salt generated by rhyolite magma systems. Norton and Friedman (1985) estimated the present *annual* geothermal chloride flux from Yellowstone as 5.8×10^{10} g. Comparing this quantity with the 4.6×10^{14} g estimated chloride content of the Searles basin (Friedman et al., 1982), shows that rhyolitic eruptions of the scale of Yellowstone or Long Valley could account for all the salt now collected in the Searles basin.

At about 100 ka (fluids now present at core depth ~150 m), Owens Lake rapidly shrank, resulting in a ~6.5-m layer of concentrated brine, which produced a 13-m thick layer of fluid in the sediment of 50% porosity. This episode was quickly followed by a return to the previous regime of outflow. The desiccation and refilling of the lake occurred so rapidly that no evidence of these events is preserved in the sediments. Note that following this event, the δD values in Owens Lake increased from about –100‰ to about –75‰. This increase in δD could have been the result of a decrease in outflow relative to inflow to the lake. This might have been caused by increased evaporation or decreased inflow.

Model of the Holocene lake

Early in the Holocene, inflow to Owens Lake decreased, whereas evaporation increased as a result of warmer and drier climatic conditions. Enough overflow to maintain Searles as a perennial lake stopped at ca. 10 ka, allowing it to desiccate (Smith, 1979). The latest date that Owens Lake could have overflowed is based on a ^{14}C-dated core from Little Lake that consisted of sediments representing uninterrupted lacustrine deposition since 5 ka (Mehringer and Sheppard, 1978). The small lake lies in Owens Lake's paleochannel and 30 km downstream from its spillway, and the alluvial fan that dams Little Lake would have easily been breached by any subsequent overflow from Owens. This date is further confirmed by the disconformity between 8 ka and 5 ka in the Owens core sediments (Bischoff et al., this volume, Chapter 8).

From >5 ka to 1872 Owens Lake shrank in size and increased in chlorinity to 40,000 ppm. Beginning in 1870, the diversion by irrigation of 25% of inflow resulted in the further shrinking of the lake. Total diversion of inflow water into the Los Angeles Aqueduct in 1913 caused the lake to desiccate by about 1921.

The δD value in the top 6 m of core is –50‰. During the desiccation following the 1969 flooding of the lake, the δD increased by Rayleigh evaporation to +3‰, but precipitation on the lake quickly diluted the brine with low δD water, yielding values in the –80‰ range (Friedman et al., 1976).

Another similar chain-of-lakes

It is interesting to compare the Owens pore water salinity and δD with that reported for the lakes of the lower Grand Coulee in Washington State (Friedman and Redfield, 1971). The series of five lakes that occupied the abandoned gorge of the Columbia River below Dry Falls, Washington, was fed by ground water and precipitation and lost water solely by evaporation; the lake series terminated with Soap Lake, which became saline. The inflow water to the system had a δD of –128‰ and a salinity of 6 ppm. In 1945–1946, before irrigation inflow disturbed the system, Soap Lake had an estimated δD of –65‰ and a salinity of 5,000 ppm. It is estimated that it required about 2,350 years of no outflow from Soap Lake for it to accumulate sufficient salt to reach this salinity level. With the exception of the somewhat lower salinity inflow to the system, these δD and salinity values are almost identical to those for the lower portions of the Owens Lake core.

CONCLUSIONS

1. Both vertical and horizontal movement of the fluids within the sediments have been demonstrated, therefore the interstitial fluids extracted from the sediments sampled in the Owens Lake Core OL-92 were not deposited contemporaneously with the sediments.

2. This fluid movement might result in fluid transport out of the basin. The loss of interstitial fluid from the sub-bottom sediments of another "closed" basin (Searles Lake) was proposed by Friedman et al. (1982).

3. For an extended period following the eruption of the 760 ka Bishop ash, the concentration of chloride in the inflow to Owens Lake was ~100 times greater than the present concentration.

4. During most of the Pleistocene, the outflow from Owens Lake was sufficient to allow the accumulation of large quantity of saline minerals in downstream Searles Lake. During this period only minor evaporative enrichment of deuterium occurred in Owens Lake.

5. Based on a diffusion model of the chlorinity-depth profile, a period of desiccation followed by flooding of the lake and outflow occurred at approximately 100 ka. Although no evidence for this event can be found in core OL-92, another well, drilled in

the Owens River delta region of the lake, contains evidence of low lake level at this time. Because no evidence of this event is found in the sediments of Searles Lake, this episode must have been of short duration.

6. Outflow ceased about 6 ka.

7. The lake slowly decreased in size until 1870, at which time major diversion of inflow by irrigation upstream resulted in the lake rapidly shrinking in size. Total diversion commenced in about 1913, causing the lake to completely desiccate by 1921. The δD of the "final" brine was ~–50‰. This relatively light δD value may have been the result of dilution by precipitation and spring discharge within the Owens Lake basin during the final desiccation. The effectiveness of precipitation in influencing the δD of these final brines is illustrated in the study of the 1969–1971 desiccation of Owens Lake (Friedman et al., 1976).

ACKNOWLEDGMENTS

The authors thank Richard Ku and George I. Smith for their helpful reviews which greatly improved this manuscript.

REFERENCES CITED

Allison, G. B., Brown, R. M., Fritz, P., 1979, Evaluation of water balance parameters from isotopic measurements in evaporation pans, in Isotopes in lake studies: Proceedings, Advisory Group Meeting, Vienna, August 29–September 2, 1977: Vienna, Austria, International Atomic Energy Agency, p. 21–32.

Benson, L., and Bischoff, J. L., 1992, Isotope geochemistry of Owens Lake drill hole OL-92: U.S. Geological Survey Open-File Report 93-683, sec. 4.3.1.

Craig, H., and Gordon, L. I., 1965, Deuterium and oxygen 18 variations in the ocean and marine atmosphere, in Stable isotopes in oceanographic studies and paleotemperatures, a series of lectures presented in Spoleto, Italy, July 1965: Pisa, Italy, Consiglio Nazionale delle Ricerche, Laboratorio di Geologia Nucleare, p. 9–130.

Craig, H., Gordon, L. I., and Horibe, Y.,1963, Isotopic exchange effects in the evaporation of water: Journal of Geophysical Research, v. 68, p. 5079–5087.

Crank, J., 1975, The mathematics of diffusion (second edition): Oxford, England, Clarendon Press, 414 p.

Dub, G., 1947, Owens Lake—Source of sodium minerals [Calif.]: American Institute of Mining, Metallurgical, and Petroleum Engineers Technical Publication 2235, p. 1–13.

Ehhalt, D. H., and Knott, K., 1965, Kinetisches isotopentrennung bei der verdampfung von wasser: Tellus, v. 17, p. 118–130.

Ericksson, E., 1965, Deuterium and oxygen-18 in precipitation and other natural waters: Some theoretical considerations: Tellus, v. 17, p. 498–512.

Fontes, J. C., and Gonfiantini, R., 1967, Comportement isotopique au cours de l'evaporation de deux bassins Sahariens: Earth and Planetary Science Letters, v. 3, p. 258–266.

Friedman, I., and Hardcastle, K., 1988, Deuterium in interstitial water from deep-sea cores: Journal of Geophysical Research, v. 93, p. 8249–8263.

Friedman, I., and O'Neil, J. R., 1977, Fractionation factors of geochemical interest: Data of Geochemistry (sixth edition), Chapter KK: U.S. Geological Survey Professional Paper 440, p. 1–12, 48 figures.

Friedman, I., and Redfield, A. C., 1971, A model of the hydrology of the lakes of the Grand Coulee, Washington: Water Resources Research, v. 7, p. 874–898.

Friedman, I., Smith, G. I., and Hardcastle, K. G., 1976, Studies of Quaternary lakes—II. Isotopic and compositional changes during desiccation of the brines in Owens Lake, California, 1969–1971: Geochimica et Cosmochimica Acta, v. 40, p. 501–511.

Friedman, I., Smith, G. I., and Matsuo, S., 1982, Economic implications of the deuterium anomaly in the brine and salts in Searles Lake, California: Economic Geology, v. 77, p. 694–702.

Gale, H. S., 1914, Salines in the Owens, Searles, and Panamint basins, southeastern California: U.S. Geological Survey Bulletin 580-L, p. 251–323.

Gat, J. R., 1970, Environmental isotope balance of Lake Tiberias, in Isotope hydrology 1970: Vienna, International Atomic Energy Agency, p. 109–127.

Gat, J. R., 1970, Isotopes in Lake Studies: Proceedings, Advisory Group meeting, Vienna, August 29–September 2, 1979: Vienna, Austria, International Atomic Energy Agency, 290 p.

Jacobson, E. A., Cochran, G. F., Lyles, B. F., and Mihevc, T. M., 1992, River site upper aquifer, Owens Dry Lake: Analysis of long-term aquifer and pumping tests: Reno, Nevada, Desert Research Institute, University of Nevada System, Water Resources Center Publication No. 41135, 37 p.

Kendall, C., and Coplen, T. B., 1985, Multisample conversion of water to hydrogen by zinc for stable isotope determination: Analytical Chemistry, v. 57, p. 1437–1440.

Li, Yuan-Hui, and Gregory, S., 1974, Diffusion of ions in sea water and in deep-sea sediments: Geochimica et Cosmochimica Acta, v. 38, p. 703–714.

Lyons, W. B., Tyler, S. W., Gaudette, H. E., and Long, D. T., 1995, The use of strontium isotopes in determining groundwater mixing and brine fingering in a playa spring zone, Lake Tyrrell, Australia: Journal of Hydrology, v. 167, p. 225–239.

Manheim, F. T., 1966, A hydraulic squeezer for obtaining interstitial water from consolidated and unconsolidated sediments: U.S. Geological Survey Professional Paper 550-C, p. 256–261.

Matsuo, S., Kusakabe, M., Niwano, M., Hirano, T., Oki, Y., 1979, Water budget in the Hakone caldera using hydrogen and oxygen isotope ratios, in Isotopes in lake studies: Proceedings, Advisory Group meeting, Vienna, August 29–September 2, 1977: Vienna, Austria, International Atomic Energy Agency, p. 131.

Mehringer, P. J., and Sheppard, J. C., 1978, Holocene history of Little Lake, Mojave Desert, California, in The ancient Californians, Rancholabrean hunters of the Mojave Lakes Country: Natural History Museum of Los Angeles County, Science Series 29, p. 153–166.

Merlivat, L., 1970, L'étude quantitative de bilans de lacs à l'aide des concentrations en deutérium et oxygène-18 dans l'eau, in Isotope hydrology 1970: Proceedings, symposium, Vienna, March 9–13, 1970: Vienna, Austria, International Atomic Energy Agency, p. 89–106.

Merlivat, L., and Nief, G., 1967, Fractionnement isotopique lors des changements d'état solide-vapour et liquide-vapour de l'eau des temperatures inférieures à 0 °C: Tellus, v. 19, p. 122–127.

Norton, D. R., and Friedman, I., 1985, Chloride flux out of Yellowstone National Park: Journal of Volcanology and Geothermal Research, v. 26, p. 231–250.

Smith, G. I., 1979, Subsurface stratigraphic and geochemistry of the late Quaternary evaporites, Searles Lake, California: U.S. Geological Survey Professional Paper 1043, p. 1–130.

Smith, G. I., Friedman, I., Gleason, J. D., and Warden, A., 1992, Stable isotope composition of waters in Southeastern California: 2. Groundwaters and their relation to modern precipitation: Journal of Geophysical Research, v. 97, p. 5813–5823.

Stewart, M. K., 1975, Stable isotope fractionation due to evaporation and isotopic exchange of falling waterdrops: Applications to atmospheric processes and evaporation of lakes: Journal of Geophysical Research, v. 80, p. 1133–1146.

Stoessell, R. K., and Hanor, J. S., 1975, A nonsteady state method for determining diffusion coefficients in porous media: Water Resources Research, v. 80, p. 4979–4982.

Zimmermann, U., 1979, Determination by stable isotopes of underground inflow and outflow and evaporation of young artificial groundwater lakes, *in* Isotopes in lake studies: Proceedings, Advisory Group Meeting, Vienna, August 29–September 2, 1977: Vienna, Austria, International Atomic Energy Agency, p. 87–94.

Zimmermann, U., and Ehhalt, D. H., 1970, Stable isotopes in study of water balance of Lake Neusiedl, Austria, *in* Isotope hydrology 1970: Proceedings, Symposium, Vienna, March 9–13, 1970: Vienna, Austria, International Atomic Energy Agency, p. 129–138.

MANUSCRIPT ACCEPTED BY THE SOCIETY JUNE 17, 1996

Geological Society of America
Special Paper 317
1997

Paleomagnetism and magnetic susceptibility of Pleistocene sediments from drill hole OL-92, Owens Lake, California

Jonathan M. Glen* and Robert S. Coe
Earth Sciences Department, Earth and Marine Sciences Building, Room A232, University of California, Santa Cruz, California 95064

ABSTRACT

Paleomagnetic measurements performed on samples from Owens Lake cores OL-92-1, 2, and 3 reveal that the composite core (OL-92), extending to a depth of 323 m, spans nearly the past 800 k.y. The Matuyama/Brunhes transition is identified near the base of the composite section yielding an average sedimentation rate of ~40 cm/k.y. The record displays several shallow inclination features that are interpreted as within-Brunhes excursions (or short polarity events) and correlated with existing dated records to provide additional age control. Variations in susceptibility, caused by changes in the concentration, mineralogy, and/or size of magnetic grains, are found to reflect paleoclimate change. Susceptibility has been used to correlate OL-92 with other cores drilled from within the Owens basin and to identify a horizon we interpret as a brief dry period that occurred just prior to the onset of the Younger Dryas.

INTRODUCTION

Closed Quaternary basins of the western United States provide a valuable opportunity to investigate past climate because they are high-deposition-rate environments whose sediments record, nearly continuously, changes in lake levels that mainly reflect changes in precipitation. In the spring of 1992, three cores (collectively referred to as OL-92) were taken from Owens Lake, California, to investigate middle to late Pleistocene climate in the western United States (Fig. 1). The composite core contains mostly fine- to medium-grained lake deposits and extends to a depth of 323 m, representing an age of nearly 800 ka (Smith and Bischoff, this volume). Because of the high deposition rate, the cores have potential for providing excellent records of climate and paleomagnetic field changes spanning the late Quaternary. Paleo- and rock-magnetic studies were performed on a suite of more than 500 discrete samples taken at roughly ~0.5–1.5-m intervals from throughout the composite section. The goal of the present study was to resolve the age and duration of climate events, to study detailed aspects of Pleistocene geomagnetic field behavior, and to obtain magnetic proxies for climate.

PROCEDURES

Sampling techniques

Samples were taken from cores OL-92-1 and OL-92-2 both on site during drilling and subsequently at the U.S. Geological Survey in Menlo Park, where the cores are presently stored. Eight-cubic-centimeter (cc) samples were taken approximately every meter, with additional samples taken in some intervals to resolve specific features of the rock- and paleomagnetic record. Continuous U-channel samples were taken through most of core OL-92-3, and then later subdivided into 8-cc specimens spaced every 2 cm.

The cores were split longitudinally and, where the sediment was soft, samples were extracted by pressing plastic boxes into the surface of the working half. In regions where the sediment was more lithified, samples were carved from the core using non-magnetic Cu-Be tools and then fit into plastic boxes. Samples were taken from the center of the core to avoid sediment contaminated or disturbed by drilling or remagnetized by the drill string. Some of the samples were later impregnated

---

*Present address: Berkeley Geochronology Center, 2455 Ridge Road, Berkeley, CA 94709.

Glen, J. M., and Coe, R. S., 1997, Paleomagnetism and magnetic susceptibility of Pleistocene sediments from drill hole OL-92, Owens Lake, California, in Smith, G. I., and Bischoff, J. L., eds., An 800,000-Year Paleoclimatic Record from Core OL-92, Owens Lake, Southeast California: Boulder, Colorado, Geological Society of America Special Paper 317.

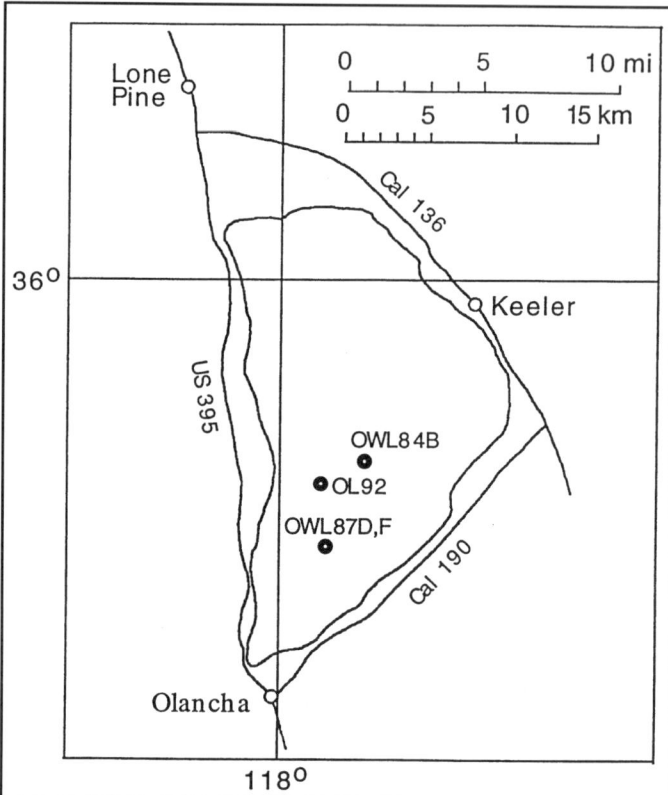

Figure 1. Site locality map showing positions of OL-92 and OWL cores within Owens playa.

with a solution of 10% sodium silicate, a nonmagnetic, high-temperature cement, to prevent them from falling apart.

Measurements

Remanent magnetization. Remanent magnetizations were measured with a 2G-Enterprises superconducting rock magnetometer housed in a field-free room. Intensity of natural remanent magnetization (J_{NRM}) varied by 3 orders of magnitude (10^{-4} to 10^{-1} A/m), with a large fraction of J_{NRM} values residing in the mid-10^{-4} range. The majority of samples were subjected to alternating field (AF) demagnetization using a Schonstedt AC tumbling sample demagnetizer. Thermal demagnetization was applied to samples for which corresponding (stratigraphically twinned) AF demagnetized samples existed. Of the paired samples, only seven have been thermally demagnetized, because the sediment constitution rendered the samples difficult to remove from their plastic boxes (which cannot be heated above 100 °C), and the samples tended to disintegrate with heating.

Generally, at least fifteen steps were applied in order to define accurately the stable direction of magnetization and to isolate any overprint direction (see Fig. 2 for representative demagnetization diagrams). A least-squares best fit is usually applied to 4–6 points on the demagnetization diagrams to yield "characteristic" paleomagnetic directions.

Magnetic susceptibility. In addition to determining remanence directions, low-field magnetic susceptibility (χ) measurements were performed on discrete samples using a 5-cm diameter Sapphire Instruments χ coil. Pass-through χ measurements (taken every 3 cm) were performed on the entire OL-92-3 core prior to its being split and on the archive halves of selected slugs from cores OL-92-1 and OL-92-2 using a 10-cm diameter coil. Susceptibilities varied by three orders of magnitude, but were generally low (~2.5×10^{-5} in SI volume units) with several measurements near or below the noise level of the instruments (5-cm coil ~1×10^{-6} SI, 10-cm coil ~1×10^{-5} SI).

RESULTS AND DISCUSSION

Remanent magnetization

Because cores OL-92-1 and OL-92-2 were rotary drilled, azimuthal orientation of individual core segments has been lost, resulting in a loss of absolute declinations. Therefore, field behavior can be inferred from inclination alone. Results from the inclination record (Fig. 3) show the presence of several low-inclination features within the Brunhes in addition to a reversal of directions at the base of the composite core. Since sediment magnetization can be affected by a variety of influences, caution should be taken when interpreting individual features as true field behavior. Further work (chemical analysis, and various rock magnetic measurements) is being performed to determine if some features are artifacts caused by disruption of the sediment, remagnetization, authigenesis, etc.

In general, however, the observed directional features are defined by multiple samples and do not show a clear relationship to lithology or a variety of other properties such as magnetic susceptibility (Fig. 4; see also Glen et al., 1993). In addition, AF and thermal demagnetization techniques applied to samples from the same horizon (Figs. 2a, 2b) reveal essentially the same directions (approximately within 5°), indicating that AF demagnetization is effectively resolving the characteristic components of magnetization. For these reasons, the directions, aside from specific cases (Glen et al., 1993), are here interpreted as indicative of real changes of the paleomagnetic field.

It should be noted however, that because transitional fields (and some excursions) are characterized by depressed intensities, some have argued that sediments deposited during such times would be incapable of accurately recording field behavior. If so, the anomalous fields would probably be reflected in the record as sequences of erratic directions, which—although not useful in deciphering details of field behavior—would still indicate the presence and duration (given an estimate of sedimentation rate) of the anomalous fields.

Matuyama/Brunhes polarity reversal

At the base of core OL-92-2, the sediments record a 12-m thick section of alternating normal, reversed, and intermediate directions (Fig. 5). This sequence of directions is distinct from any other in the core; it contains the only occurrence of fully reversed inclinations (the single exception being one sample

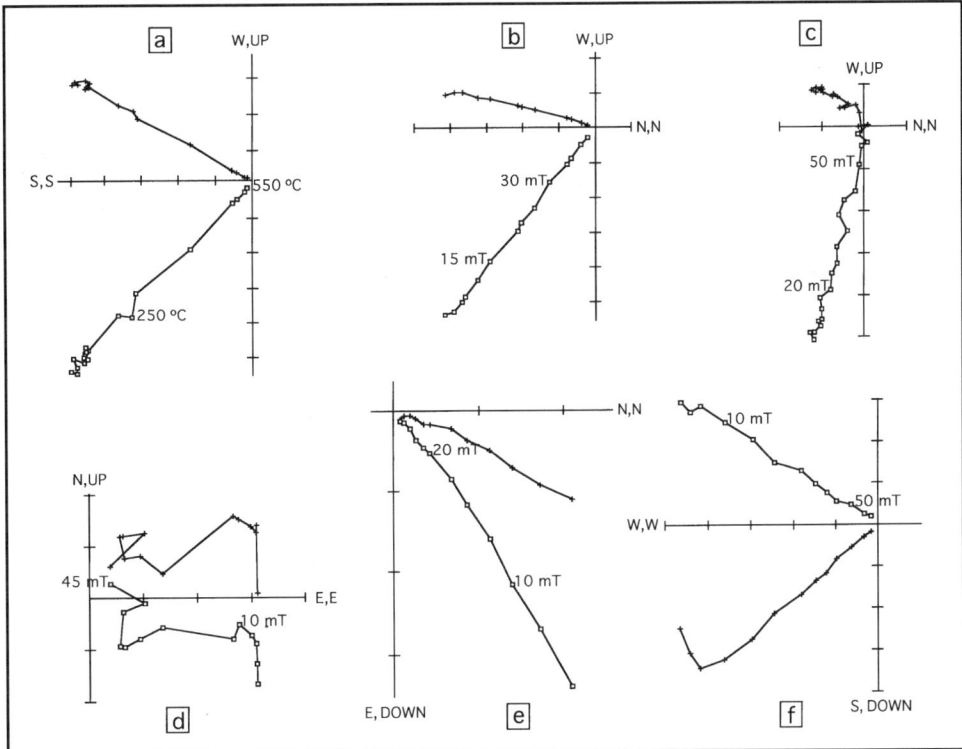

Figure 2. Typical vector endpoint plots for samples from core OL-92. Open symbols represent projections onto the vertical plane. Scale divisions represent 5×10^{-3}, 1×10^{-2}, 2.5×10^{-4}, 5×10^{-5}, 5×10^{-4}, 2.5×10^{-3} A/m on plots a, b, c, d, e, and f respectively. (a) Sample O71B, 83.54 m, has been thermally demagnetized. (b) Sample O71A, 83.54 m, has been AF demagnetized and was taken from the same stratigraphic horizon as sample O71B (See Fig. 3a). (c) Sample O5A, 10.62 m, is from a section of core spanning an excursion. Note that steep positive inclinations persist through progressive demagnetization and that the NRM intensity is an order of magnitude weaker than 071 specimens. (d) Sample O86A, 99.80 m, is taken from a section of core spanning an excursion. This is an example of a poorly resolved characteristic component. Note that the sample's NRM intensity is weak, which makes it difficult to resolve the magnetization. (e) Sample OB25A, 317.25 m, is from within the M/B transition. (f) Sample OC35A, 79.71 m, is from a section of core spanning an excursion that displays reverse inclination.

associated with the inclination anomaly at a depth near 100 m) and spans a thickness of section several times greater than any of the inclination anomalies found throughout the remaining 310 m of core. We therefore interpret these directions as part of the Matuyama/Brunhes (M/B) transition.

Although stable reversed directions were not reached by the bottom of the core (indicating that the record of the reversal is incomplete), inclinations throughout the lowermost 12 m of core undergo several large swings in polarity, suggesting that if the directions reflect true field behavior, a fairly large segment of time is represented and that the bulk of the transition is spanned. This is supported by the apparently long duration of the reversal record (>14 ka) based on the average sedimentation rate (note that the average reversal duration is ~5 k.y.; Bogue and Merrill, 1992). Thus, the M/B transition may be both unusually long and complex, consisting of several rapid swings between fully reversed and normal directions. Although such inferences depend on the reliability of remanence directions and on the applicability of the average sedimentation rate through this lithologically variable section, they are consistent with observations from other studies (e.g., Sun et al., 1993; Okada and Niitsuma, 1989).

Assuming that most of the transition has been recovered and that the directions are reliable, we consider the middle of the reversal to lie midway (at 320.2 m) through the recovered section of the main sequence of transitional directions (within the lowermost 5.5 m of core). It should be noted, however, that many of the directions in this section of core are poorly resolved and lithology varies significantly. Thus, if we disregard all outlier, or moderately to poorly resolved directions, the most probable depth to place the transition is at 317.6 m (see discussion that follows) where well-defined normal (315.4–317.3 m) and reversed (317.8–318.8 m) segments are juxtaposed.

Nonetheless, it seems unlikely that the directions occurring below 311 m, which span 12 m of section and a wide range of lithologies, are entirely the result of varying degrees of normal overprinting. It should be noted that this would imply that the "lock-in" of remanence occurred over an unusually large zone, in contrast to what is generally believed to be a relatively narrow region within rapidly deposited sediments (Lund, 1989). In addition, if directions below 311 m were overprinted, one would anticipate, unless the overprinting was extremely efficient, inclinations to be shallower than expected. On the contrary, long stretches of both normally (315.3–317.3 m) and reversely (317.8–318.8 m) magnetized samples yield well-defined inclinations that are consistent with those expected for a geocentric axial dipole field.

In addition, the quality of the demagnetization behavior,

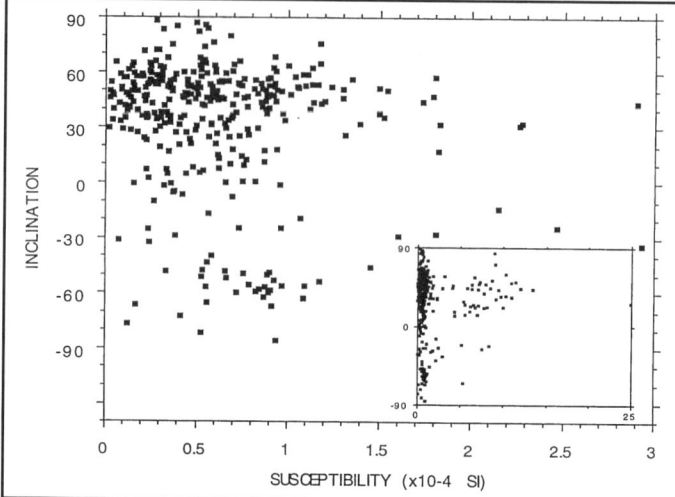

Figure 4. Plot of susceptibility (χ) versus inclination for samples spanning the entire core OL-92. The lack of correlation evident in this plot suggests that inclinations are not influenced by effects tied to changes in χ. Axes labels on the inset are the same as for the primary figure.

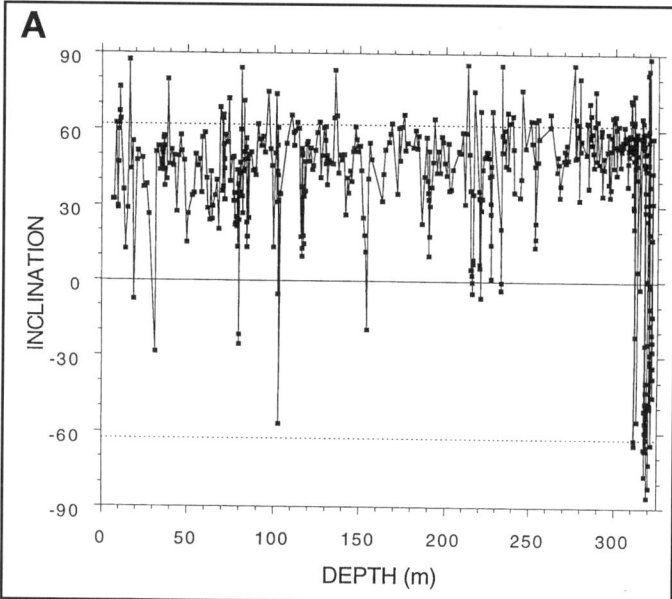

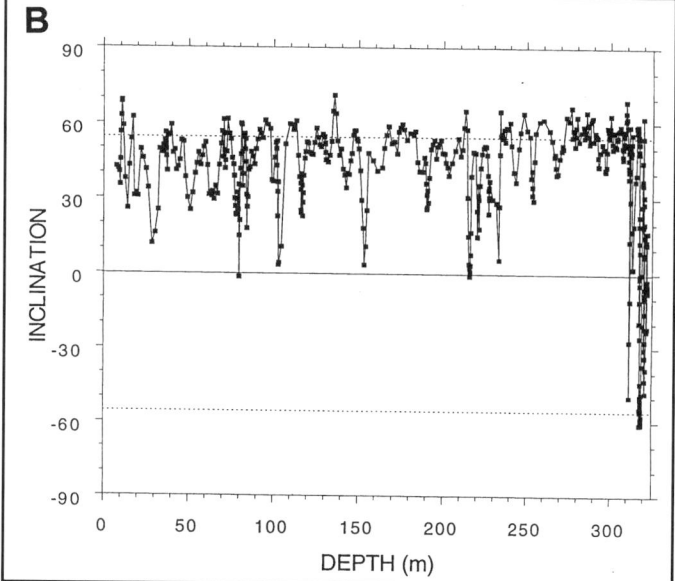

Figure 3. Plot of (A) inclination; and (B) smoothed inclination (3-sample sliding average) through the composite core spanning 0–323 m. Dotted lines show the expected inclinations for a geocentric axial dipole field at Owens Lake.

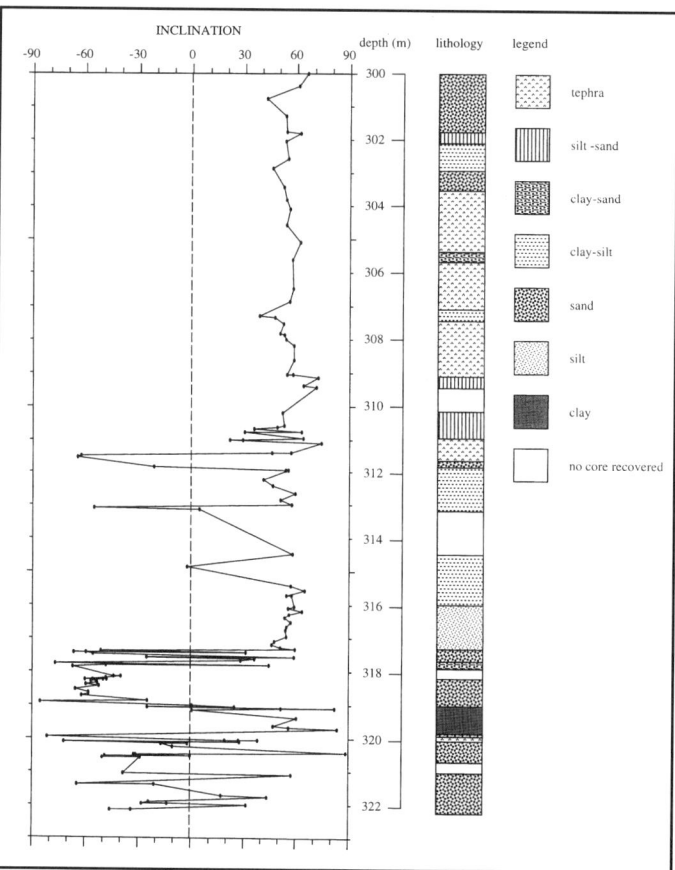

Figure 5. Plot of inclination and lithology spanning the lowermost 23 m of core OL-92-2 (Smith, 1993) that record the M/B transition. Note that the transition is punctuated by rapid fluctuations and that stable reversed directions were not reached by the bottom of the core indicating the transition record is incomplete. The Bishop ash has been identified in the interval 309.15 m to 298.60 m (Sarna-Wojcicki et al., this volume).

agreement of directions from stratigraphically twinned samples, coherent trends in directions, and agreement of directions defined by AF and Thermal demagnetization techniques (Glen et al., 1993) all suggest that at least some of the directions are reliable. There are sections below 311 m that display well-resolved (normal and transitional) directions and that are contained in fine-grained muds and silts—generally the best recording material.

Thus for the mentioned reasons, we consider the best estimate of the middle of the transition to lie at or below 317.6 m. This, together with the age of the transition, yields an estimate of the average deposition rate throughout the core (discussion in a following section). Recently, several different workers have derived ages for the M/B transition (Johnson, 1982; Spell and McDougall, 1992; Mankinen and Dalrymple, 1979; Izett and Obradovich, 1992; Baksi et al., 1992). Baksi et al. (1992) derived an age of 783 ± 11 ka from $^{40}Ar/^{39}Ar$ incremental heating studies of lava flows erupted during the transition; since this is the only study involving flows from within the transition, this date is used here for the age of the reversal. Nonetheless, the particular choice of age (or ages to average) or precise position of the transition, has little effect on the estimated average deposition rate.

Within-Brunhes excursions and secular variation

There are several (16) large inclination anomalies, distinct from those interpreted as part of the Matuyama/Brunhes transition, that deviate significantly (>35°) from the expected inclination of 55° resulting from a geocentric axial dipole field. Although shallowing of inclination can occur from a variety of sedimentological effects (King and Channell, 1991), in a visual comparison no clear correlation was found between inclination and concentration (inferred from χ and Anhysteretic Remanent Magnetization—ARM) or grain size (inferred from χ / χ_{ARM}) of magnetic constituents, clay mineralogy, or concentration of clay minerals (e.g., Fig. 4; Glen et al. 1993). Thus, the features could represent true field behavior, either field excursions (global and/or local) or short reversals. Such anomalous directional features within the Brunhes have been reported by many workers in both sedimentary and volcanic rocks from around the globe (Champion et al., 1988; Herrero-Bervera et al., 1994; Negrini et al., 1987; Bleil and Gard, 1989; Harland et al., 1990; Nowaczyk, 1990; Meynadier et al., 1992; Schnepp 1992; Valet and Meynadier, 1993).

We tentatively correlate eleven inclination anomalies to previously identified events in the existing literature (Table 1, Figs. 3, 6). Because of the lack of declination and intensity data, correlations rest entirely on inclination. They were based on the depth of the inclination anomalies in the core, allowing as well for the effects of sediment compaction (see e.g., Bischoff et al., this volume, Chapter 8), and the ages of events cited in existing literature. Time was principally constrained by the age and position of the M/B transition (discussed in the previous section) at the base of the core, and ^{14}C ages located within the upper 30 m of core. Additional consideration involved matches of the $CaCO_3$ and smectite records (see Menking, this volume) with the deep-sea sediment record of $\delta^{18}O$.

Because there are more anomalies in the OL-92 record than "known" events, there is inherently some ambiguity as to which anomalies correlate with which (if any) existing events. Nonetheless, the inclination anomalies picked (particularly Laschamp, Blake, Jamaica, Levantine, and Emperor events) are the best defined, large swings that are not surrounded by a crowd of similar swings. In light of this, it is surprising how closely the events match the mass accumulation curve of depth versus age (Fig. 7). Figures 6 and 7 give our best estimate of event correlations, barring additional information on the details of the transition. Further diagnosis is underway to characterize the inclination, relative intensity, and reconstructed declination waveforms of the anomalies and compare them to existing records.

Just as excursions may be used to match records globally, secular variation can be used to tie records on a regional scale. Negrini et al. (1987) describe a record from the Humbolt River Canyon (HRC), Nevada, that displays a long-term trend (10 k.y. to ~100 k.y.) of shallow inclinations much like that observed in core OL-92, in addition to the Lava Creek and Dibekulewe ashes. By correlating the two records, we were able to infer the correct position of the Dibekulewe ash in core OL-92 prior to our knowledge of its position (Fig. 8). In addition, the position of the Lava Creek ash in OL-92, that we deduced from the correlation, corresponds to a segment of core loss, which would explain why the ash was not found in the core despite its expected deposition based on the ash's regional distribution. Notably, this inferred position of the Lava Creek Ash (given its age of 665 ka, Izett et al., 1992) is quite consistent with the position predicted from magnetostratigraphy (Fig. 7) and mass accumulation (Bischoff et al., this volume, Chapter 8) models of age versus depth.

Interestingly, the comparison of OL-92 and HRC records indicates that the HRC section spans the timing of one of the largest changes of lithology seen in core OL-92 (at a depth of 223 m), but unlike OL-92, the HRC section is relatively uniform (fine grained) throughout (R. Negrini, 5/1993, oral communication) and does not show a pronounced lithologic break. Because the two basins are relatively close to each other, the lithologic change in OL-92 is either the result of a very localized shift in the climate pattern or more likely a tectonic/geologic event that altered the basin geometry, perhaps tectonic uplift of the spillway or modification by slides and/or debris- or lava-flows.

Age versus depth

The depth and age of the M/B reversal yield an average sedimentation rate of 40.9 cm/k.y. A 5-m segment of ash near the base of the composite section (at depths between 298.6 m and 309.2 m) has been identified as the Bishop ash (Sarna-Wojcicki et al., this volume). Removing the segment of core logged as ash, and assuming the ash was effectively deposited as an instantaneous pulse, yields an average sedimentation rate of 39.5 cm/k.y.

TABLE 1. CORRELATION OF EXCURSIONS WITHIN THE BRUNHES CHRON DETECTED IN OL-92

Event	Event Depth (m)	Event Age (ka)	Age Error* (k.y.)	Age Source†	Age Control	Corrected Event Age§ (ka)
Mono Lake	20	28	2	1	Isotopic	28
Laschamp	29	42	10	1	Isotopic	42
Norwegian-Greenland Sea	50	77	12	2	Magnetostratigraphic	77
Fram Strait	65	100	...	2	Magnetostratigraphic	100
Blake	78	114	10	1	Magnetostratigraphic	122
Jamaica (BI)	103	182	31	1	Isotopic	182
Pringle Falls	116	218	10	3	Isotopic	218
Levantine (BII)	154	289	19	1	Magnetostratigraphic	310
Biwa III	190	389	9	1	Magnetostratigraphic	417
Emperor	215	443	19	1	Magnetostratigraphic	475
West Eifel	227	510	14	1	Isotopic	510
Matuyama/Brunhes	320	783	11	4	Isotopic	783

*Age errors represent either (1) error bands given for isotopically derived ages, (2) standard deviations of ages derived from a set of sedimentary records, or (3) the full range of ages if only a few sedimentary records were available to provide an average.
†Sources of age control: 1 = Champion et al., 1988; 2 = Nowaczyk, 1990; 3 = Herrero-Bervera et al., 1994; 4 = Baksi et al., 1992. Event ages derived from sedimentary records given by Champion et al., 1988, which were calibrated to the M/B age of 730 ka, have been adjusted to the newest estimate of 783 ka.
§Ages used in this chapter (e.g., Fig. 7).

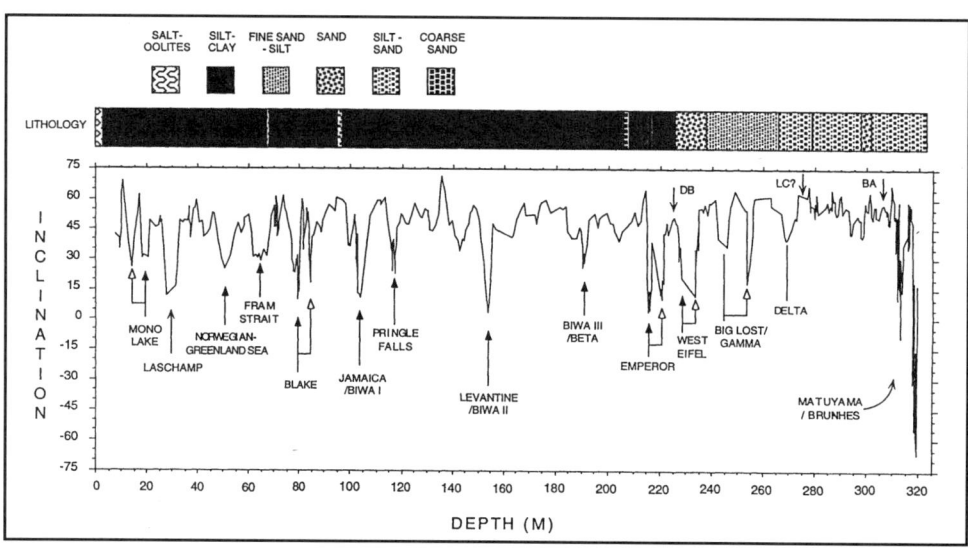

Figure 6. Plot of excursions (discussed in text and listed in Table 1) seen in inclination (smoothed) throughout core OL-92. Also shown is a generalized stratigraphy of the composite core. Solid arrows identify excursions that may correlate with events described by Champion et al. (1988). Open arrows identify alternative or less well constrained correlations. Short solid arrows above the inclination curve identify positions (or expected positions based on correlations of secular variation) of ashes (DB = Dibekulewe, LC = Lava Creek, BA = Bishop ash). Also indicated are features that may correspond to the Big Lost and Delta excursions, but because these features are both poorly constrained and less pronounced (<35° from "expected"), they are not included in the magnetostratigraphic age depth curve (Fig. 7, Table 1).

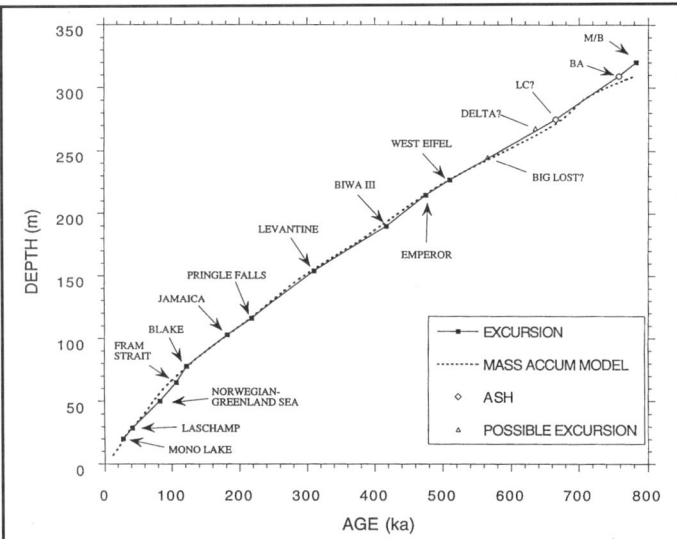

Figure 7. Age versus depth curve for OL-92 based on a correlation of inclination anomalies with excursions. Depths and correlation ages of shallow inclination events (Table 1) are plotted (solid line) against an age versus depth curve based on sediment bulk density and pore water measurements (dotted line) derived from sediment and pore water measurements (Bischoff et al., this volume, Chapter 8). The two curves agree remarkably well. Included in the "excursion curve" are the Lava Creek (LC), Bishop ashes (BA), and Matuyama/Brunhes (M/B). Also shown are possible occurrences of the Big Lost and Delta events (see also Fig. 6).

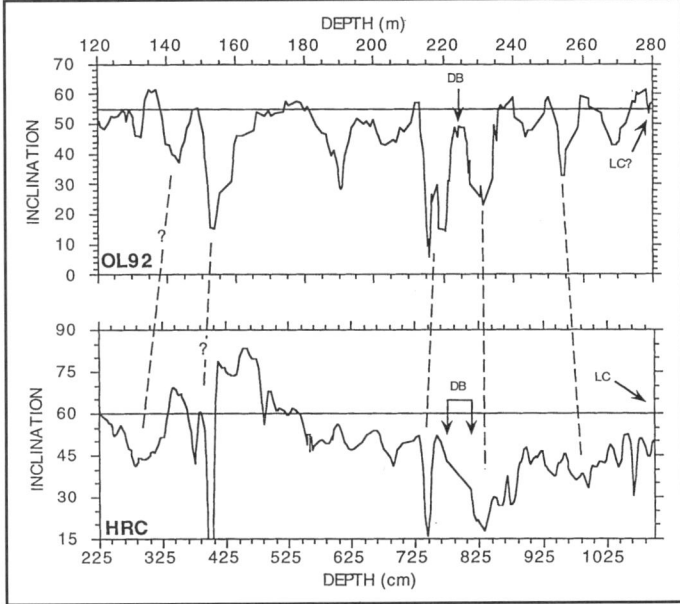

Figure 8. Correlation of inclination records from the Humbolt River Canyon (HRC) section (smoothed using a 7-point running average; Negrini et al., 1987) and Owens Lake core OL-92-2 (smoothed using a 5-point running average). The position of the Dibekulewe (DB) ash is shown in both plots by arrows. The Lava Creek (LC) ash lies immediately to the right of the HRC plot and is interpreted to lie just to the right of the OL-92 plot. Dashed lines show an interpretation of correlative features.

Depths and correlation ages of shallow inclination events (Table 1) are plotted (Fig. 7) against an age versus depth curve based on sediment bulk density, pore water, and salinity measurements (Bischoff et al., this volume, Chapter 8). The curves agree remarkably well and reveal that sediment accumulated relatively steadily throughout OL-92.

Susceptibility and climate

Variations in χ (Fig. 9), which depend on the amount and types of magnetic material present in a sample, may reflect changes in climate (e.g., mediated through changes in the input of soil or airborn, biogenic, or lithogenic sources of magnetic material) or geology (e.g., through volcanic input or changes in source areas). Because χ depends on the concentration and mineralogy of magnetic grains, they may potentially reflect variations in climate via changes in the source material (e.g., soil production), the rates of detrital input, the production of diagenetic/authigenic phases, or the destruction of existing material (e.g., via reduction diagenesis) in the lake arising from changing climate conditions. Several studies spanning a variety of sedimentary environments have successfully identified correlations between magnetic records and traditional climate indicators (for a review of works, see King and Channell, 1991).

A comparison of χ with a variety of other chemical and lithologic indicators, including several traditionally interpreted as reflecting climate change (e.g., $CaCO_3$, $\delta^{18}O$, and clay mineralogy; Menking, this volume) reveals several clearly defined correlatable features (Fig. 10; see also Glen et al., 1993). Correlation refers here to the coincidence of variations in different parameters identified through visual comparison. A quantitative measure of correlation is precluded by the facts that the (1) depths and spacing between samples used for various measurements differ;

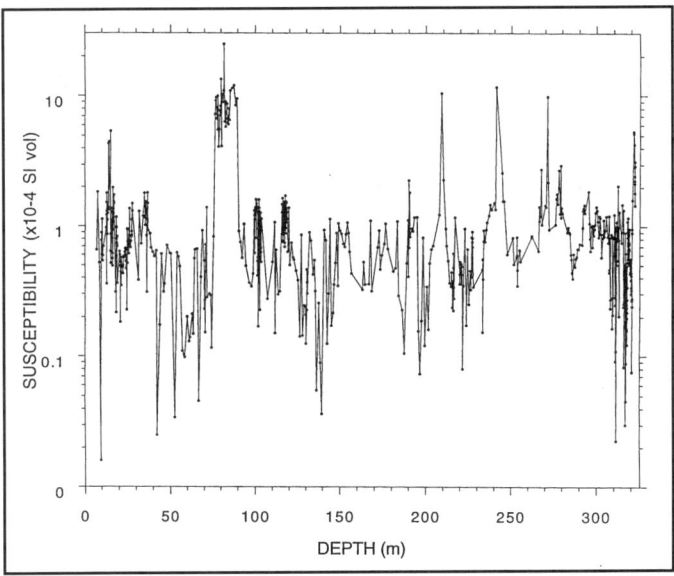

Figure 9. Plot of susceptibility (χ, in SI units) through the composite core spanning 0–323 m.

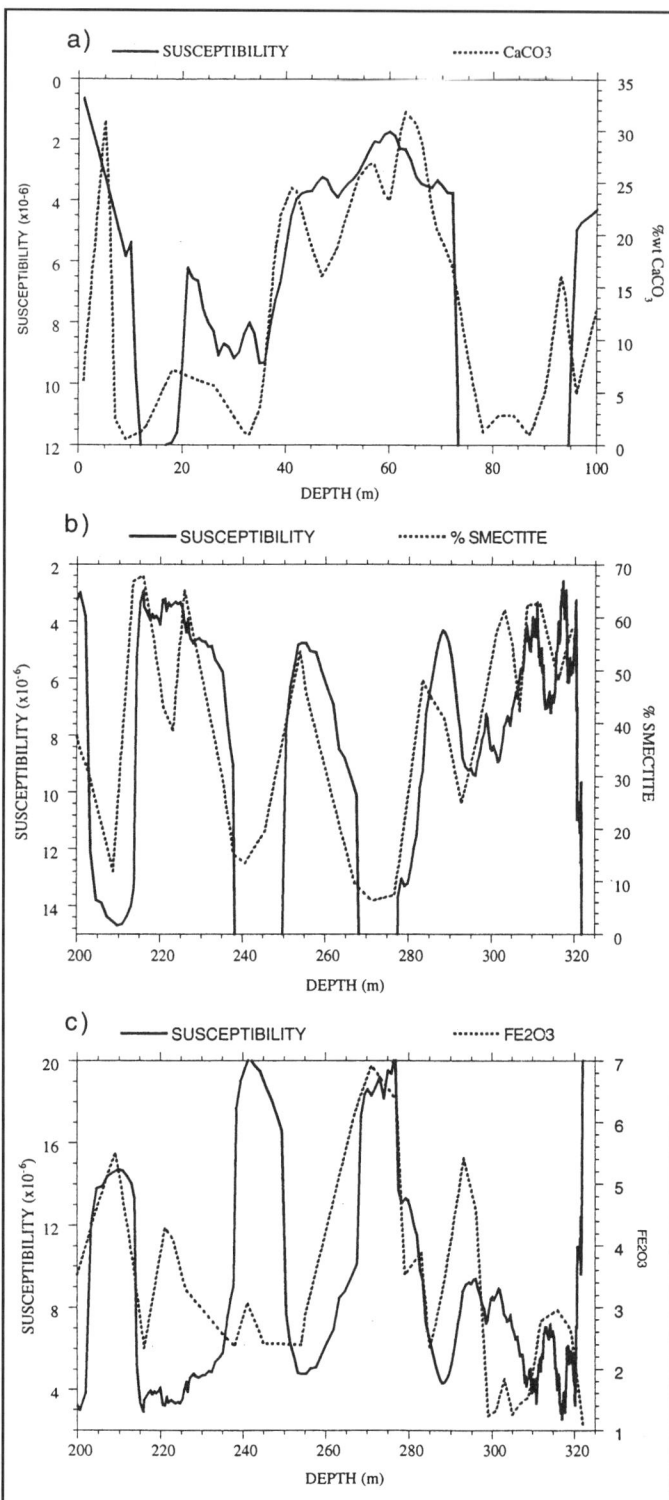

Figure 10. Comparison of susceptibility with (a) CaCO$_3$ (Bischoff et al., this volume, Chapter 4), (b) % smectite (Menking, this volume), and (c) Fe$_2$O$_3$ (Bischoff et al., this volume, Chapter 4). Susceptibility (χ) is smoothed using a 10-point running average. Correlations between these variables suggest that χ may reflect climate change.

(2) relative changes between parameters appear to be nonlinear; (3) correlations can be intermittent; (4) parameters can either be directly or inversely proportional; and (5) successive parameter records are commonly variable.

Susceptibilities in core OL-92 vary by three orders of magnitude (Fig. 9). The most prominent change occurs in the depth range 75–90 m. The drop near 75 m is believed to correspond to the Termination II boundary between glacial Stage 6 and the subsequent interglacial Stage 5e, based on its inferred age and correlation to other climate proxies (Figs. 9, 10). There were no comparable advances for nearly 300 k.y. prior to glacial stage 6. The fact that the χ anomaly is so prominent could be because of the long-term production and storage of weathering products behind terminal moraines (R. Anderson, 4/1992, oral communication). On the other hand, the χ high may be the result of the presence of greigite, commonly formed during reduction diagenesis in lake environments (Roberts and Turner, 1993) and produced in the deep-lake conditions prevailing during this glacial stage. Reduction diagenesis can result in the preferential loss, via dissolution, of fine-grained magnetite particles and therefore a net increase in the average grain size of magnetic grains. The section of core between depths 75–90 m displays larger grain sizes inferred from χ/ARM (Glen et al., 1993). By either of these mechanisms (i.e., an increase in production of weathering products or reduction diagenesis), the other susceptibility highs may also be tied to prominent glacial stages. For example, the highs at depths 210, 242, and 275 m may correspond to glacial stages at 440, 560, and 630 ka respectively.

If χ variations actually reflect climate change, they should correspond to fluctuations in other lithologic/chemical climate indicators. Plots of χ, CaCO$_3$, Fe$_2$O$_3$, and smectite versus depth show distinct correlations among these variables (Fig. 10; Glen et al., 1993). Note, however, that the relative amplitudes of change vary. For example, χ and CaCO$_3$ correlations cannot be entirely ascribed to the dilution of a constant fraction of magnetic constituents by the addition of CaCO$_3$—some other mechanism such as reduction diagenesis, as discussed earlier, must also be active. In addition, χ generally varies nonlinearly and inversely proportionally with CaCO$_3$; however, during the most recent drying event, χ and CaCO$_3$ covary (Fig. 11). As this event is both lithologically and climatically unique to the 800-k.y.-long OL-92 record, one might expect distinctly different processes that control sediment characteristics to prevail then. Such variations in the scaling and sign of correlation of proxies, undoubtedly caused by the action of multiple processes, may be useful to identify distinct lithologic zones reflecting major climate/lake conditions (e.g., χ versus CaCO$_3$ suggests perhaps five such regimes: 0–70, 70–95, 95–145, 145–185, 185–320 m).

Late Pleistocene and Holocene

Understanding the history of the late Pleistocene and Holocene climate recorded at Owens Lake was a primary goal of the present study. Not only do χ measurements contribute to

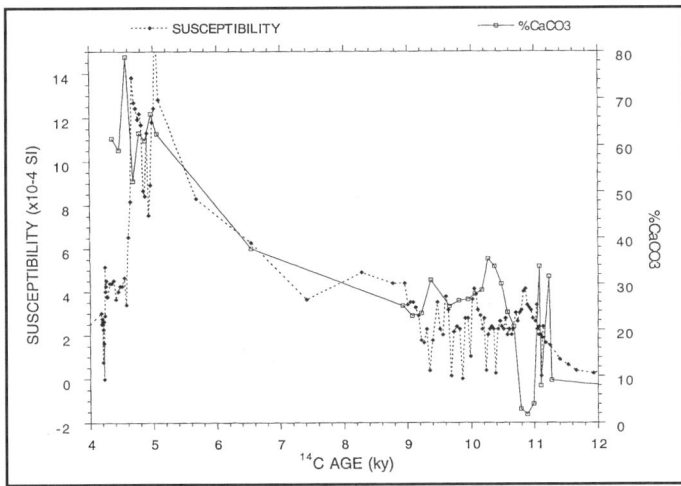

Figure 11. Comparison of susceptibility (χ) with $CaCO_3$ (Bischoff et al., this volume, Chapter 4) spanning 4–12 k.y. Age is estimated from ^{14}C dates (Bischoff et al., this volume, Chapter 8). Note that through this span of time χ and $CaCO_3$ covary.

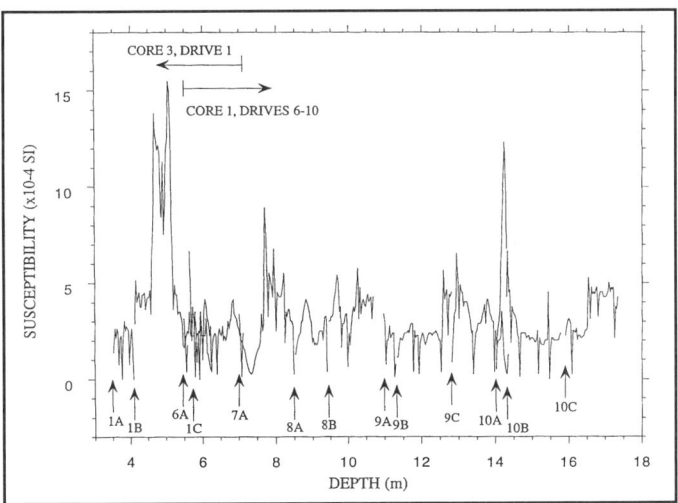

Figure 12. Pass-through susceptibility versus depth (measurements made every 3 cm) in cores OL-92-3 and OL-92-1. Drive tops are shown by arrows, and alphanumeric labels identify drive and slug numbers (Smith, 1993, supplemental data). Prominent features at 5, 8, and 14 m are interpreted as reflecting paleoclimatic change—see text for discussion.

this goal by providing a proxy for climate change, they can also be used to correlate OL-92 with cores taken from elsewhere within the basin. Recently, several shallow cores were taken from nearby locations on the playa surface (Newton, 1991). We have attempted to correlate these cores (referred to as OWL cores) with OL-92 based on a comparison of χ in OL-92 and magnetization (J) in the OWL cores.

The OL-92 record of pass-through susceptibility (Fig. 12) spanning the OWL cores, shows several large and distinct fluctuations (near depths 5, 8, 14.5 m). The χ anomaly at 8 m is remarkably similar in form to the J anomaly across a boundary seen in the OWL cores (Fig. 13). Correlating these features has allowed us to tie the two records together (Fig. 14). Doing so has revealed several other aspects of the cores that are quite consistent, and has thus resolved ambiguities over how the cores relate.

Radiocarbon ages of oolites and underlying humates from the OL-92 cores (Bischoff et al., this volume, Chapter 8) agree reasonably well with those from bulk sediment organic carbon from the OWL cores (Lund et al., 1991; Newton, 1991), and evidence exists in all cores for a shallowing of the lake at this oolite/clay boundary. In addition, the top of the χ anomaly in OL-92 (at ~8 m) is marked by a thin sand layer like the horizon in the OWL cores, and the sediment color above the sand in both cores is described as olive brown and below as gray. The sediments below the sand layer were found to contain abundant diatoms—the species present (*Stephaodiscus niagarae*) is identical to those noted in the OWL cores (Bradbury, 1993). Furthermore, the J anomaly in OWL-84B and χ anomaly in OL-92 span approximately the same thickness of section as do the intervening sediments between the oolite clay boundary and the sand layers in both cores.

Finally, the cores grade from OWL87D and 87F (the most southerly and thickest) to OL-92 (intermediate in thickness and geographic position) to OWL84B (the thickest and most northerly) (Fig. 1).

In part, it was differences between OL-92 and OWL log descriptions that had originally led to confusion over how the cores related (that issue arose during a meeting titled: "Ongoing paleoclimatic studies in the Northern Great Basin," Reno, Nevada, May 1993). After correlating χ and J anomalies, we reexamined OL-92 and found that the mud below the sand was marked by an orange speckled color that manifested only after the cores has sufficiently oxidized during storage. This pattern, which is a distinct characteristic of Newton's description of the unit, was not indicated in the OL-92 logs that were made immediately after core recovery. Thus, contrary to what was thought earlier, it seems that there are no significant discrepancies between OL-92 and OWL cores.

However, there are still some interesting differences between OL-92 and OWL cores. Newton (1991) identified the sand layer as an aeolian deposit and found mudcracks at the top of the unit below the sand, indicating the lake had desiccated. The sand in OL-92 is apparently more poorly sorted than that seen in the OWL cores, and no mudcracks are visible. The lack of mud cracks and the probable fluvial or shallow lacustrine deposition of the sand layer in core OL-92, in contrast to OWL cores, suggest that either the surface was partially deflated or that the lake did not desiccate.

Whether this boundary (at ~8 m) marks the passage from the Holocene to the Pleistocene remains unresolved. Lund et al. (1991) estimate its age to be 12.5 ka—consistent with OL-92 radiocarbon ages. A similar dry period, which is of the same age, is found recorded in the Searles Basin (Smith et al., 1983). How-

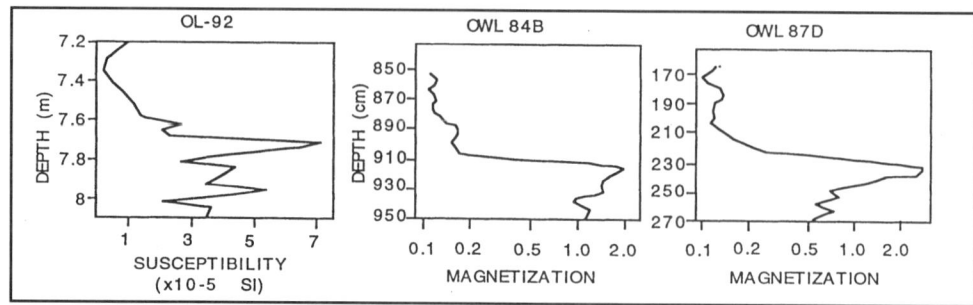

Figure 13. Correlation of magnetization (given in relative units) from OWL cores and susceptibility from OL-92-1 across the horizon identified by Newton (1991) as the Holocene/Pleistocene boundary (adapted from Newton 1991, fig. 4-7).

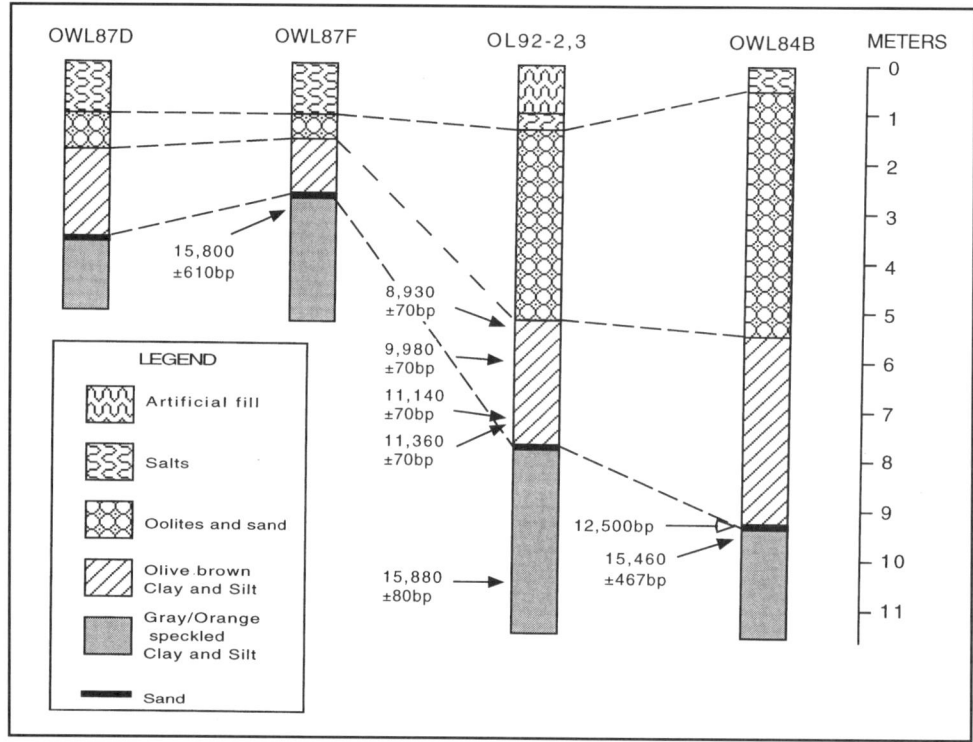

Figure 14. Generalized stratigraphic columns and correlations of OWL and OL-92 cores (OWL lithologies adapted from Newton, 1991, fig. 4-4). Radiocarbon ages are located with black arrows, Lund et al.'s (1991) estimate for the Holocene/Pleistocene unit boundary with a white arrow.

ever, it was not the last of the deep-water lakes in that basin. The horizon in OL-92 may therefore represent an earlier stage of drying than the Holocene/Pleistocene boundary, such as the onset of the Late Glacial Interstadial, that occurred at ca. 12,500 B.P., just prior to the Younger Dryas Stadial (Roberts et al., 1993).

Interestingly, the boundary is close in age to the last of the rapid transitions (Dansgaard-Oeschger events) from periods of extremely cold stadials (Heinrich events) to markedly warm interstadials that are observed in North Atlantic sediments and Greenland ice cores, dated at 13–14 ka (e.g., Bond et al., 1993). Links between western North America and North Atlantic climates have been proposed by Phillips et al. (1994) who claim to find a striking similarity in the numbers and ages of mid-Wisconsin salt deposition episodes at Searles Lake, California (during Pleistocene time, part of the same hydrologic system as Owens Lake), and the ages of stadial-interstadial cycles in the Greenland Summit ice core. Thus, such extreme cold events, or their associated transitions to warmer conditions, may be responsible for many of the features in susceptibility found further downcore. Figure 15 presents a possible correlation of χ anomalies and Heinrich events, and compares the age versus depth curve based on this correlation with those determined from magnetostratigraphy and mass-accumulation models.

CONCLUSIONS

The M/B polarity reversal occurs at or just below a depth of 317.6 m in a composite of three cores taken from Owens Lake, California. The age and depth of the reversal yield an average sedimentation rate of 40 cm/k.y. Several excursions of magnetic field directions, found within the Brunhes epoch, have been correlated with features seen and dated in other records to provide timing throughout the OL-92 core. The results are in excellent agreement with an age versus depth curve based on bulk density calculations from sediment and pore water data. Both indicate that the sedimentation rate is relatively steady throughout the core. Correlations of the upper 10 m of the OL-92 core with gravity cores from adjacent sites on the playa have been achieved using pass-through χ measurements and corroborated by reexamining lithology. This correlation led to our identifying a distinct, but brief, dry period that probably marks a temporary shift from glacial conditions just prior to the onset of the Younger Dryas which preceded the close of pluvial conditions in both Owens and Searles Lakes. Correlations of secular variation in the OL-92 and HRC sections provide additional age control and help constrain the cause of lithologic changes in the OL-92 core. Work is being continued in order to fill in gaps, improve the resolution of field excursions, and refine rock magnetic records.

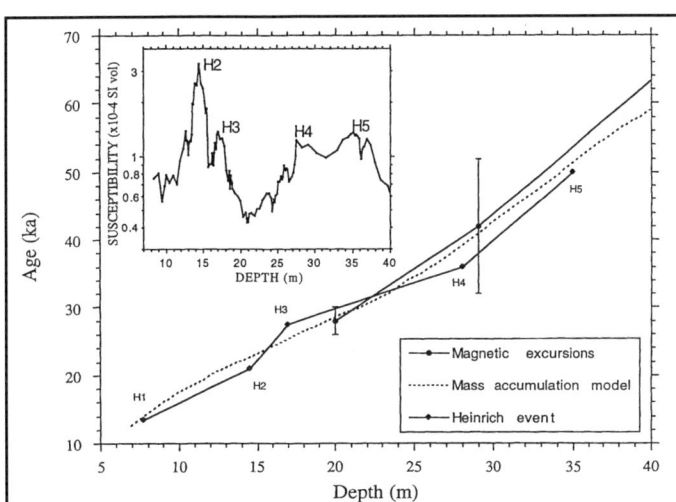

Figure 15. Age versus depth curves spanning the upper 40 m of core OL-92 based on inclination excursions, a mass accumulation model (Bischoff et al., this volume, Chapter 8), and a possible correlation of peaks in susceptibility with Heinrich events observed in North Atlantic sediments and Greenland ice cores. (Inset) Smoothed (5-sample running average) pass-through susceptibility (χ) measurements through the upper 40 m of core OL-92. Labels H2 through H5 identify peaks in χ which may correspond to Heinrich or Dansgaard-Oeschger events observed in North Atlantic sediments and Greenland ice cores (Bond et al., 1993).

ACKNOWLEDGMENTS

We thank Robert Anderson, Jim Bischoff, Duane Champion, Kirsten Menking, Joe Rosenbaum, Andrei Sarna-Wojcicki, and George Smith for helpful comments and discussions. We also thank Shannon Boughn for assistance in performing laboratory measurements.

REFERENCES CITED

Baksi, A. K., Hsu, V., McWilliams, M. O., and Farrar, E., 1992, $^{40}Ar/^{39}Ar$ dating of the Brunhes-Matuyama geomagnetic field reversal: Science, v. 256, p. 356.

Bleil, U., and Gard, G., 1989, Chronology and correlation of Quaternary magnetostratigraphy and nannofossil biostratigraphy in Norwegian-Greenland Sea sediments: Geologische Rundschau, v. 78, p. 1173–1187.

Bogue, S. W., and Merrill, R. T., 1992, The character of the field during geomagnetic reversals: Annual Review of Earth and Planetary Sciences, v. 20, p. 181–219.

Bond, G., Broecker, W., Johnsen, S., McManus, J., Labeyrle, L., Jouzel, J., and Bonani, G., 1993, Correlations between climate records from North Atlantic sediments and Greenland ice: Nature, v. 365, p. 143–147.

Bradbury, J. P., 1993, Diatoms in sediments, in Smith, G. I., and Bischoff, J. L., eds., Core OL-92 from Owens Lake, southeast California: U.S. Geological Survey Open-File Report 93-683, p. 261–302.

Champion, D. E., Lanphere, M. A., and Kuntz, M. A., 1988, Evidence for a new geomagnetic reversal from lava flows in Idaho: Discussion of short polarity reversals in the Brunhes and late Matuyama polarity chrons: Journal of Geophysical Research, v. B93, p. 11667–11680.

Glen, J. M., Coe, R. S., Menking, K., Boughn, S. S., and Altschul, I., 1993, Rock- and paleo-magnetic results from core OL92, Owens Lake, CA, in Smith, G. I., and Bischoff, J. L., eds., Core OL-92 from Owens Lake, southeast California: U.S. Geological Survey Open-File Report 93-683, p. 127–183.

Harland, B. W., Armstrong, R. L., Cox, A. V., Craig, L. E., Smith, A. G., and Smith, D. G., 1990, A geologic timescale 1989: Cambridge, U.K., Cambridge University Press, 263 p.

Herrero-Bervera E., Helsley, C. E., Sarna-Wojcicki, A. M., Lajoie, K. R., 1994, Age and correlation of a paleomagnetic episode in the western United States by Ar-40/Ar-39 dating and tephrochronology—The Jamaica, Blake, or a new polarity episode: Journal of Geophysical Research—Solid Earth, v. B99, p. 24091–24103.

Izett, G. A., and Obradovich, J. K., 1992, $^{40}Ar/^{39}Ar$ dating of the Jaramillo Normal Polarity Subchron and the Matuyama-Brunhes geomagnetic boundary: U.S. Geological Survey Open-File Report 92-699, 22 p.

Izett, G. A., Pierce, K. L., Naeser, N. D., and Jaworowski, C., 1992, Isotopic dating of Lava Creek B tephra in terrace deposits along the Wind River, Wyoming: Implications for post 0.6 Ma uplift of the Yellowstone hotspot: Geological Society of America Abstracts with Programs, v. 24, no. 7, p. A102.

Johnson, R. J., 1982, Brunhes-Matuyama magnetic reversal dated at 790,000 yr BP by marine-astronomical correlations: Quaternary Research, v. 17, p. 135.

King, J. W., and Channell, J. E. T., 1991, Sedimentary magnetism, environmental magnetism, and magnetostratigraphy, in Contributions in geomagnetism and paleomagnetism, U.S. national report to IUGG 1978-1990: Reviews of Geophysics, Supplement, p. 358–370.

Lund, S. P., 1989, Paleomagnetic secular variation, in James, D., ed., Encyclopedia of Solid Earth geophysics: New York, Van Nostrand Reinhold, p. 876–888.

Lund, S. P., Newton, M., Hammond, D., and Stott, L., 1991, Paleohydrology of the Owens River/Lake system as a proxy indicator of late Quaternary glacial variations within the Sierra Nevada: Geological Society of America Abstracts with Programs, vol. 23, no. 5, p. A61.

Mankinen, E. A., and Dalrymple, G. B., 1979, Revised geomagnetic polarity time scale for the interval 0–5 m.y.B.P.: Journal of Geophysical Research, v. B84, p. 615–626.

Meynadier, L., Valet, J. P., Weeks, R., Shackleton, N. J., and Hagee, V. L., 1992, Relative geomagnetic intensity of the field during the last 104 ka: Earth and Planetary Science Letters, v. 114, p. 39–57.

Negrini, R. M., Verosub, K. L., and Davis, J. O., 1987, Long-term nongeocentric axial dipole directions and a geomagnetic excursion from the middle Pleistocene sediments of the Humbolt River Canyon, Pershing County, Nevada: Journal of Geophysical Research, v. B92, p. 10617–10627.

Newton, M., 1991, Holocene stratigraphy and magnetostratigraphy of Owens and Mono Lake, eastern California [Ph.D. thesis] Los Angeles, University of Southern California.

Nowaczyk, N. R., 1990, Hochauflösende magnetostratigraphie spätquartärer Sedimente arktischer Meeresgebiete [Ph.D. thesis]: Bremen, Germany, Bremen University.

Okada, M., and Niitsuma, N., 1989, Detailed palaeomagnetic records during the Brunhes-Matuyama geomagnetic reversal, and a direct determination of depth lag for magnetisation in marine sediments: Physics of the Earth and Planetary Interiors, v. 56, p. 133–150.

Phillips, F. M., Campbell, A. R., Smith, G. I., Bischoff, J. L., 1994, Interstadial climate cycles: A link between western North America and Greenland? Geology, v. 22, p. 1115–1118.

Roberts, A. P., and Turner, G., M., 1993, Diagenetic formation of ferrimagnetic iron sulphide minerals in rapidly deposited marine sediments, South Island, New Zealand: Earth and Planetary Science Letters, v. 115, p. 257–273.

Roberts, N., Taeb, M., Barker, P., Damnati, B., Icole, M., and Williamson, D., 1993, Timing of the Younger Dryas event in East Africa from lake-level changes: Nature, v. 366, p. 146–148.

Schnepp, E., 1992, Paleointensity in the Quaternary West Eifel volcanic field, Germany: Preliminary results: Physics of the Earth and Planetary Interiors, v. 70, p. 231–236.

Smith, G. I., 1993, Field log of Core OL-92, in Smith, G. I., and Bischoff, J. L., eds., Core OL-92 from Owens Lake, southeast California: U.S. Geological Survey Open-File Report 93-683, p. 4-57.

Smith, G. I., Barczak, V. J., Moulton, G. F., and Liddicoat, J. C., 1983, Core KM-3, a surface-to-bedrock record of late Cenozoic sedimentation in Searles Valley, California, U.S. Geological Survey Professional Paper 1256, 24 p.

Spell, T. L., and McDougall, I., 1992, Revisions to the age of the Brunhes-Matuyama boundary and the Pleistocene geomagnetic polarity timescale: Geophysical Research Letters, v. 19, p. 1181–1184.

Sun, D., Shaw, J., An, Z., and Rolph, T., 1993, Matuyama/Brunhes (M/B) transition recorded in chinese loess: Journal of Geomagnetism and Geoelectricity, v. 45, p. 319–330.

Valet, J. P., and Meynadier, L., 1993, Geomagnetic field intensity and reversals during the last four million years: Nature, v. 366, p. 234–238.

MANUSCRIPT ACCEPTED BY THE SOCIETY JUNE 17, 1996

Age and correlation of tephra layers, position of the Matuyama-Brunhes chron boundary, and effects of Bishop ash eruption on Owens Lake, as determined from drill hole OL-92, southeast California

Andrei M. Sarna-Wojcicki, Charles E. Meyer, and Elmira Wan
U.S. Geological Survey, 345 Middlefield Road, MS 975, Menlo Park, California 94025

ABSTRACT

Tephra layers in the ~323-m-deep Owens lake drill hole OL-92 correlate to tephra layers that have been identified and dated elsewhere in the western United States. Tephra layers identified are the Bishop ash bed (758 ka) at 309.2–298.6 m; the Dibekulewe (ash) bed (ca. 470 ka to ca. 610 ka) at ~224 m; and one of several ash beds in Walker Lake (ca. 60 ka to ca. 80 ka) at ~50.7 m. Other tephra layers, the ages of which are poorly constrained, have also been identified. Age constraints from a sedimentation-rate curve based on dry bulk density and independently derived magnetostratigraphy provide new age constraints to the undated or poorly dated tephra layers: ca. 740 ka for the ash of Thermal Canyon, and ca. 510 ka for the Dibekulewe (ash) bed.

Bishop tephra fell into a deep Owens Lake, but the lake shallowed as ash was rapidly reworked by wind and water within the Owens Lake basin. The shallowing of the lake was the result in part to filling with the large volume of ash that was deposited in the basin and then reworked into the lake, but the filling was also an effect of the onset of a moderately warm interstadial period of hemispheric or global extent corresponding to oxygen-isotope stage 19. The lake deepened again as the last several meters of the 10-m-thick, composite ash bed were deposited in the lake. Despite its great thickness, reworking of the light ash must have been rapid.

The position of the Matuyama-Brunhes paleomagnetic boundary is estimated to be between 311.4 m and 314.8 m, and most likely between 311.4 m and 312.9 m, in the core, based on (1) the pattern of magnetic inclinations in the Owens Lake core as compared with those at other sites in the region; (2) estimates of the time elapsed between the magnetic reversal and the deposition of the Bishop ash bed; and (3) the probable range of sediment-deposition rates in Owens Lake during this time.

INTRODUCTION

Tephra (volcanic ash) layers in cores from the composite ~323-m-deep core hole (OL-92), drilled in 1992 in Owens Lake, east-central California, were sampled and chemically analyzed to provide correlation and age control for a paleolimnologic and paleoclimatic study of upper Quaternary lake sediments in this depositional basin. The primary focus of this larger study is to characterize the history of climatic fluctuations during the latter part of Quaternary time for this region.

Volcanic glass shards from macroscopic tephra layers were separated and chemically analyzed using an electron micro-

---

Sarna-Wojcicki, A. M., Meyer, C. E., and Wan, E., 1997, Age and correlation of tephra layers, position of the Matuyama-Brunhes chron boundary, and effects of Bishop ash eruption on Owens Lake, as determined from drill hole OL-92, southeast California, *in* Smith, G. I., and Bischoff, J. L., eds., An 800,000-Year Paleoclimatic Record from Core OL-92, Owens Lake, Southeast California: Boulder, Colorado, Geological Society of America Special Paper 317.

probe. Glass compositions were then compared to those of previously analyzed tephra layers, some of known age. Correlations were made on the basis of petrographic characteristics, chemical composition, and stratigraphic sequence in the core, as compared to other localities in the western United States where the age, chemical composition, and stratigraphic sequence of potentially correlative tephra layers had been previously documented.

We also examined in detail the lithology of the basal 27 m of the core, the segment containing the Bishop ash bed and the Matuyama Reversed Polarity Chron–Brunhes Normal Polarity Chron boundary, to determine the environments of deposition that existed during the deposition of the ash bed and to better define the positon of the boundary within a basal zone of scattered magnetic inclinations (Glen and Coe, this volume).

METHODS

Approximately 1–3 cm^3 of tephra were sampled from macroscopic tephra layers in the drill cores. Volcanic glass shards from these were separated and analyzed using methods described by Sarna-Wojcicki et al. (1984). In brief: samples were wet-sieved with water in plastic sieves fitted with nylon screens, retaining the 200–100-mesh-size fraction (~80 µm to ~150 µm, respectively) for separation of glass shards. This fraction was placed in an ultrasonic vibrator in water, then treated with a 10% solution of HCl for a few minutes, to remove authigenic carbonate adhering to the glass particles, and with an 8% solution of HF for about 30 sec to 1 min, to remove other coatings or altered rinds that may have been present on the glass shards. The glass shards were then separated from other components of the tephra sample using (1) a magnetic separator and (2) heavy liquids of variable density made from mixtures of methylene iodide and acetone.

The glass separate was mounted in epoxy resin in shallow holes drilled into plexiglass slides, and the slides were grounddown and polished with diamond paste to expose the shards and prepare a smooth, uniform surface for analysis. The polished sample was coated with carbon, and individual shards were analyzed by electron-microprobe. See Sarna-Wojcicki et al. (1984), and Sarna-Wojcicki et al. (1985), for specifics of analytical conditions.

The polished glass shards were analyzed for Si, Al, Fe, Mg, Mn, Ca, Ti, Na, and K, and results were compared with our data base of approximately 3,200 previously analyzed samples of volcanic glasses from upper Neogene tephra layers collected within the conterminous western United States and from bottom sediments of the adjacent Pacific Ocean. The best matches were identified using numerical and statistical programs (SIMAN; Borchardt, 1974; Sarna-Wojcicki et al., 1984). The best matches were then examined for petrographic similarities and for stratigraphic position and sequence (Sarna-Wojcicki and Davis, 1991). Correlative layers were identified on the basis of the following four main criteria: (1) chemical composition of volcanic glass, (2) petrographic characteristics, (3) stratigraphic position and sequence, and (4) magnetostratigraphy (when applicable). Comparisons of tephra layers for the purpose of correlation were made with several different combinations of elements, excluding those elements that are present in concentrations close to the detection limit when comparing sample groups.

RESULTS

The depth of each tephra layer, its position, and type of intercalated sediment is given in Table 1. A summary of results of electron-microprobe analysis is presented in Table 2, together with comparative compositions of shards from tephra layers from other sites. Detailed lithologic and petrographic descriptions of the samples analyzed are given in Sarna-Wojcicki et al. (1993). Individual comparisons for each analyzed tephra sample showing the 30 closest matches on the basis of glass chemical composition are also given in Sarna-Wojcicki et al. (1993), together with the locations and age information of tephra layers that match most closely with those in the Owens Valley cores.

The following widespread, dated tephra layers are present in the Owens Lake bore hole (OL92-1 and OL92-2): the Bishop ash bed (309.15–298.60 m; 758 ± 1 ka; Sarna-Wojcicki and Pringle, 1992); the Dibekulewe (ash) bed (224.15 m; >400 ka, <665 ka; Sarna-Wojcicki et al., 1991), and a tephra layer at 50.66–50.69 m attributed to early eruptions of the Mono Craters, a correlative of which has been found in core samples from Walker Lake, Nev., estimated to be between ca. 60 ka and ca. 80 ka in age based on age control at the latter site (Sarna-Wojcicki et al., 1988) (Tables 1, 2; Fig. 1). Several tephra layers that have been identified at other locations but not dated at those sites are also present in the Owens Lake core. Estimates of the ages of these layers are obtained from the sedimentation-rate curve presented by Bischoff et al. (this volume, chapter 8). In addition, zones of reworked Bishop tephra are present at several levels above the Bishop ash bed in the Owens Lake core, suggesting that the Bishop ash bed was repeatedly reworked within the Owens Lake drainage basin.

Glass Mountain set of ash beds and the Bishop ash bed

Abundant coarse- to fine-sand-sized tephra layers are present within the lowermost ~23 m of the Owens Lake core (Fig. 2). Several zones within this interval are massive, and the entire interval consists dominantly of material that was probably both direct airfall tephra and tephra reworked by streams and wind, and eventually deposited within Owens Lake. These layers are intercalated with clays, silts, and sands that contain little or no pyroclastic material. Compositionally, the volcanic glass shards of most of these layers are relatively homogenous, and represent the products of several eruptions from the same eruptive source area spaced closely in time.

Determining the base of the Bishop ash bed in the core:
The lowermost three tephra layers in the core, in the depth inter-

TABLE 1. DATA ON TEPHRA SAMPLES ANALYZED FROM THE 1992 OWENS LAKE CORES

Sample	Core	Depth (m)	Thickness of Layer (cm)	Sediment in Which Tephra Is Present
OL92-1	92-1	50.6	Spot sample	Silty clay
OL92-1001	92-1	50.66-50.69	3	Silty clay
OL92-1003	92-1	52.20-52.25	5	Silty clay
OL92-2	92-2	131.1	Spot sample	Silty clay and sand
OL92-1015	92-2	216.54	2	Sand, fine to very coarse
OL92-1016	92-2	224.15	3	Sand, fine to medium
OL92-1018	92-2	241.14	5	Sand, mostly fine to medium, some silt
OL92-1019	92-2	266.43	4	Clay, some silt
OL92-1020	92-2	296.04	10 to 15	Clay and silt
OL92-1021	92-2	303.94	90	Tephra, air fall and/or reworked
OL92-1022	92-2	304.44	90	Tephra, air fall and/or reworked
OL92-1023	92-2	304.91	4	Tephra, air fall and/or reworked
OL92-1024	92-2	305.38	10	Tephra, air fall and/or reworked
OL92-1025	92-2	305.93	60	Tephra, air fall and/or reworked
OL92-1026	92-2	306.89	35	Tephra, air fall and/or reworked
OL92-1027	92-2	308.46	65	Tephra, air fall and/or reworked
OL92-1028	92-2	311.08	80	Tephra, air fall and/or reworked
OL92-1029	92-2	311.48	20 to 30	Tephra, air fall and/or reworked
OL92-1030	92-2	320.01	57	Tephra, air fall and/or reworked

val from about 320.8–310.6 m, are fine grained compared to the 10-m-thick zone of tephra extending from a depth of 309.15–298.6 m. The tephra layer at 320.86 is impure, 57 cm thick, and consists of medium- to fine-sand- and silt-sized pumiceous shards, with pyroclastic and detrital crystalline material. The second layer, at about 311.9 m depth, is 20 cm thick and consists of loose, structureless, well-sorted, pure, medium- to coarse-sand-size pumiceous shards. Because this unit was found at the top of a drive, and because of its loose condition, it may be material slumped down from higher in the core during coring. The third layer, at about 311.4 m depth, is 80 cm thick, and consists of well-sorted, medium- to fine-grained-sand-size tephra containing pumiceous, bubble-wall, and bubble-wall junction shards, as well as pyroclastic and detrital clastic crystalline material. This layer grades in its upper half into fine-silt- and clayey-silt-size tephra or ashy silt and clay.

At 309.15 m in the core is the base of a 1-m-thick, compound tephra layer that contains coarse pumice lapilli to as much as 3 cm in long diameter in its upper part. This layer marks the base of a 10-m-thick zone of tephra layers and intercalated ashy sediments containing rounded pumice lapilli, extending upward to a depth of 298.6 m (Fig. 2). Above 298.2 m, the massive silts contain little or no tephra, except for a 10–15-cm-thick layer close to the top of this interval, at about 296 m.

All the tephra samples (OL92-1021 through OL92-1030) in this basal 27-m-thick interval of the core except the uppermost one at ~296 m (OL92-1020), contain shards that are chemically and morphologically similar or identical to those of a set of Quaternary tephra layers that had their source in the Glass Mountain–Long Valley Caldera area of east-central California, about 155 km north of Owens Lake (Tables 1, 2; Fig. 1).

These layers are referred to as the Glass Mountain set of ash beds and the Bishop ash bed, which erupted from the adjacent Long Valley Caldera (Gilbert, 1938; Bailey et al., 1976; Izett, 1981; Izett et al., 1988; Sarna-Wojcicki et al., 1984, 1991; Izett et al., 1970, 1988) (Table 2; Fig. 1). These layers were erupted during the period extending from 1.8–0.76 Ma. Older tephra layers that had their source at or near Glass Mountain (for example, the tuff of Taylor Canyon), can be distinguished readily from the younger set by their glass chemical composition.

The youngest, coarsest, and by far the most voluminous tephra layer of the younger set of layers derived from the Glass Mountain-Long Valley Caldera complex is the Bishop ash bed, which erupted at ca. 0.76 Ma. This eruption caused the subsidence that formed Long Valley caldera. The Bishop ash bed is the distal air-fall deposit and reworked air-fall ash equivalent of the proximal air-fall pumice and ash flows of the Bishop Tuff. We infer that the base of the Bishop ash bed in the Owens Lake core is at the first appearance of thick, coarse air-fall ash, at 309.15 m in the core, and that the finer-grained ash layers below are the Glass Mountain set of ash beds. Similar stratigraphic relationships have been observed at many localities throughout this region, for example at the southern end of the Volcanic Tableland north of Bishop (Izett, 1981; Sarna-Wojcicki et al., 1984; Liddicoat, 1993), in Fish Lake Valley (Reheis et al., 1993), and at Lake Tecopa (Sarna-Wojcicki et al., 1984).

Other ash beds in the Owens Lake core

Ash bed of sample OL92-1020, the ash of Thermal Canyon. The ash bed at 296.04 m in OL-92 (Fig. 2) contains 90% glass shards. Its glass chemical composition matches well

TABLE 2. RESULTS OF ELECTRON-MICROPROBE ANALYSIS OF VOLCANIC GLASS SHARDS FROM TEPHRA LAYERS IN OWENS VALLEY CORE OL-92 AND COMPARATIVE DATA FOR SIMILAR OR CORRELATIVE TEPHRA LAYERS*

Sample	SiO_2	Al_2O_3	Fe_2O_3	MgO	MnO	CaO	TiO_2	Na_2O	K_2O	Total, T(o)
TEPHRA LAYER WITH AFFINITY TO THE MONO CRATERS TEPHRA SUITE, CA. 75 KA[§]										
OL92-1 (J)	76.85	12.82	1.09	0.03	0.03	0.44	0.06	4.11	4.58	94.41
OL92-1 (J)	76.94	12.99	1.06	0.01	0.00	0.41	0.04	4.07	4.47	94.80
GS-90 (Mt. Jefferson) (S)	76.95	12.93	1.06	0.01	0.03	0.44	0.07	4.11	4.41	n.a.[†]
TEPHRA LAYER WITH AFFINITY TO THE MONO CRATERS TEPHRA SUITE, BETWEEN CA. 60 KA TO CA. 80 KA[**]										
OL92-1001 (S)	77.42	12.67	0.78	0.06	0.05	0.77	0.08	3.61	4.56	97.24
Walker L. 4-26 67 m (S)	77.27	12.59	0.84	0.04	0.05	0.74	0.06	3.73	4.70	95.91
Walker L. 5-19 79 m (S)	77.29	12.54	0.79	0.06	0.05	0.84	0.08	3.62	4.73	95.19
TEPHRA LAYER WITH AFFINITY TO MAMMOTH MOUNTAIN VOLCANIC ROCKS, CA. 80 KA[§]										
OL92-1003 (S)	73.85	14.38	1.07	0.09	0.11	0.47	0.22	5.19	4.62	96.10
Owens R. UC8966B (S)	73.00	14.93	1.11	0.11	0.10	0.49	0.21	5.20	4.86	94.04
Negit KRL71082(II-3) (S)	72.65	14.86	1.27	0.12	n.a.[†]	0.49	0.24	5.26	5.11	94.23
DISSEMINATED GLASS SHARDS, REWORKED BISHOP ASH (?)										
OL92-1015 (S)	78.01	12.19	0.77	0.03	0.04	0.47	0.05	3.85	4.57	97.75
THE DIBEKULEWE ASH BED, BETWEEN 400 KA AND 665 KA[**]**, ESTIMATED TO BE 510 KA**[§]										
OL92-1016 (J)	76.12	13.54	1.29	0.05	0.04	0.59	0.08	4.22	4.06	93.98
Tulelake 318 (S)	76.32	13.31	1.31	0.05	0.03	0.59	0.07	4.28	4.05	93.73
SPARSE, DISSEMINATED GLASS SHARDS; REWORKED BISHOP ASH (?)										
OL92-1019 (S)	77.20	12.83	0.76	0.04	0.08	0.42	0.06	4.20	4.42	95.65
ASH OF THERMAL CANYON, MECCA HILLS (MH), AND IN OCATILLO CONGLOMERATE (RVS), SOUTHERN CALIFORNIA, CA. 740 KA[§]										
OL92-1020 (J)	74.27	14.09	1.39	0.17	0.03	0.91	0.21	3.83	5.10	93.77
MH9402 (J)	73.89	14.55	1.38	0.18	0.04	0.91	0.22	3.71	5.12	91.99
RVS-BL-1 (J)	74.42	14.43	1.37	0.18	0.03	0.94	0.21	3.53	4.88	94.42
BISHOP ASH BED, IN THE OWENS LAKE CORE AND AT THE SOUTH END OF THE VOLCANIC TABLELAND (BT-8), 758 KA[**]										
OL92-1021 (J)	76.91	13.00	0.72	0.03	0.03	0.45	0.06	4.00	4.81	95.46
OL92-1022 (J)	76.95	12.95	0.74	0.03	0.03	0.45	0.07	3.99	4.79	95.53
OL92-1023 (J)	76.96	12.93	0.70	0.03	0.04	0.46	0.05	4.01	4.83	95.23
OL92-1024 (J)	76.86	13.00	0.70	0.03	0.03	0.44	0.05	4.04	4.85	95.24
OL92-1025 (S)	76.86	12.99	0.68	0.04	0.02	0.45	0.08	3.92	4.96	95.02
OL92-1026 (J)	76.96	12.93	0.69	0.03	0.03	0.45	0.08	3.97	4.87	95.25
OL92-1027 MAJ (J)	76.94	13.00	0.70	0.03	0.03	0.46	0.05	3.99	4.80	94.88
OL92-1027 MIN (J)	76.52	12.90	0.80	0.04	0.05	0.44	0.06	4.10	5.11	92.49
BT-8 (S)	77.48	12.68	0.71	0.03	0.02	0.45	0.06	3.85	4.72	94.28
ASH BEDS OF GLASS MOUNTAIN (MONO COUNTY), IN OWENS LAKE CORE, AND AT SOUTH END OF VOLCANIC TABLELAND										
OL92-1028 MAJ (J)	76.89	12.97	0.73	0.04	0.04	0.45	0.06	3.97	4.85	93.37
OL92-1028 MIN (J)	76.73	13.16	0.71	0.04	0.01	0.45	0.07	4.08	4.75	91.43
OL92-1029 (J)	76.92	13.06	0.71	0.03	0.04	0.42	0.05	3.92	4.85	93.61
OL92-1030 (J)	76.94	12.91	0.74	0.04	0.03	0.44	0.07	3.90	4.93	94.61
BT-1A (S)	77.53	12.65	0.74	0.03	0.04	0.42	0.06	3.71	4.81	95.44
HOMOGENOUS GLASS STANDARD, ANALYZED BY THE SEMQ (S) AND JEOL (J) ELECTRON-MICROPROBES										
RLS 132 (S)	75.37	11.26	2.12[‡]	0.06	0.16	0.11	0.19	4.88	4.42	98.55
±1 σ (18)	0.57	0.16	0.04	0.01	0.01	0.01	0.01	0.13	0.06	0.78
RLS 132 (J)	74.40	11.51	2.13[‡]	0.05	0.16	0.10	0.19	5.23	4.41	98.08
±1 σ (28)	0.34	0.10	0.06	0.004	0.01	0.01	0.01	0.13	0.04	0.48

*Oxides in wt %, recalculated to 100% (fluid-free basis). Original values for each oxide can be obtained by multiplying by T(o), the original total on analysis and dividing by 100. (J) - JEOL 5-channel electron microprobe; (S) - SEMQ 9-channel electron-microprobe. Results of replicate analyses of a homogenous glass standard, RLS 132, on both types of microprobes are given at the end of the table, as an approximation of analytical precision. C. E. Meyer, analyst, U.S. Geological Survey, Menlo Park, California. Data for sample GS-90 from J. O. Davis, Desert Research Institute, Reno, Nevada.
[†]No data available.
[§]Estimate from sedimentation-rate curve of Bischoff (this volume, chapter 8).
[**]Estimate or isotopic age from other localities in the western United States (see text).
[‡]Reported as FeO for the standard.

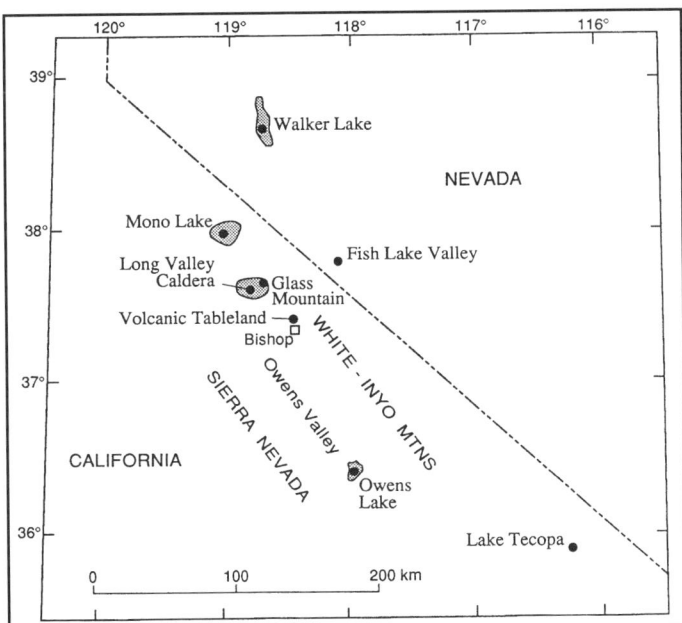

Figure 1. General location map of Owens Lake and localities mentioned in the text.

with a tephra layer (RVS-BL-1) from the Borrego Badlands of southern California, west of the Salton Sea (R. V. Sharpe, 1988, written communication). This ash bed has no independent age constraint, nor is its eruptive source known. In the Mecca Hills in the same general region, this tephra layer is found in deformed fine- to medium-grained sands 40 m stratigraphically above the Bishop ash bed (M. Rymer and A. Sarna-Wojcicki, 1994, unpublished data). We informally assign the name "ash of Thermal Canyon" to this ash bed, based on its occurence in that canyon at the Mecca Hills locality. Using the sedimentation rate curve of Bischoff et al. (this volume) for the Owens Lake core, the age of this tephra layer is estimated to be 740 ka.

Ash of sample OL92-1019. Diatomaceous clay at 266.43 m in the bore hole contains about 25% glass shards, chemically identical to the Bishop ash bed and the younger ash layers of the Glass Mountain set of ash beds. This sample is most likely the Bishop ash, reworked from within the Owens Valley drainage basin. Alternatively, it could be tephra of a later eruption that is compositionally indistinguishable from the Bishop ash bed. The existence of such a bed, however, has not been documented; similar occurrences of these shard types occur at other levels in the core.

Dibekulewe (ash) bed. This clayey, very fine grained ash layer, containing 55% glass shards, is present at a depth of 224.15 m in the core. Chemically, the volcanic glass shards of sample OL92-1016 match well with those of the Dibekulewe Bed of Davis (1978), found at localities in Oregon, California, and Nevada, where it overlies the Lava Creek B ash bed, and is overlain, in turn, by the Rockland ash bed (Davis, 1978; Izett, 1981; Sarna-Wojcicki et al., 1985, 1991). The age of the Dibekulewe (ash) bed is not known, but the associated tephra layers bracket it between 400 ka and 665 ka (Meyer et al., 1991; Izett et al., 1992; Alloway et al., 1992). An age estimate based on a curve derived from dry bulk densities in this study (Bischoff et al., this volume, chapter 8) and magnetostratigraphy (Glen and Coe, this volume) is 510 ka.

Sample OL92-1015. Disseminated glass shards in diatomaceous and calcareous, fine-grained sediment, at a depth of 216.54 m in the core, comprised only about 5% of the sample. These shards are chemically identical to the Bishop ash bed and the Glass Mountain set of ash beds, and are most likely reworked Bishop tephra.

Sample OL92-1003. This sample, at a depth of 52.25 m in the core, consists of 50–55% glass shards. Of these, most are clear, colorless, but 2–3% are brown. The chemical composition of shards from this sample is generically similar to rhyolite flows erupted from Mammoth Mountain (155 km north of Owens Lake) that are dated between 50 ka and 150 ka by the conventional K-Ar method (Bailey et al., 1976). None of the latter units, however, are sufficiently similar to OL92-1003 to be considered correlative. Iron in particular is higher in the Mammoth Mountain source rocks. The similarity does, however, suggest common provenance. OL92-1003 is most similar to a tephra layer (UCSB-FS-89-6-6BC) present in river terrace sediments of Fish Slough, a tributary of Owens River at the south edge of the Volcanic Tableland (Fig. 1), but we have no independent age or stratigraphic control on this latter layer.

Sample OL92-1003 is also similar to several tephra layers intercalated with lake sediments that were exposed along a natural causeway connecting Negit Island to the north shore of Mono Lake during a lowstand of the lake in 1982. Two chemical types of layers are present in this section: one set that is similar to OL92-1003 and the Mammoth Mountain rhyolites, and a second set that is generically similar to tephra erupted from the Mono Craters, situated south of Mono Lake. The lake beds of the Negit Island causeway unconformably underlie the informally named Wilson Creek beds (Lajoie, 1968), exposed north of Mono Lake. The latter contain numerous younger tephra layers erupted from the Mono Craters, ranging in age from ca. 12 ka to ca. 36 ka (Lajoie, 1968; Benson et al., 1990). The beds of the Negit Island causeway are younger, however, than lake beds exposed on Paoha Island south of Negit Island, containing tephra layers with correlated ages ranging between ca. 140 ka and ca. 210 ka (Herrero-Bervera et al., 1994). On the basis of these data and correlations to Walker Lake (Sarna-Wojcicki et al., 1988), we infer that the Negit Island causeway section is about 60–85 ka in age, equivalent to deep-sea oxygen-isotope stage 4 (Imbrie et al., 1984). Using the sedimentation-rate curve of Bischoff et al. (this volume, chapter 8) for the Owens Lake core, we obtain an age estimate of about 80 ka for the ash bed of sample OL92-1003, in reasonably good agreement with previous age data.

Sample OL92-1001. This sample, obtained from a depth of 50.69 m in the core, contains about 75% glass shards. Chemically, this sample matches well with the group of early, Mono

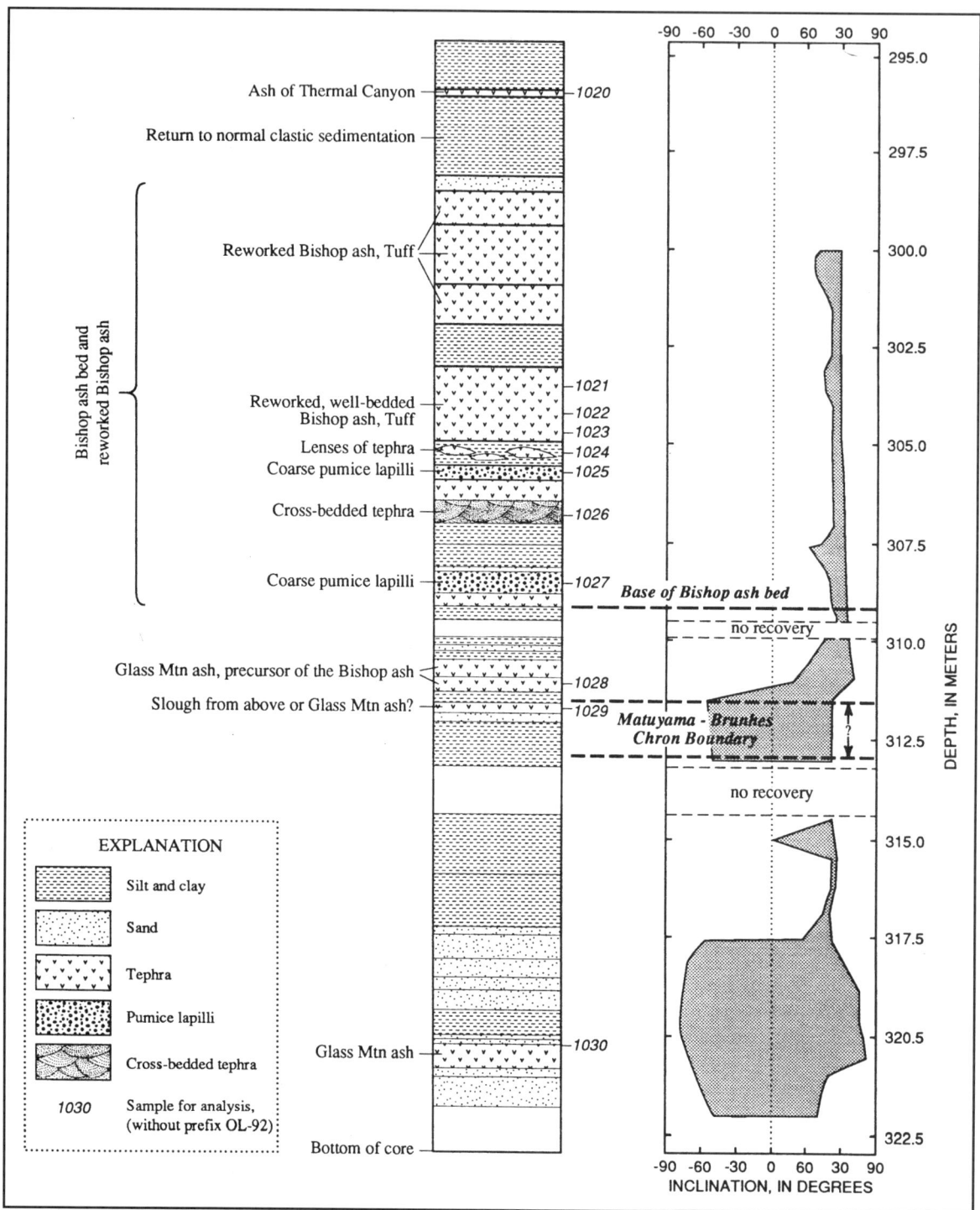

Figure 2. Lithology of the basal 27 m in Owens Lake core OL92-2 and generalized magnetostratigraphy of the basal 23 m (from Glen and Coe, this volume). Left side: lithology in core; intervals marked with "v"s are tephra layers or ashy sediments containing 50% or more glass shards. Right side: envelope curve of all magnetic inclinations in the basal 23 m of OL-92, determined by Glen and Coe (this volume). The position of the base of the Bishop ash bed is inferred from grain size and other lines of evidence (see text). A depth range within which the Matuyama-Brunhes boundary is probably situated is calculated from an estimated duration of 14.1 k.y. between the base of the Bishop ash bed and the boundary, the age of the Bishop ash bed (Sarna-Wojcicki and Pringle, 1992; A. Sarna-Wojcicki, 1992, unpublished data), and probable sedimentation rates of between 16 and 26.5 cm/k.y. in the core (see text). This range is indicated by the two dashed lines at 311.4 m and 312.9 m.

Craters–like tephra layers found in the Negit Island causeway section (see above), as well as with tephra layers at depths of about 63–79 m in a bore hole in Walker Lake, Nev., 260 km south of Owens Lake. At Walker Lake, these beds are estimated to be between 60 ka and 85 ka in age, based on radiocarbon ages in the upper parts of the core, several ages in the core obtained by tephra correlations, and uranium series ages in the middle part of the core (Sarna-Wojcicki et al., 1988; John Rosholt, cited in Benson, 1988). Using the sedimentation-rate curve of Bischoff et al. (this volume, chapter 8), we obtain an age estimate of ca. 75 ka on the tephra layer of sample OL92-1001.

Sample OL92-1. This sample, obtained from a depth of 50.6 m in core OL-92, contained about 75% glass shards, the remainder being crystalline material, both clastic detrital and pyrogenic minerals. Two replicate analyses were run on this sample. Chemically, the glass shards of this tephra layer match most closely with tephra layers obtained from depths of ~54 m to ~62 m in the Tulelake core, in northern California (Rieck et al., 1992), that have interpolated ages ranging from 160–550 ka. Because the latter ages are incompatible with chronostratigraphic data in core OL-92, and because the distance between Tulelake and Owens Lake is great (~700 km), it is unlikely that OL92-1 correlates with any of the Tulelake tephra layers. The tephra of this layer is also generically similar to tephra erupted from the Mono Craters, specifically to tephra layers collected by J. O. Davis in the vicinity of Walker Lake, Nevada (J. O. Davis's BO-series of samples), and from terrace alluvium in the area of Mt. Jefferson, in the Toquima Range, Nevada (J. O. Davis's sample GS-90) (J. O. Davis, written communications, 1980–1990). These layers are latest Pleistocene to Holocene in age, but have no precise age control. An estimate of the age of this ash layer from the sedimentation rate curve of Bishoff et al. (this volume, chapter 8) is also ca. 75 ka.

DISCUSSION AND INTERPRETATIONS

Stratigraphic relationship of the Bishop ash bed and the Matuyama-Brunhes boundary in the Owens Lake core

The ~10-m-thick interval of tephra at depths of ~309.2–298.6 m in the Owens Lake core is the Bishop ash bed, based on several independent lines of evidence discussed in previous sections of the chapter. Magnetostratigraphic studies of the Owens Lake core (Fig. 2, and Glen and Coe, this volume), and previous studies of exposed, temporally equivalent sections in the Volcanic Tableland (Liddicoat, 1993) and at Lake Tecopa (Hillhouse and Cox, 1976; Valet et al., 1988), support our interpretation that the base of the Bishop ash bed is at 309.2 m in the Owens Lake core. Liddicoat (1993) identified the transition between the Matuyama Reversed Polarity Chron and the Brunhes Normal Polarity Chron at the southern end of the Volcanic Tableland about 2 m below the Bishop Tuff. He also showed that there was a zone of highly scattered, transitional inclination and declination directions several meters thick between the stable, coherent reversed directions below the transition, and the stable normal directions above the transition. The pattern of inclinations reported by Liddicoat is similar to the one obtained by Glen and Coe (this volume) from a depth of about 300 m down to the bottom of the Owens Lake core (Fig. 2). In both instances, the Bishop ash bed is within a normally polarized interval with stable normal inclinations, a short stratigraphic distance above scattered inclination directions associated with the Matuyama-Brunhes transition. Thus, our pick for the base of the Bishop ash bed at ~309.2 m is supported by the *general* pattern of magnetic inclination directions observed in the Owens Lake core and at other sites in the region.

Stable reversed inclinations of the Matuyama Reversed Polarity Chron that are observed elsewhere in the region below the Matuyama-Brunhes boundary, however, were not observed within OL-92 below the zone of scattered inclinations. The core bottomed-out within the scattered zone of inclinations. Thus, the actual position of the Matuyama-Brunhes boundary is not determined in the core. Glen and Coe (this volume) propose that the Matuyama-Brunhes boundary is at about 320.2 m, on the assumption that most of the transitional interval was probably recorded in the recovered core, that stable reversed inclinations are probably present a short distance below the bottom of the recovered core, and thus that the Matuyama-Brunhes boundary should lie at about the midpoint of the oberved transitional directions, analogous to inclination directions and the associated low in magnetic intensity observed at Lake Tecopa by Hillhouse and Cox (1976). For reasons that follow, we prefer the interpretation that the Matuyama-Brunhes boundary is present higher in the core, between 311.4 m and 312.9 m depth:

(1) As mentioned, Liddicoat (1993) proposed that the Matuyama-Brunhes boundary was close to the top of the interval of scattered, transitional inclinations that he observed at the south end of the Volcanic Tableland (fig. 1), and about 2 m below the base of the Bishop Tuff. Moreover, Liddicoat identified a short interval of stable, normal-polarity inclinations within the zone of scattered transitional inclinations, below the Matuyama-Brunhes boundary, that is similar to the interval found at 317.5 m to 315.2 m in the Owens Lake core (Fig. 2).

(2) In the studies by Hillhouse and Cox (1976), Liddicoat (1993), and Glen et al. (1993), most of the samples were demagnetized in an alternating field (AF) to obtain remanent magnetic directions. Subsequent to the work by Hillhouse and Cox (1976), Valet et al. (1988) restudied the section at Lake Tecopa using both AF and thermal demagnetization. By using the latter method, they found that the transitional zone of scattered inclinations was considerably compressed and situated somewhat higher in the section (50 ± 5 cm below the Bishop ash bed) than what they and Hillhouse and Cox (1976) had determined from the AF demagnetization. From this, Valet et al. (1988) concluded that much of the scatter in the transitional zone as defined by AF demagnetization was a result of varying degrees of overprinting of stable, reversed paleomagnetic directions, a conclusion similar to that reached by Liddicoat (1993), who showed

that replicate samples obtained at the same depths within the transition showed widely different inclination directions, and thus probably did not record the directions of the actual ambient magnetic field. These conclusions again support the idea that the Matuyama-Brunhes boundary is situated closer to the top of the zone of scattered directions in the Owens Lake core.

(3) The Bishop ash bed has been precisely redated by the laser-fusion $^{40}Ar/^{39}Ar$ method (Sarna-Wojcicki and Pringle, 1992), yielding an age of 758.3 ± 1.2 ka. The Bishop ash bed and the Matuyama-Brunhes boundary are in stratigraphic superposition at five well-documented sites in the western United States where other well-dated chronostratigraphic horizons are present above and below both the Bishop and the Matuyama-Brunhes boundary (Eardley et al., 1973; Davis et al., 1977; Hillhouse, 1987; Sarna-Wojcicki et al., 1984, 1987; Colman et al., 1986; Williams, 1993). The average duration of the stratigraphic interval between the base of the Bishop ash bed and the Matuyama-Brunhes boundary, calculated from sedimentation rates at these five sites, is 14.1 ± 1.7 k.y. Adding this duration to the age of the Bishop ash bed yields an age of 772.4 ± 2 ka for the Matuyama-Brunhes boundary (A. Sarna-Wojcicki, unpublished data, 1992). Thus, if we can derive reasonable sedimentation rates for the base of the Owens Lake core and apply the 14.1-k.y. duration for the time interval between the Bishop and the Matuyama-Brunhes boundary, we can estimate where the Matuyama-Brunhes boundary should fall within the Owens Lake core.

Although the long-term sedimentation rate for the Owens Lake core is about 40 cm/k.y., the sedimentation rates appear to decrease systematically downward in the core; part of that decrease is the result of sediment compaction by overlying sediment (Fig. 3, Bischoff et al., this volume, chapter 8). Given the above constraints on the position of the base of the Bishop ash bed, a 14.1 k.y. duration for the stratigraphic interval between the base of the Bishop ash bed and the Matuyama-Brunhes boundary, and the pattern of magnetic inclinations in the core observed by Glen and Coe (this volume), the sedimentation rate for this interval cannot be less than 15.6 cm/k.y., a rate that would place the Matuyama-Brunhes boundary at 311.4 m, at the top of the zone of scattered inclinations, coinciding with the level of the first fully reversed inclinations (Fig. 2). Any slower rate given the above constraints would place the Matuyama-Brunhes boundary within the overlying normal zone of inclinations. Using a rate of 26.5 cm/k.y. based on the slope of the curve in the basal part of the core (Fig. 3), a rate that appears to be most appropriate for these depths given the sedimentation rate curve of Bischoff et al. (this volume, chapter 8), would place the Matuyama-Brunhes boundary at 312.9 m in the core. The long-term average rate, 40 cm/k.y., probably represents a maximum rate for compacted sediments at the base of the Owens lake core. This rate would place the boundary at 314.8 m. There is no lithologic evidence, such as the presence of thick sand beds or gravels, that would argue for more rapid sedimentation rates than those in the range given above. Although the deposition of the ~10-m-thick Bishop ash bed was undoubtedly rapid, this event does not affect the above calculations, because they are made on the cored interval from the *base* of the ash bed downward. Thus, we propose that the Matuyama-Brunhes boundary lies between 311.4 m and 314.8 m in the core, and most likely within the 1.5-m-thick interval between 311.4 m and 312.9 m depth.

An alternative interpretation of the magnetostratigraphic pattern obtained by Glen and Coe (this volume) for the base of OL-92 is that the brief normal interval at 315.7–317.5 m represents the Kamikatsura Normal Polarity Subchron (of the Matuyama) (Maenaka, 1983; Champion et al., 1988) or an older normal-polarity magnetostratigraphic subchron. Such an interpretation, given the aforementioned constraints, would require that sedimentation rates were even slower than 15.6 cm/k.y., placing the Matuyama-Brunhes boundary even higher above the first fully reversed directions, unless a hiatus, or hiatuses, are present within this interval. No evidence of a hiatus, such as a paleosol or truncation of bedding, was observed in this part of the core. However, there are two zones from which core was not recovered in this interval of the hole, and the nature of these unrecovered sediments is unknown.

The highly scattered inclinations in the lowermost 11–12 m of the Owens Lake core may result from different degrees of overprinting of stable reversed-polarity directions within these basal silts and sands, a conclusion also reached by Liddicoat (1993) in his study of the Matuyama-Brunhes transition at the south end of the Volcanic Tableland, near Bishop. These observations are compatible with independent isotopic studies of the lower Owens Lake sediments, which suggest that very young, interstitial water has been circulating within the lower part of the Owens Lake core (Friedman, this volume). Based on the evidence we have presented, we conclude that the Matuyama-

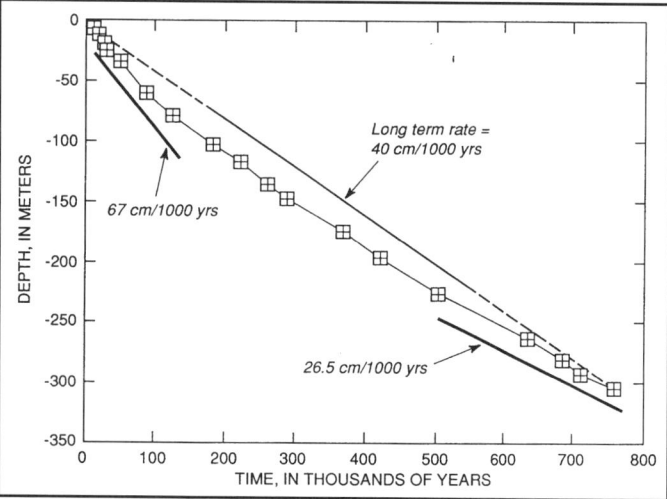

Figure 3. Sedimentation-rate curve for the Owens Lake drill core (from Bischoff, 1993, and this volume, chapter 8), showing selected sedimentation rates used for calculations discussed in the text. Boxes with crosses = estimates of age and depth errors for data points from which curve is drawn.

Brunhes boundary is probably situated between 311.4 m and 314.8 m in OL-92, most likely between 311.4 m and 312.9 m (Fig. 2). However, see Glen and Coe (this volume) for a different interpretation.

Environment of Owens Valley during and after deposition of the Bishop ash bed

Level of Owens Lake during and after eruption of the Bishop ash.
Textures of the Bishop ash bed found in the Owens Lake core suggest that the air-fall ash fell into standing water, and that the ash was also subsequently reworked by water currents. Within the Owens Lake core, we did not find a massive bed of coarse, well-sorted, angular pumice lapilli that is typical of the inital, subaerial fallout of the Plinian phase of the Bishop ash bed (or, for that matter, typical of other ash-fall beds that have been observed after historic Plinian eruptions). There are well-sorted medium- to coarse-sand-size and finer tephra beds with abundant but scattered, rounded, finer-matrix–supported pumice lapilli of coarser size, as much as 2–3 cm in long diameter. We interpret these textures to result from ash fallout deposited into standing water.

The basal, sandy ash intervals in the core, beginning at 309.2 m and above, are composed dominantly of solid-glass bubble-wall and bubble-wall junction shards. The coarser, well-vesiculated pumice granules and lapilli are dominantly well rounded. The latter are commonly found in the upper parts of the more massive tephra beds (that is, these beds tend to be reversely graded) and as scattered, occasional pumice pebbles in well-bedded matrices of silts and clays. These observations suggest that the pumices floated at the lake surface and were also abraded along the shores of the lake and during transport by streams before they became water-logged and sank. Pumice is lighter (often less than ~1 g/cm^3) than the unvesiculated solid glass (2.35–2.40 g/cm^3) of which it is composed. Tubular vesicles that extend through the length of a pumice clast will fill up with water by capillary action, and such pumice clasts will become waterlogged and sink more rapidly than pumice clasts with unconnected vesicles. Pumice clasts with unconnected vescicles may float indefinitely. Such clasts may eventually be worn down into their component solid-glass shards by long-shore drift and abrasion, or they may eventually become hydrated through the solid body of the glass so that their vesicles begin to fill with water, a much slower process than capillary filling.

The sedimentary textures seen in the Bishop ash bed argue for the existence of a standing body of water at the time of eruption. The even bedding and laminations, reversed grading, and absence of ripple marks, cross-bedding, or other current structures in the basal part of the Bishop ash bed suggest that initial deposition occurred below the effective depth of wave action.

Although there was standing water within Owens Lake at the time of the eruption of the Bishop ash, the lake may have shallowed with time. The initially well-bedded and laminated ashy sediments in the basal interval at 309.2–307 m are succeeded upward by an interval of ashy sediments that are cross-bedded at low angles (~10–20°) at 307–306.6 m, and in turn are succeeded upward at 305.8–305.2 m by asymmetrical lenses of pumiceous tephra in an ashy mud matrix, suggestive of currents and shallow water. We don't know if this shallowing of Owens Lake after the eruption of the Bishop ash was a consequence of the rapid filling of the lake with about 4 m of tephra, or the consequence of a climate change—or both. Evidence from correlative lake beds at Lake Tecopa to the east, where the Bishop ash bed is only about 30 cm thick in the central parts of the basin (Hillhouse, 1987) indicates that the Bishop ash bed was deposited during a minor regression but not during a completely dry lake. This regression correlates with deep-sea oxygen-isotope interstadial stage 19 (Shackleton and Opdyke, 1973; Imbrie et al., 1984; Sarna-Wojcicki et al., 1987, 1991). Thus, the regression of the lakes at this time was not just a local phenomenon. The bedding in the upper part of the Bishop ash bed and overlying sediments, from about 305 m to 295 m in the core, consists again of well-bedded silts and sands, suggesting that lake level had risen once again.

Deposition and reworking of Bishop tephra.
The Owens Lake drainage basin and its tributary basins were probably completely covered with tephra after the eruption of the Bishop Tuff, because this basin and its tributaries are situated wholly within the fallout region of the Bishop ash bed. The streams within these basins would not be transporting anything but tephra for some time after the eruption because their basins would be overloaded with loose, easily erodable tephra of low density. The underlying clastic load would not begin to move again until most of the tephra had been cleared from the channels and slopes of the drainage basin, both because the tephra is lighter, and because it covered the normal clastic load.

Reworking of Bishop tephra was probably not uniform over the entire drainage basin tributary to Owens Lake. The tephra was most likely eroded fastest from the higher, steeper terrane, where average runoff was highest, or from broad, flat, dry areas exposed to wind. For the Owens Lake drainage basin, reworking would have been most rapid from the eastern Sierra Nevada east of the drainage divide. Erosion of ash from the more arid Inyo Range and White Mountains east of Owens Valley was more likely slower and more sporadic.

The winds undoubtedly moved the tephra from place to place, building large drifts and dunes of the material at sheltered sites, only to rework them as winds shifted. Persistent dust veils probably reduced insolation, perhaps modifying the regional climate. The local sink for the eolian ash and dust would be a wet Owens Lake, for once the ash and dust fell into standing water it could not be picked up again by wind. Bishop tephra was reworked sporadically from thicker, more protected areas within the Owens Lake drainage basin during exceptionally heavy storms, thus accounting for the zones of reworked Bishop tephra that are present higher in the core.

How rapidly was tephra of the Bishop ash bed eroded from the landscape of the Owens Lake drainage basin after eruption?

Only the basal, 1-m-thick part of the Bishop ash bed represents direct fallout over Owens Lake that settled directly to the bottom. The rest of the ~10-m-thick interval most likely represents repeated episodes of reworking and redeposition of Bishop tephra within the Owens Lake drainage basin. At ~298.4 m, a change to finer grain size, clay and silt, and a change in sediment composition mark a return to normal clastic sedimentation in the basin. The change is gradual, with a progressive decrease in tephra content and increase in detrital material. Above this transition, at a depth of ~296 m, a younger tephra layer (the ash of Thermal Canyon, OL92-1020; Table 2) of different chemical composition than the Bishop ash bed, derived from an unknown eruptive source outside the Owens Valley drainage basin, is present in the sediments. These sediments and exotic tephra layer indicate that the drainage system of Owens Valley had largely cleansed itself of the easily erodable pyroclastic material. What was left was protected by overlying indurated welded tuff, or by capping gravels, as can now be seen from erosional remnants of the Bishop Tuff and ash bed exposed in the northern part of Owens Valley.

Reworking of Bishop tephra from the landscape probably was quite rapid, despite the large mass of material that was deposited, because the tephra is loose, light, and consequently presents little resistance to erosion. Eolian deposition in Owens lake was probably also accelerated at this time, because a very large surface area of loose ash was exposed to winds. An estimate of the *maximum* time that it might take to rework most of the ash from the surrounding countryside would be that calculated from the fastest observed long-term sedimentation rate in the core, equivalent to about 69 cm/k.y. in the uppermost part of core OL-92 (Fig. 3), or ca. 13 k.y. for the ~9-m-thick interval above the air-fall layer. The actual period of reworking may have been much shorter, perhaps only several hundred to several thousand years.

Estimates of ages of undated tephra layers in OL-92

The dry-bulk-density sedimentation-rate curve derived by Bischoff et al. (this volume, chapter 8) for core OL-92 is calibrated to the age and position of the Bishop ash bed in the core. A depth of ~304 m, near the top of the ash bed, was chosen by Bischoff et al. to eliminate the effect of a rapid sedimentation "spike" from the initial air fall and reworking of the ash bed. The curve appears to agree reasonably well with independent age data presented here, and generally with magnetostratigraphy determined by Glen et al. (1993), and Glen and Coe (this volume). Consequently, the curve can be used for providing age estimates for undated units in the core, such as tephra layers or paleoclimatic events. The correlated age of the Walker lake tephra identified in the Owens Lake core (sample OL92-1001), ca. 60 ka to ca. 85 ka, is compatible with the 75 ka age estimated from the curve. The age estimate for the Dibekulewe (ash) bed of 510 ka falls within the known age constraints of >400 ka and <665 ka for the ash bed. The age of the ash of Thermal Canyon, OL92-1020 at 296.04 m, is estimated to be 740 ka.

Glass Mountain set of ash beds at the base of the core.

The three layers below the Bishop ash bed in OL-92 are derived from the older Glass Mountain source adjacent to Long Valley (Fig. 1), as is inferred from their very similar chemical composition and mineralogy to that of the Bishop ash bed and the similar stratigraphic relationships among these ash beds within the region. Although the Glass Mountain set of ash beds is below the Bishop ash bed, which serves as the calibration point for Bischoff et al.'s (this volume, chapter 8) sedimentation-rate curve, we can estimate the ages of these ash layers using the age and position of the Bishop ash bed and a sedimentation rate of about 26.5 cm/k.y., which appears reasonable for this part of the core (Fig. 3). The age estimates are ca. 802 ka for the lowest of the three ash beds, ca. 769 ka for the middle bed (providing that this is indeed an ash bed and not slough from above), and ca. 767 ka, for the uppermost bed. The positions of the latter two ash beds are close to the estimated position of the Matuyama-Brunhes boundary; direct isotopic ages of these layers may better constrain the age of the boundary.

Absence of some widespread tephra layers in OL-92

Fallout patterns of some widespread middle to upper Quaternary ash layers indicate that they fell, or might have fallen, into Owens Lake but have not been observed in the drill cores. These ash layers are (1) the Rye Patch Dam Bed of Davis (1978), derived from the central Cascade Range west of Bend, Oregon, at about 700 ka (Sarna-Wojcicki et al., 1991); (2) the Lava Creek B ash bed, derived from the Yellowstone National Park area of northwestern Wyoming, at about 665 ka (Izett et al., 1992); (3) the Rockland ash bed, derived from the southern Cascade Range near the present site of Lassen Peak, in northern California, at about 400–470 ka (Sarna-Wojcicki et al., 1991; Alloway et al., 1992); and (4) the "Orange" ash beds, a set of five tephra layers, derived from several central and southern Cascade Range volcanic sources; these layers range in age from about 160–218 ka. Although Owens Valley may be too far south to have received fallout from (1) and (3), two to four of the "Orange" ash beds have been traced as far south as Paoha Island, in Mono Lake and Long Valley, in eastern California, and to Walker Lake in western Nevada, and one has been traced as far south as Lake Tecopa (Fig. 1; Sarna-Wojcicki et al., 1991; Herrero-Bervera et al., 1994; A. M. Sarna-Wojcicki and R. M. Morrison, unpublished data). The Lava Creek B ash bed has been found throughout much of the western and central U.S., and Owens Lake is well within the fallout area of this ash bed. Indeed, this ash bed has been identified from a previous bore hole drilled into Owens Lake (Smith and Pratt, 1957; A. M. Sarna-Wojcicki, 1984, unpublished data). Our inability to find these ash beds in OL-92 may be caused by incomplete core recovery, by bioturbation of thinner ash beds, or by rapid clastic sedimentation that accompanied tephra deposition, which could dilute and mask the tephra so that the layers could not be recognized by visual inspection. Lack of core recovery at about 266 m (3.7 m), 270 m (1.6 m), and 276 m (1.9 m) may explain the absence of the Lava Creek B ash bed in OL-92. Using the curve of Bischoff et al. (this volume, chapter 8), these gaps in recovery are close in age to the Lava Creek B ash bed.

At Owens Lake, just as found at several other modern and ancient lakes such as Lake Tecopa, the initial air-fall layer often comprises a small fraction of the total thickess of a tephra layer. The reworked tephra is usually thickest near the shores in the stream deltas, and it thins outward into the depositional center of the lake. For small or moderate eruptions, or eruptions that did not have a significant air-fall component over a lake, such ash layers may not leave an obvious record within the lake's depositional center.

Sedimentation rates in Owens Lake drill holes

The previously drilled hole in Owens Lake, 3.5 km to the northeast of OL-92 (Smith and Pratt, 1957), encountered the Lava Creek B ash bed at a depth of ~234 m in the hole, and disseminated shards of presumably the Bishop ash bed, at ~262 m (A. M. Sarna-Wojcicki and C. E. Meyer, 1984, unpublished data). According to the sedimentation rate curve of Bischoff et al. (this volume, chapter 8), corresponding units are deeper in the 1992 hole than in the older one. The Lava Creek B ash bed should be about 40 m deeper in OL-92 than in the older core. Calculated sedimentation rates down to the position of the Lava Creek B ash bed are 35 cm/k.y. for the old hole, and 41 cm/k.y. to the estimated position of this ash bed in OL-92.

CONCLUSIONS

Presence of tephra layers in Owens Lake drill holes provides a method by which the stratigraphy of the lake can be correlated from site to site within the lake basin, as well as within the larger drainage basin tributary to Owens Lake. Moreover, because many tephra layers are widespread, this stratigraphy and associated climate-proxy parameters can be correlated in turn with other lake and drainage basins within the western United States, as well as with sediments of the northeastern Pacific Ocean. As such studies progress in the future, these correlations will help to define the rates and nature of climate change within this large region, and provide information on the types of hydrologic and sedimentologic responses that result from specific climatic changes. In addition, presence of the thick Bishop ash bed within the stratigraphic record of Owens Lake allows us to study the regional climatic and other ecological impacts from large-magnitude volcanic eruptions. These data define the important baseline conditions from which we can predict future climate change.

ACKNOWLEDGMENTS

We thank James L. Bischoff, Frank H. Brown, Jonathan M. Glen, Brian P. Housback, James R. Lecompte, and George I. Smith for helpful discussions during this study and for reviews of the manuscript. We also thank Stan Soles and Greg Swanson for assistance in the laboratory, and Kathy Nimz for drafting the figures.

REFERENCES CITED

Alloway, B. V., Westgate, J. A., Sandhu, A. S., and Brught, R. C., 1992, Isothermal plateau fission-track age and revised distribution of the widespread mid-Pleistocene Rockland tephra in west-central United States: Geophysical Research Letters, v. 19, no. 6, p. 569–572.

Bailey, R. A., Dalrymple, G. B., and Lanphere, M. A. 1976, Volcanism, structure, and geochronology of Long Valley caldera, Mono County, California: Journal of Geophysical Research, v. 81, p. 725–744.

Benson, L., 1988, Preliminary paleolimnologic data for the Walker Lake subbasin, California and Nevada: U.S. Geological Survey Water-Resources Investigations Report 87-4258, 50 p.

Benson, L. V., and 7 others, 1990, Chronology of expansion and contraction of four Great Basin lake systems during the past 35,000 years: Paleogeography, Palaeoclimatology, Palaeoecology, v. 78, p. 241–286.

Bischoff, J. L., 1993, Age-depth relations for the sediment column at Owens Lake, California: OL-92 drill hole: U.S. Geological Survey Open-File Report 93-683, p. 250–260.

Borchardt, G. A., 1974, the SIMAN coefficient for similarity analysis: Classification Society Bulletin, v. 3, p. 2–8.

Champion, D. E., Lanphere, M. A., and Kuntz, M. A, 1988, Evidence for a new geomagnetic reversal from lava flows in Idaho: Discussion of short polarity reversals in the Brunhes and late Matuyama polarity chrons: Journal of Geophysical Research, v. 93, p. 11,667–11,680.

Colman, S. M., Choquette, A. F., Rosholt, J. N., Miller, G. H., and Huntley, D. J., 1986, Dating the upper Cenozoic sediments in Fisher Valley, southeastern Utah: Geological Society of America Bulletin v. 97, p. 1422–1431.

Davis, J. O., 1978, Quaternary tephrochronology of the Lake Lahontan area, Nevada and California: Nevada Archeological Survey Research Paper No. 7, 137 p.

Davis, P., Smith, J., Kukla, G. J., and Opdyke, N. D., 1977, Paleomagnetic study at a nuclear power plant site near Bakersfield, California: Quaternary Research, v. 7, p.380–397.

Eardley, A. J., Shuey, R. T., Gvosdetsky, V., Nash, W. P., Dane Picard, M., Grey, D. C., and Kukla, G. J., 1973, Lake cycles in the Bonneville basin, Utah: Geological Society of America, v. 84, p. 211–216.

Gilbert, C. M., 1938, Welded tuff in eastern California: Geological Society of America Bulletin, v. 49, p. 1829–1862.

Glen, J. M., Coe, R. S, Meking, S. S., Boughn, S. S., and Altschul, I., 1993, Rock- and paleo-magnetic results from core OL-92, Owens Lake, CA: U.S. Geological Survey Open-File Report 93-683, p. 127–183.

Herrero-Bervera, Emilio, and 9 others, 1994, Age and correlation of a paleomagnetic episode in the Western United States by $^{40}Ar/^{39}Ar$ dating and tephrochronology: The Jamaica, Blake or a new polarity episode?: Journal of Geophysical Research, v. 99, p. 24,091–24,103.

Hillhouse, J. W., 1987, Late Tertiary and Quaternary geology of the Tecopa basin, southeastern California: U.S. Geological Survey Miscellaneous Investigations Series Map I-1728, scale 1:62,500.

Hillhouse, J. W., and Cox, A., 1976, Brunhes-Matuyama polarity transition: Earth and Planetary Science Letters, v. 29, p. 51–64.

Imbrie, J., and 8 others, 1984, The orbital theory of Pleistocene climate; Support from a revised chronology of the marine ^{18}O record, in Berger, A. L., et al., eds., Milankovich and climate, Part I: Dordrecht, The Netherlands, D. Reidel, p. 269–305.

Izett, G. A., 1981, Volcanic ash beds; Recorders of upper Cenozoic silicic pyroclastic volcanism in the western United States: Journal of Geophysical Research, v. 86, no. B11, p. 10,200–10,222.

Izett, G. A., Wilcox, R. E., Powers, H. A., and Desborough, G. A., 1970, The Bishop ash bed, a Pleistocene marker bed in the western United States: Quaternary Research, v. 1, p. 121–132.

Izett, G. A., Obradovich, J. D., and Mehnert, H. H., 1988, The Bishop ash bed (middle Pleistocene) and some older (Pliocene and Pleistocene) chemically similar ash beds in California, Nevada, and Utah: U.S. Geological

Survey Bulletin 1675, 37 p.

Izett, G. A., Pierce, K. L., Naeser, N. D., and Jaworowski, C., 1992, Isotopic dating of Lava Creek B tephra in terrace deposits along the Wind River, Wyoming—Implications for post 0.6 Ma uplift of the Yellowstone hotspot: Geological Society of America Abstracts with Programs, v. 24, no. 7, p. A102.

Lajoie, K. R., 1968, Late Quaternary stratigraphy and history of Mono Basin, eastern Calif. [Ph.D. thesis]: Berkeley, University of California, 373 p.

Liddicoat, J. C., 1993, Matuyama/Brunhes polarity transition near Bishop, California. Geophysics Journal International, v. 112, p. 497–506.

Maenaka, K., 1983, Magnetostratigraphic study on the Osaka Group, with special reference to the existence of pre- and post-Jaramillo episodes in the late Matuyama Polarity Epoch: Memoirs of the Hanazono University, v. 14, p. 1–65.

Meyer, C. E., Sarna-Wojcicki, A. M., Hillhouse, J. W., Woodward, M. J., Slate, J. L., and Sorg, D. H., 1991, Fission-track age (400,000 yr) of the Rockland tephra, based on inclusion of zircon grains lacking fossil fission tracks. Quaternary Research, v. 35, p. 367–382.

Reheis, M. C., Slate, J. L., Sarna-Wojcicki, A. M., and Meyer, C. E., 1993, A late Pliocene to middle Pleistocene pluvial lake in Fish Lake Valley, Nevada and California: Geological Society of America Bulletin, v. 105, p. 953–967.

Rieck, H. J., Sarna-Wojcicki, A. M., Meyer, C. E., and Adam, D. P., 1992, Magnetostratigraphy and tephrochronology of an upper Pliocene to Holocene record in lake sediments at Tulelake, northern California: Geological Society of America Bulletin, v. 104, p. 409–428.

Sarna-Wojcicki, A. M., and Davis, J. O., 1991, Quaternary tephrochronology, in Morrison, R. B., ed., Quaternary nonglacial geology: Conterminous U.S.: Boulder, Colorado, Geological Society of America, The Geology of North America, v. K-2, p. 93–116.

Sarna-Wojcicki, A. M., and Pringle, M. S., Jr., 1992, Laser-fusion $^{40}Ar/^{39}Ar$ ages of the Tuff of Taylor Canyon and Bishop Tuff, E. California–W. Nevada [abs.]: Eos (Transactions, American Geophysical Union), v. 73, p. 633.

Sarna-Wojcicki, A. M., and 9 others, 1984, Chemical analyses, correlations, and ages of upper Pliocene and Pleistocene ash layers of east-central and southern California: U.S. Geological Survey Professional Paper 1293, 40 p.

Sarna-Wojcicki, A. M., Meyer, C. E., Bowman, H. R., Hall, N. T., Russell, P. C., Woodward, M. J., and Slate, J. L., 1985, Correlation of the Rockland ash bed, a 400,000-year-old stratigraphic marker in northern California and western Nevada, and implications for middle Pleistocene paleogeography of central California: Quaternary Research, v. 23, p. 236–257.

Sarna-Wojcicki, A. M., Morrison, S. D., Meyer, C. E., and Hillhouse, J. W., 1987, Correlation of upper Cenozoic tephra layers between sediments of the western United States and eastern Pacific Ocean and comparison with biostratigraphic and magnetostratigraphic age data: Geological Society of America Bulletin, v. 98, p. 207–223.

Sarna-Wojcicki, A. M., Lajoie, K. R., Meyer, C. E., Adam, D. P., Robinson, S. W., and Anderson, R. S., 1988, Tephrochronologic studies of sediment cores from Walker Lake, Nevada: U.S. Geological Survey Open-File Report 88-548, 25 p.

Sarna-Wojcicki, A. M., Lajoie, K. R., Meyer, C. E., Adam, D. P., and Rieck, H. J., 1991, Tephrochronologic correlation of upper Neogene sediments along the Pacific margin, conterminous United Sates, in Morrison, R. B., ed., Quaternary nonglacial geology: Conterminous U.S.: Boulder, Colorado, Geological Society of America, The Geology of North America, v. K-2, p. 117–140.

Sarna-Wojcicki, A. M., Meyer, C. E., Wan, E., and Soles, S., 1993, Age and correlation of tephra layers in Owens Lake drill cores OL-92-1 and -2, in Smith, G. I., and Bischoff, J. L., eds., Core OL-92 from Owens Lake, southeast California: U.S. Geological Survey Open-File Report 93-683, p. 184–245.

Shackleton, N. J., and Opdyke, N. D., 1973, Oxygen isotope and paleomagnetic stratigraphy of equatorial Pacific core V28-238; Oxygen isotope temperatures and ice volumes on a 10^5 and 10^6 year scale: Quaternary Research, v. 3, p. 39–55.

Smith, G. I., and Pratt, W. P., 1957, Core logs from Owens, China, Searles, and Panamint basins, Calif.: U.S. Geological Survey Bulletin 1045-A, 62 p.

Valet, Jean-Pierre, Tauxe, L., and Clark, D. R., 1988, The Matuyama-Brunhes transition recorded from Lake Tecopa sediments (California): Earth and Planetary Science Letters, v. 87, p. 463–472.

Williams, S. K., 1993, Tephrochronology and basinal correlation of ash deposits in the Bonneville basin, northwest Utah [M.S. thesis]: Salt Lake City, University of Utah, 104 p.

MANUSCRIPT ACCEPTED BY THE SOCIETY JUNE 17, 1996

A time-depth scale for Owens Lake sediments of core OL-92: Radiocarbon dates and constant mass-accumulation rate

James L. Bischoff
U.S. Geological Survey, 345 Middlefield Road, MS 910, Menlo Park, California 94025
Thomas W. Stafford, Jr.
INSTAAR, University of Colorado, Boulder, Colorado 80309
Meyer Rubin
U.S. Geological Survey, National Center, MS 971, Reston, Virginia 22092

ABSTRACT

Results of radiocarbon analyses of carbonates and humates from the upper 31 m of OL-92 indicate coherent and linear progression of dates with depth down to about 24 m and 30 ka. Scatter of results below this depth indicates that the practical limit of radiocarbon dating in this core is about 30 ka.

The average mass-accumulation rate (MAR) for the top 24 m is 52.4 g/cm^2/k.y. calculated from radiocarbon dates and bulk density reconstruction. A similar calculation for the entire core down to the Bishop ash bed (304 m, 759 ka) gives a MAR of 51.4 g/cm^2/k.y., suggesting a constant MAR throughout the past 760 k.y. A time-depth curve for the entire core was then constructed from this constant value and using pore-water content to correct for sediment compaction. The resulting curve is remarkably coincident with a similar plot independently derived from 10 within-Brunhes paleomagnetic events identified by Glen et al. (this volume). That MAR remained relatively constant through the glacial/interglacial cycles may imply that the increased sediment supply during the high runoff of glacial times was balanced by increased size of the depositional area and vice-versa during interglacials. Surface area of the lake is approximately proportional to river discharge, but only up to the spill point of the lake. The constant MAR suggests, therefore, that spilling of Owens Lake was relatively infrequent during the past 800 k.y., as independently inferred from the sediment geochemistry (Bischoff et al., this volume, chapter 4).

INTRODUCTION

Uncertainty in chronology is the major barrier in the correlation of paleoclimate records throughout the world, whether from polar ice cores, marine sediments, or from lacustrine sections. The present study attempts to construct an age-depth curve for Owens Lake drill hole OL-92 in order to place oscillations of climate-controlled sediment parameters in chronological context. Chronological control on the core consists of radiocarbon dates for the top 25 meters, presented herein, and the identification of the Bishop ash bed extending from 309–300 m (Sarna-Wojcicki et al., this volume), the identification of the Brunhes-Matuyama paleomagnetic reversal at 320 m and a series of 11 paleomagnetic excursion events between 20 m and 269 m (Glen et al., this volume). These results show that the age versus depth is not linear. We outline below our finding that the mass-accumulation rate (MAR, g/cm^2/k.y.), however, was essentially constant for the part of the core with radiocarbon control and for the entire core down to the Bishop ash bed. A constant MAR allows construction of a continuous age-depth curve down to the top of the Bishop ash bed. The age to any given depth can be calculated by estimating the mass of solids for various segments progressively down the core, using the measured pore-water content to correct for compaction. The

Bischoff, J. L., Stafford, T. W., Jr., and Rubin, M., 1997, A time-depth scale for Owens Lake sediments of core OL-92: Radiocarbon dates and constant mass-accumulation rate, *in* Smith, G. I., and Bischoff, J. L., eds., An 800,000-Year Paleoclimatic Record from Core OL-92, Owens Lake, Southeast California: Boulder, Colorado, Geological Society of America Special Paper 317.

resulting age-depth relationship is derived independently of the paleomagnetic results, but when the two records are compared, the agreement is remarkable. In what follows, we detail the radiocarbon results and then show how the *MAR* age-depth curve was calculated and present a table of age-depth data useful for placing geochemical, sedimentological, and biological parameters of the core in a chronological context.

RADIOCARBON DATING

Stratigraphy and sampling procedures

The occurrence near the bottom of the core of distal air-fall deposits of the Bishop Tuff (= Bishop ash bed, Sarna-Wojcicki et al., this volume), which is well dated at 759 ka (Sarna-Wojcicki and Pringle, 1992), indicates an overall average sedimentation rate of about 40 cm/k.y. Correcting for less compaction in the upper part of the core, we estimated that the practical limit for radiocarbon dating of about 35 ka occurs at about 30 m depth, and thus we limited our sampling down core to this level. Samples were taken from cores 1 and 3 of OL-92 series. Core OL-92-1 was drilled by rotary drilling down to 61 m. Because the top 5.5 m were disturbed by the drilling, OL-92-3, essentially a pushed-in gravity core, was taken immediately adjacent to the drilling pad to provide a high-resolution sampling of the topmost 7 m of the section. This latter core (OL-92-3) was driven from a back hoe–excavated 3.5-m pit to a total depth of 7.16 m below the surface. Depth of the uppermost stratigraphy as exposed in the pit and in the two cores (Smith, 1993, and this volume) is as follows (present surface = 0 m): 0–0.94 m, artificial fill; 0.94–1.3 m, historic salt bed; 1.32 m–5.16 m, oolitic sand; 5.16 m to bottom of core, silty clay with occasional thin sand beds. The salt bed was deposited between 1912 and 1921 as a consequence of the desiccation of the lake as a result of the diversion of the Owens River by the City of Los Angeles (Smith, this volume). Thus, the natural section in the time range of interest is characterized by a 3.8-m bed of oolitic sand, presumably forming until historic time, overlaying silty clay. The contact between the oolitic sand and silty clay is sharp and represents the most striking lithologic change in the entire core. In fact, oolitic sands are absent throughout the rest of the section.

We took 8 samples of the oolites from depths between 3.72 m and 5.12 m from OL-92-3, and 12 samples of silty clay from depths between 5.23 m and 7.11 m from OL-92-3, and from depths between 7.21 m and 31.13 m from OL-92-1. Correlation of beds between OL-92-1 and OL-92-3 based on lithology and on measured depth was straightforward and unambiguous. Each sample consisted of about 50 cc of wet sediment and represented about 3 cm of section. In addition, we took two samples of silty-clay sediment from considerably deeper in the core, beyond the limit of radiocarbon dating, in order to assess the ^{14}C background and contamination limits. These samples were taken at depths of 36.1 m (ca. 50 ka) and 55.1 m (ca. 80 ka).

Extractions and target preparation

The humate fraction of the organic material was extracted from the silty clays by suspending the samples in 200 ml of 1N NaOH solution in capped polyethylene bottles held at 70 °C overnight. The coffee-colored supernate was then separated from the residual solids by filtering. The supernate was then titrated to pH 5 with 3N HCl and the resulting precipitate ("humate fraction") was collected on filter paper, rinsed with distilled water, and air-dried. Air-dried humate yields ranged from 12–200 mg, more-or-less in proportion to the bulk organic content of each sediment sample. Humates were combusted to CO_2 in sealed evacuated silica tubes using CuO, Cu, and Ag.

Oolite samples were cleaned, weighed, and acidified with 10% HCl in vacuum. The resulting CO_2 was dried by passing through an oxygen flow–combustion train, using hot platinum as a catalyst and trapping water and SO_2. The CO_2 gas volumes for oolites and humates were measured and admitted to graphitizers designed for accelerator-mass-spectrometer (AMS) target preparation (Vogel et al., 1984). These small-volume units were charged with 400 torr of CO_2 and 900 torr of H_2. Heating overnight in the presence of iron and zinc catalysts at 675 °C produced sufficient graphite to be pressed into targets for accelerator analysis. All reactions were carried to completion to eliminate isotope fractionation caused by processing. Replicate targets were made of four samples of oolites and one of humate to evaluate contamination during target preparation and to evaluate precision.

For the two "infinitely old" control samples, targets were prepared of the "total organic," "humate" and "humin" fractions to determine the viability of the CuO-Cu-Ag procedure described above. For the total-organic fraction, a decalcified-sediment sample was combusted. The humate fraction was the NaOH extractable fraction from another aliquot, separated out as above, and the humin fraction was the residue. Graphite targets were prepared as above. Targets were analyzed at the Center for Accelerator Mass Spectroscopy (CAMS) of the Lawrence Livermore National Laboratory.

Radiocarbon results

Results for the samples from the top 31 m are given in Table 1 and Figure 1. Dates reported in this section are as uncalibrated radiocarbon years before present and are based on the Libby half-life (5,568 yr), using an assumed $\delta^{13}C$ of –5‰ for the oolites and –15‰ for the organic fractions. Counting errors on individual samples are generally on the order of ±2% or less. Replicates on the five samples for which duplicate targets were prepared agreed within counting error for two of the oolite samples (4.8 m and 5.02 m) and differed by between 5% and 6% for the other two oolite samples and the single humate (3.82 m, 4.29 m, and 31.13 m). Results on the "infinite" samples indicate a probable limiting age of between 30 ka and 35 ka. For the 36-m sample, three replicate analyses of the total organic frac-

TABLE 1. AMS RADIOCARBON DATES (UNCALIBRATED) FROM CARBONATES AND ORGANIC MATTER FROM SEDIMENTS, DRILL HOLE OL-92, OWENS LAKE, CALIFORNIA

Depth* (m)	Material	CAMS† Number	^{14}C Date (B.P.)
3.72	Oolites	4662-B	3,320 ± 70
3.82	Oolites	6317-R	4,080 ± 70
"	Oolites	6318-R	4,280 ± 70
4.02	Oolites	4663-B	3,390 ± 70
4.29	Oolites	6319-R	4,130 ± 60
"	Oolites	6320-R	4,400 ± 60
4.65	Oolites	6316-R	4,660 ± 100
4.80	Oolites	6314-R	5,300 ± 70
"	Oolites	6315-R	5,310 ± 70
5.02	Oolites	6326-R	5,010 ± 80
"	Oolites	6327-R	5,010 ± 100
5.12	Oolites	4664-B	5,090 ± 80
5.23	Humate	6310-R	8,280 ± 120
5.27	Humate	4672-B	8,930 ± 70
5.99	Humate	4675-B	9,980 ± 70
7.11	Humate	4657-B	11,140 ± 70
7.21	Humate	4671-B	11,360 ± 70
10.45	Humate	4659-B	15,880 ± 80
12.63	Humate	4674-B	13,490 ± 80
12.97	Humate	6313-B	11,990 ± 150
15.32	Humate	4658-B	12,570 ± 70
18.31	Humate	4661-B	25,370 ± 160
23.27	Humate	4673-B	30,310 ± 310
31.13	Humate	6311-R	32,320 ± 1,780
"	Humate	6312-R	30,670 ± 1,420

*Ditto marks (") under sample depth refer to analyses of replicate targets made from the same sample.
†Analyzed at the Center for Accelerator Mass Spectroscopy (CAMS) of the Lawrence Livermore Laboratory by John Southon. B = graphite target prepared by Institute of Arctic and Alpine Research Boulder ^{14}C Laboratory. R = graphite target prepared at U.S. Geological Survey, Reston ^{14}C Laboratory.

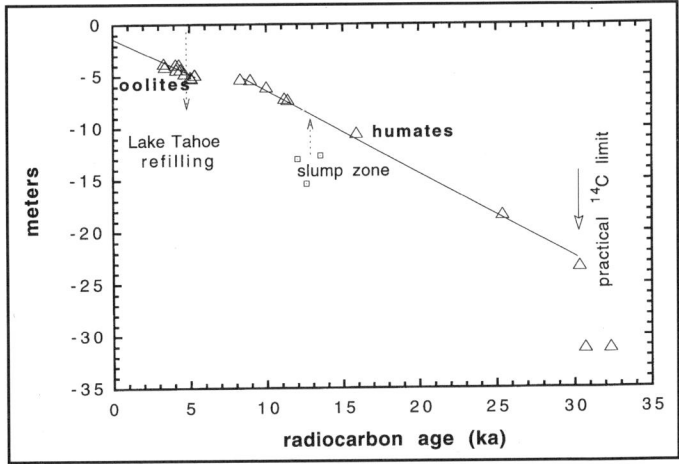

Figure 1. AMS radiocarbon dates (uncalibrated) on sediments (oolites and humate extractions) from Owens Lake drill hole OL-92. Triangle symbol is about twice the size of the actual counting error for each sample. Disconformity between lowest oolite and uppermost humate sample represents a middle Holocene erosion event. The end of this event appears to correspond to radiocarbon dates of submerged tree trunks (5 m below present sill) at Lake Tahoe (dashed arrow), interpreted as representing a prolonged 1,500-yr drought in the Tahoe basin (Furgurson, 1992). This gap also corresponds to a postglacial climate optimum at 6.8 ka manifest in hydrogen isotopic composition in tree rings of Bristlecone pine trees from the nearby White Mountains (Feng and Epstein, 1994). Square symbols refer to radiocarbon dates from sediments from 12.97–15.32 m believed to have been slumped by drill re-entry from a horizon at 8–9 m (dashed arrow).

tion yielded dates ranging from 25–27 ka, whereas the corresponding humate yielded a date of 30 ka, and the humin yielded a date of 32 ka. Similarly, the total organics for the 55-m sample yielded dates from 35.0–35.7 ka, the humate 38 ka, and the humin 37 ka. Targets made from infinitely old calcite yielded a date of 43.5 ka whereas that of infinitely old coal yielded a date of 35 ka (unpublished data, 1992, U.S. Geological Society Reston radiocarbon laboratory). Thus, we conclude that the practical limit for dating humates from the Owens core is about 30 ka, about the same as encountered in radiocarbon analyses of organic material extracted from other lake sediments (i.e., Robinson et al., 1988; Thompson et al., 1990).

Results on samples younger than 30 ka are relatively coherent and represent two linear trends with an apparent hiatus between 5.1 ka and 8.3 ka (Fig. 1). The twelve dates on the oolites, which span only 1.25 m of section (3.72–5.02 m) define a linear trend which projects to a zero age at 1.5 m depth. This depth is essentially that of the base of the historic salt layer (–1.3 m), a result which adds considerable confidence to the dates. The magnitude of the reservoir effect on the dates (low initial ^{14}C/C ratios in lake water), based on the detailed analysis of Benson (1993) for nearby Walker Lake, should be on the order of 200 yr and certainly less than 500 yr.

The dates on the humate extracts likewise result in a linear trend, except for the three samples from –12.97 m to –15.32 m. Reexamination of this interval in the core indicated a slump. The slump was probably cave-in from the interval of 8–9 m above caused by hole re-entry during drilling. The texture, mottling, color, and bedding of this slumped sediment is identical to that found above between 8 m and 9 m where the dates would plot exactly on the trend. The deepest samples at 23.27 m and at 31.13 m all give dates of 30–32 ka, essentially at the practical limit as indicated by the deeper control samples. Thus, there exists a coherent trend in the humates from the base of the oolites at 8.2 ka to 23.27 m at 25.4 ka. The apparent sedimentation rate (SR) of the oolites is 70 cm/k.y. and that of the silty clay is 83 cm/k.y. The contact between the oolites and the silty clay is abrupt, the bedding apparently conformable, and there is no evidence of pedogenesis at the top of the silty clay. The linear trend of the humate dates projects to zero-age at +2 m. Thus, because the true zero-age of the section is at –1.3 m (base of the salt), the amount of missing section represented by the disconformity is about 3.3 m. We interpret the 3.2-k.y. offset between the two trends as a disconformity, possibly caused by sublacustrine wave erosion resulting from a severe shallowing of the lake during a prolonged dry period. The end of the hiatus corresponds to the youngest radiocarbon dates of submerged tree trunks (5 m below present sill) at Lake Tahoe interpreted as representing a prolonged 1,500-yr drought in the Tahoe basin (Furgurson, 1992). This gap also corresponds to a postglacial climate optimum at 6.8 ka manifest in hydrogen isotopic composition in tree rings of Bristlecone pine trees from the nearby White Mountains (Feng and Epstein, 1994). The coincidence of these observations suggests there was perhaps an unusually severe and prolonged drought in the entire western Great Basin and Sierra Nevada that ended about 5,000 yr ago.

Deeper in the core, there is no other evidence of a significant change in sedimentological conditions as might be expected for the Pleistocene-Holocene transition, between 10 ka and 14 ka. Rather the significant change in lake conditions seems to have occurred only after 8.2 ka.

CONSTRUCTION OF A TIME-DEPTH CURVE

Calculation of mass-accumulation rate

The calculation of *MAR* is based on radiocarbon ages at the top of the core, the age of the Bishop ash bed near the bottom, pore-water content and salinity, and grain-density measurements of the sediments. Sarna-Wojcicki et al. (1993) describe a 10-m sequence of the Bishop ash bed occurring between 309 m and 299 m in OL-92, the age of which is 759 ± 2 ka (Sarna-Wojcicki and Pringle, 1992, and Sarna-Wojcicki et al., this volume). The top of the direct ash-fall facies of the Bishop appears to be at about 304 m. Layers above this depth contain progressively more reworked tephra, probably representing run-off of the ash from the surrounding drainage basin. We assume that the section between 309 m and 304 m was deposited very rapidly, thus 304 m represents 759 ka. This translates to an *SR* of 40.1 cm/k.y. (Fig. 2).

Using the calibration of Mazaud et al. (1991), AMS radiocarbon dates from the top 5–24 m (Fig. 2) translate to an *SR* of 83.0 cm/k.y. in radiocarbon years, or 78.8 cm/k.y. in absolute years, a rate almost twice that for the entire segment down to the Bishop ash bed. We next calculate and compare the average *MAR* for the two segments by correcting for the water content using bulk density estimates. Manheim et al. (1974) have shown that bulk densities of marine sediments are best determined from separate measurements of pore-water content, salinity, and grain density. Pore-water content and salinity were determined for 120 samples from OL-92 taken every 2–3 m (Bischoff et al., 1993b; Friedman et al., this volume). The water content varies very erratically, but generally decreases down the core, whereas the salinity varies in a smoother pattern consisting of a single cycle reflecting the overriding effects of ionic diffusion (Fig. 3). Grain densities were determined via gas-comparison pycnometer on six composite samples of sediment from various parts of the core that had been previously rinsed of pore-water salts and dried. The results (Bischoff et al., 1993a) show the grain density to be extremely uniform at 2.63 ± 0.05 g/cc. Such uniformity is

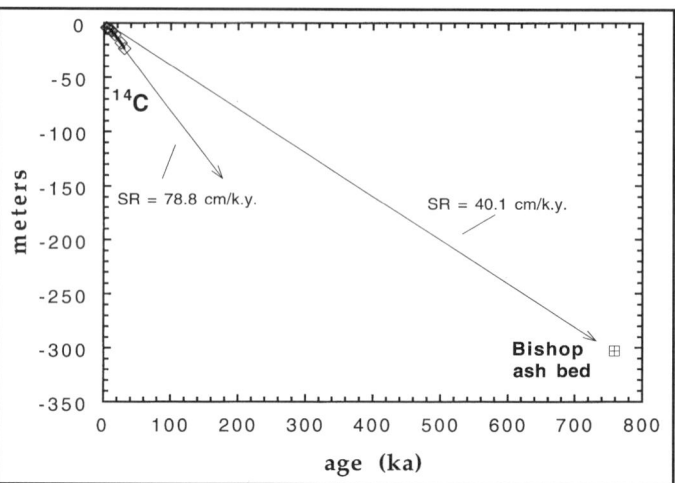

Figure 2. Radiometric age control on sediments from Owens Lake drill hole OL-92. Radiocarbon dates from 6–24 m from this work. Bishop ash bed at 304 m from Sarna-Wojcicki et al. (this volume). Apparent sedimentation rates (*SR*) for the radiocarbon segment and for the entire core down to the Bishop ash bed differ by a factor of two because of the effects of compaction.

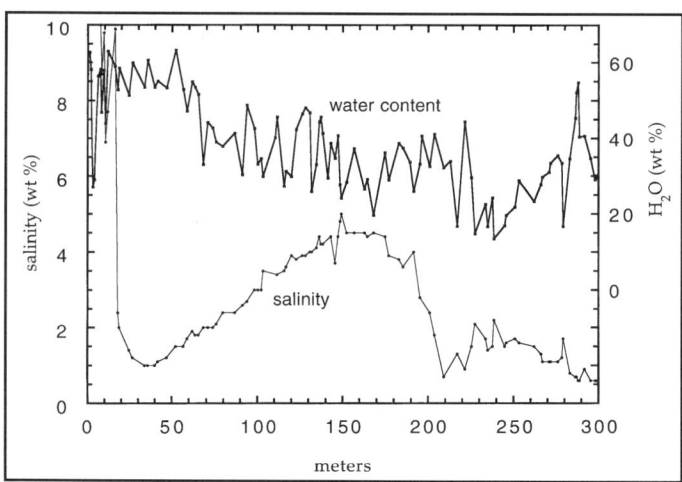

Figure 3. Pore-water content and salinity in sediments from Owens Lake drill hole OL-92 from Bischoff et al. (1993b).

expected (Manheim et al., 1974) considering the narrow density range of the constituent minerals: calcite (2.72 g/cc), quartz (2.65 g/cc), feldspar (2.6–2.75 g/cc), and clay minerals (2.6–2.9 g/cc). Thus, in what follows, the density of the solids, ρ_s, is taken as a constant 2.63 g/cc, and variations in bulk density are attributable to, and calculated from, pore-water content and salinity. The method of calculation follows. First, the mass accumulation rate is given by

$$MAR = zW_s/t = SR \cdot W_s \qquad (1)$$

where W_s is the mass of dry sediment per unit volume of wet sediment, z is the thickness of the sediment, and t is the time of accumulation. W_s is first calculated for the radiocarbon segment (6–24 m) and then for the Bishop ash bed segment (6–304 m) from bulk density considerations, and then MAR is calculated by multiplying each W_s by its respective SR. Bulk density is given by

$$\rho_b = X_s\rho_s + X_{pw}\rho_{pw} = (1 - X_{pw})\rho_s + X_{pw}\rho_{pw} \quad (2)$$

where ρ is density, X is weight fraction and subscript "b" refers to bulk, "s" to solids and "pw" to pore water. The weight of solids in 1 cc of wet sediment (W_s) is simply the weight-fraction of the solids times the bulk density:

$$W_s = X_s\rho_b = (1 - X_{pw})\rho_b. \quad (3)$$

Substituting from Equation 2,

$$W_s = (1 - X_{pw})\left[(1 - X_{pw})\rho_s + X_{pw}\rho_{pw}\right]. \quad (4)$$

Rearranging:

$$W_s = \rho_s + X_{pw}^2(\rho_s + \rho_{pw}) - X_{pw}(2\rho_s + \rho_{pw}) \quad (5)$$

To calculate W_s from parameters that were actually measured, expressions are needed for X_{pw} and ρ_{pw} in terms of measured pore-water content, X_{H_2O}, and measured salinity, x_{sal}, where X_{H_2O} is the weight fraction of H_2O in the bulk sample, and x_{sal} is the weight-fraction of salt in the pore water. X_{pw} is related to these variables by:

$$X_{pw} = X_{H_2O} + \left[X_{H_2O}x_{sal} / (1 - x_{sal})\right] \quad (6)$$

The density of the pore water is primarily a function of its salinity, which can be approximated using the salinity-density relations for seawater (Sverdrup and others, 1942):

$$\rho_{pw} = 1 + 0.8x_{sal} \quad (7)$$

Averaged values for pore-water content and salinity for the two segments are given in Table 2, from which respective average W_s is calculated using Equations 5, 6, and 7. Average MAR is then calculated for each segment from values in Table 2 as follows:

6 to 24 m: $MAR = SR \cdot W_s = 78.8$ cm/k.y. $\times$ 0.665 g/cm^3
 = 52.4 g/cm^2/k.y.

6 to 304 m: $MAR = SR \cdot W_s = 40.1$ cm/k.y. $\times$ 1.282 g/cm^3
 = 51.4 g/cm^2/k.y.

The agreement of average MAR for the two segments is remarkable. Such agreement was unexpected, particularly because the 6–24 m segment represents primarily maximum glacial conditions, whereas the 6–309 m segment encompasses several glacial and interglacial cycles through which rates of MAR might be expected to vary. The next step is to construct a time-depth curve for the core down to the Bishop ash bed based on the constant value for MAR and test for the reasonableness of the curve against independent criteria.

TABLE 2. AVERAGED PARAMETERS (WEIGHTED) FROM OWENS LAKE SEDIMENTS IN DRILL HOLE OL-92 USED FOR CALCULATING MASS-SEDIMENTATION RATES

AVERAGES FOR ENTIRE CORE TO BISHOP ASH BED (6-304 M)
 SR (sedimentation rate) = 40.1 cm/k.y.
 X_{H_2O} (pore-water content) = 0.372
 x_{sal} (pore-water salinity) = 0.0027
 ρ_s (grain density) = 2.63 g/cm^3
 ρ_b (bulk density) = 2.01 g/cm^3

 MAR (mass sedimentation rate) = 51.4 g/cm^2/k.y.

AVERAGES FOR PRE-HOLOCENE RADIOCARBON-DATED SEGMENT (6-24 M)
 SR (sedimentation rate) = 78.8 cm/k.y.
 X_{H_2O} (pore-water content) = 0.548
 x_{sal} (pore-water salinity) = 0.00093
 ρ_s (grain density) = 2.63 g/cm^3
 ρ_b (bulk density) = 1.68 g/cm^3

 MAR (mass sedimentation rate) = 52.4 g/cm^2/k.y.

Calculation of age to any given depth

The time for accumulation of any segment of sediment of thickness z with average solid content of W_s and average mass-accumulation rate of MAR is

$$t = zW_s/MAR \quad (8)$$

Taking MAR as constant and recognizing that W_s varies widely:

$$t = \frac{1}{MAR}\int_0^z W_s dz \quad (9)$$

Because, as shown in Figure 3, X_{H_2O} (and therefore X_{pw}) varies erratically down the core and is not amenable to a mathematical expression, the core was divided into small separate segments and the technique of finite sums was used to estimate the age to any given depth:

$$t = \frac{1}{MAR}\sum_0^n W_{s_i}z_i \quad (10)$$

The core was subdivided into fifteen parcels, an average W_s was calculated from the average pore-water content and salinity for each parcel, and from these values, the time of accumulation for each segment was calculated from Equation 1. The time to any given depth is then calculated from Equation 10. Fifteen segments divide the core into approximately 20-m parcels, each of which is represented by 5–10 pore-water analyses. The resolution of the process could be increased by taking progressively smaller intervals, and based on the limiting number of pore-water samples, the number of parcels for the finite sum could be extended to 120 parcels. At increasingly smaller intervals, however, the assumption of constant *MAR* becomes less tenable, and little is gained in precision. For the purposes of the present study, 15 parcels were broken out with boundaries taken at major changes in pore-water content (Fig. 3). Parameters for each of the 15 segments are listed in Table 3 with the calculated time span for each and the cumulative age to the bottom of each segment. The top of the section is at 6 m for which the radiocarbon results (10 radiocarbon ka, Table 1) yield a calibrated age of 11 ka.

Time-depth curve

The plot of cumulative age versus depth from the model (Fig. 4) is generally smooth, but the plot has some irregularities between 200 m and 300 m where the massive sandy units interbed with fine-grained sediments (Smith, 1993). The points were fit by cubic spline (moving 3-point polynomial) which reproduces each of the 15 derived ages within ±0.1 ka. An expanded table of ages versus depth (Table 4) was then generated from the cubic spline fit using, as independent variable, the mid-point depths of the 85 channel samples used in the geochemical study (Bischoff et al., 1993a). This table is the primary product of the exercise, and it is recommended for use in interpolating the age for any given depth in OL-92. The age-depth curve in Figure 4 can also be described by a polynomial, albeit less accurately:

$$t = 11.83 + 0.37425m + 0.022243m^2 - 0.0001354m^3 + 4.5397 \times 10^{-7}m^4 - 5.7878 \times 10^{-10}m^5 \quad (11)$$

where t is in ka, m is meters below the surface and is valid for $6 \leq m \leq 304$. The polynomial reproduces the derived ages within ±5 ka for 0–200 m and within ±10 ka for the lower 200–300 m.

DISCUSSION

The reasonableness of the constant *MAR* model is evaluated by comparison with the paleomagnetic data. Figure 5 shows the *MAR*-derived age-depth plot on which are superimposed the 11 paleomagnetic excursions between the Bishop ash bed and the surface, as recognized by Glen et al. (this volume). The identification of these excursions was made assuming that the section was continuous, and, therefore, was somewhat tentative. Constraints on the ages of the excursions, as shown by the error bars on Figure 5, varies widely, from very close for some to rather broad for others (see Champion et al., 1988). Nevertheless, the *MAR* plot passes close to, or exactly on, the proposed age of each and all of the excursions in what might otherwise appear to be a hand-drawn curve fit through the excursion points. In fact, the excursion data appear to provide no additional control, and therefore, cause no modification of the derived position of the age-depth curve. These results, inde-

TABLE 3. PARAMETERS OF SEDIMENT INTERVALS FROM OWENS LAKE DRILL HOLE OL-92 USED TO CALCULATE ACCUMULATION TIMES AND AGE TO BASE OF INTERVALS FROM MODEL OF CONSTANT ACCUMULATION RATE*

Depth to Base of Interval (m)	Interval Thickness (m) (z)	X_{H_2O}†	x_{sal}† (x100)	ρ_{pw}† (g/cc)	ρ_b† (g/cc)	$(z)W_s$† (kg/cm²)	k.y. in Interval	k.y. to Base of Interval
16.0	10.00	0.551	9.30	1.07	1.74	0.66	12.6	23.6
33.6	17.60	0.552	1.60	1.01	1.78	1.33	25.4	49.0
58.8	25.20	0.551	1.30	1.01	1.78	1.92	37.4	86.4
77.8	19.00	0.456	1.94	1.02	1.94	1.91	37.2	123.6
102.2	24.40	0.384	2.73	1.02	2.07	2.94	57.3	180.9
117.0	14.80	0.348	3.50	1.03	2.13	1.94	37.7	218.6
134.5	17.50	0.389	3.94	1.03	2.06	2.07	40.2	258.7
147.3	12.80	0.392	4.22	1.03	2.05	1.50	29.1	287.9
174.8	27.50	0.286	4.58	1.04	2.24	4.11	80.0	367.9
195.0	20.20	0.330	3.62	1.03	2.16	2.76	53.8	421.6
225.3	30.30	0.329	1.43	1.01	2.17	4.21	81.9	503.6
262.1	36.77	0.201	1.68	1.01	2.39	6.73	130.9	634.5
279.1	17.08	0.297	1.23	1.01	2.23	2.56	49.8	684.2
291.8	12.65	0.446	0.71	1.01	1.97	1.33	25.8	710.0
306.0	14.20	0.338	0.60	1.00	2.16	1.95	37.8	748.0

*Pore-water content and salinity are averaged for each interval from Bischoff et al., 1993b.
†X_{H_2O} = weight fraction of H_2O in bulk sample; x_{sal} = weight fraction of salt in pore water; ρ_{pw} = density of pore water; ρ_b = bulk density; and W_s = weight of solids in 1 cc of wet sediment.

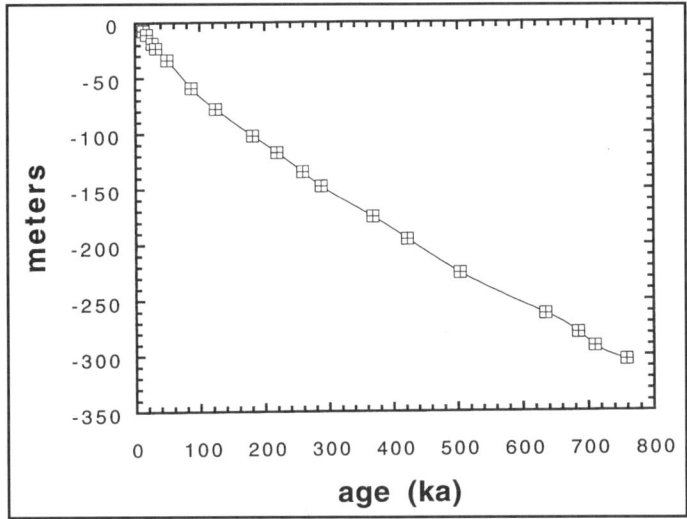

Figure 4. Age-depth plot of sediment parcels (box symbols) from Owens Lake drill hole OL-92 based on constant mass-accumulation rate (*MAR*) model. Data from Table 3. Points are fitted via cubic spline.

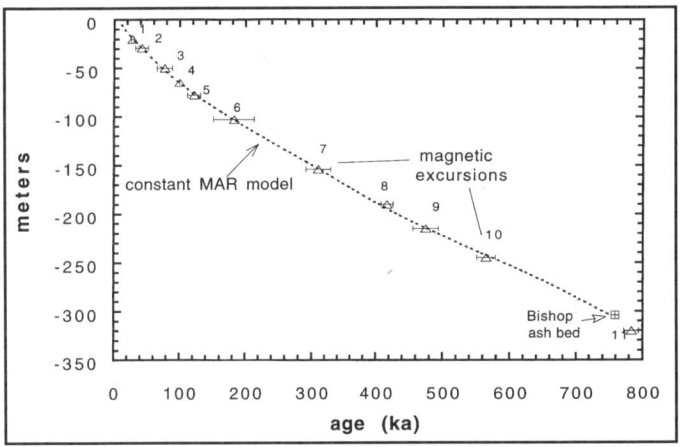

Figure 5. Age-depth plot for Owens Lake drill hole OL-92 based on constant mass-accumulation rate *(MAR)* model, shown with the 11 paleomagnetic excursions between the Bishop ash bed and the surface reported by Glen et al. (this volume) as follows: 1. Mono Lake; 2. Laschamp; 3. NGS; 4. Fram Strass; 5. Blake; 6. Jamaica-Biwa 1; 7. Levantine-Biwa 2; 8. Biwa 3; 9. Emperor; 10. Big Lost; and 11. Brunhes-Matuyama. Age-depth correspondence between *MAR* model and magnetic excursions is almost exact.

TABLE 4. AGE-DEPTH RELATIONS FOR OWENS LAKE CORE OL-92 BASED ON MODEL OF CONSTANT ACCUMULATION RATE*

Meters	ka	Meters	ka	Meters	ka
6.0	11.0	99.3	173.7	200.4	435.1
7.0	12.6	102.3	181.2	202.9	441.6
9.0	16.0	110.4	201.9	208.6	456.1
12.7	20.8	112.8	208.1	213.3	468.7
18.1	26.6	115.7	215.4	215.7	475.1
23.3	31.9	117.7	220.3	220.7	489.6
26.3	36.2	120.8	227.7	223.1	496.7
32.7	47.4	123.4	233.9	225.7	504.9
35.4	52.1	125.6	238.8	235.1	536.5
38.4	56.8	128.0	244.3	237.7	545.7
41.5	61.4	131.1	251.3	240.6	556.3
44.6	65.8	133.8	257.3	245.4	573.8
47.1	69.2	136.6	263.3	253.8	604.8
49.9	73.2	140.5	272.0	255.3	610.4
53.9	78.8	143.5	278.8	267.1	651.5
56.7	83.1	146.5	285.8	271.2	664.2
59.8	88.1	148.9	292.1	276.6	678.5
62.7	93.1	153.2	303.6	279.0	683.8
65.6	98.4	155.9	311.3	283.1	691.6
68.5	104.1	163.8	334.8	285.2	695.4
71.8	110.9	167.7	346.7	288.5	701.8
75.5	118.5	172.8	361.9	292.6	712.4
78.0	124.0	176.0	371.3	296.2	725.1
81.1	130.9	183.2	391.3	299.0	736.2
84.1	137.6	186.3	399.3	301.0	744.9
87.2	144.6	189.3	407.0	303.0	753.7
90.3	151.7	192.1	414.4	305.1	763.0
93.3	158.9	195.1	421.9		
95.8	165.1	196.9	426.5		

*Ages are calculated for each depth from the cubic spline fit to 15 age-depth points from Table 3.

pendently derived, confirm the identification of the excursions, and actually provide them with improved age control. The comparison clearly validates the constant *MAR* model and confirms the continuity and integrity of the sedimentary section. The age-depth curve can, therefore, be used as a time scale for plotting other measured/observed parameters on the core. For example, ages can be tentatively assigned to otherwise poorly dated tephra in OL-92 identified by Sarna et al. (this volume) above the Bishop ash bed, such as the Dibekulewe (ash) Bed of Davis (1978) at 224 m (*MAR* age = 500 ka), and the Walker Lake tephra layer at 50 m (*MAR* age = 74 ka).

The constant *MAR* age-depth model works surprisingly well, a result not predictable considering the variations in runoff that must have occurred over the 800-k.y. span. The transport of sediment into the lake must have varied considerably between glacial and interglacial conditions with the wide variation in river discharge. Hydrologic studies in hundreds of normal watersheds show that particulate mass is strongly correlated with water discharge. Therefore, one might expect highest sediment accumulation during the wet (glacial) periods. The fact that bulk *MAR* has remained relatively constant through the glacial-interglacial cycles may imply that the increased sediment supply during the high run-off of glacial times is balanced by increased size of the depositional area, and vice versa during interglacials. Surface area of the lake up to the spill point is proportional to river discharge, if evaporation rate remains about constant. Given the wide and relatively flat configuration of the Owens Lake basin, the area of sublacustrine sediment will be approximately equal to the area of the lake surface. Therefore, *MAR* is a ratio of sediment mass per unit time divided by the area of the sublacustrine sediment surface. Each of these two parameters is approximately proportional to river discharge. The apparent success of the *MAR* model, therefore, implies that Owens Lake was closed for the bulk of the past 800 k.y., and that periods of overflow were relatively brief. This conclusion is consistent with the inference that Owens Lake was below its spill level for about 66% of the past 800 k.y. suggested by the $CaCO_3$ content of the sediments and Owens River budget considerations (Bischoff et al., this volume, chapter 4).

ACKNOWLEDGMENTS

We thank Frank Manheim for advice on pore-water matters and for advice on how to measure bulk density of the sediments. Accelerator analyses at Lawrence Livermore Laboratory were skillfully carried out by John Southon. We are grateful to Dave Piper, Walter Dean, Andrei Sarna-Wojcicki, and Dave Adam for critically reviewing an early version of the manuscript.

REFERENCES CITED

Benson, L., 1993, Factors affecting ^{14}C ages of lacustrine carbonates: Timing and duration of the last highstand lake in the Lahontan Basin: Quaternary Research, v. 39, p. 163–174.

Bischoff, J. L., Fitts, J. P., Fitzpatrick, J. A.. and Menking, K., 1993a, Sediment Geochemistry of Owens Lake Drill Hole OL-92: U.S. Geological Survey Open-File report 93-683, p. 83–99.

Bischoff, J. L., Fitts, J. P., and Menking, K., 1993b, Sediment pore-waters of Owens Lake drill hole OL-92: U.S. Geological Survey Open-File report 93-683, p. 100–105.

Champion, D. E., Lanphere, M., and Kuntz, M., 1988, Evidence for a new geomagnetic reversal from lava flows in Idaho: Discussion of short polarity reversals in the Brunhes and Late Matuyama polarity chrons: Journal of Geophysical Research, v. 93, p. 11,667–11,680.

Davis, J. O., 1978, Quaternary tephrochronology of the Lake Lahontan area, Nevada and California: Nevada Archeological Survey Research Paper 7, 137 p.

Feng, X., and Epstein, S., 1994, Climatic implications of an 8000-year hydrogen isotope time series from Bristlecone pine trees: Science, v. 265, p. 1079–1081.

Furgurson, E. B., 1992, Lake Tahoe, playing for high stakes: National Geographic, v. 181, no. 3 p. 113–132.

Manheim, F. T., Dwight, L., and Belastock, R. A., 1974, Porosity, density, grain density, and related physical properties of sediments from the Red Sea drill cores, *in* Whitmarsh, R. and 12 others, Initial Reports of the Deep Sea Drilling Program, v. 23, p. 887–907.

Mazaud, A., Laj, C., Bard, E., Arnold, M., and Tric, E., 1991, Geomagnetic field control of ^{14}C production over the last 80 ky: Implications for the radiocarbon time-scale: Geophysical Research Letters, v. 18, p. 1885–1888.

Robinson, S. W., Adam, D. P., and Sims, J. D., 1988, Radiocarbon content, sedimentation rates, and a time-scale for core CL-73-4 from Clear Lake Ca, *in* Sims, J. D., ed., Late Quaternary climate, tectonism, and sedimentation in Clear Lake, northern California Coast Ranges, Geological Society of America Special Paper 214, p. 151–160.

Sarna-Wojcicki, A. M., and Pringle, M. S., 1992, Laser-fusion $^{40}Ar/^{39}Ar$ ages of the tuff of Taylor Canyon and Bishop ash bed, E. California–W. Nevada [abs.]: Eos (Transactions, American Geophysical Union), v. 73, p. 146.

Sarna-Wojcicki, A. M., Meyer, C., and Wan, E., 1993, Tephra in Owens Lake drill hole OL-92: U.S. Geological Survey Open File Report 93-683, p. 184–245.

Smith, G. I., 1993, Field log of core OL-92: U.S. Geological Survey Open-File report 93-683.

Sverdrup, H. U., Johnson, M. W., and Fleming, R. H., 1942, The oceans: Englewood Cliffs, New Jersey, Prentice-Hall, 1087 p.

Thompson, R., Toolin, L., Forester, R., and Spencer, R., 1990, Accelerator-mass spectrometer (AMS) radiocarbon dating of Pleistocene lake sediments in the Great Basin: Palaeography, Palaeoclimatology, Palaeoecology, v. 78, p. 301–313.

Vogel, J. S., Southon, J. R., Nelson, D. E., and Brown, T. A., 1984, Performance of catalytically condensed carbon for use in accelerator mass spectrometry: Nuclear Instruments and Methods, v. 223, p. 289–293.

Manuscript Accepted by the Society June 17, 1996

A diatom-based paleohydrologic record of climate change for the past 800 k.y. from Owens Lake, California

J. Platt Bradbury
U.S. Geological Survey, Federal Center, Box 25046, MS 980, Denver, Colorado 80225

ABSTRACT

A 323-m (~800 k.y.) core of lake deposits beneath Owens Lake playa, Inyo County, California, contains a nearly continuous paleolimnological record based on diatom assemblages. The core chronology is anchored by the Matuyama/Brunhes magnetostratigraphic boundary and the Bishop ash near the base of the record and by radiocarbon dates near the top.

Throughout most of its history, Owens Lake was characterized by fresh-water diatoms, indicating a positive hydrologic input from the Owens River and overflow to lake systems downstream. Both benthic and planktic freshwater diatoms dominate in ashy and sandy sediments between 800 ka and 440 ka and suggest shallow, open-water environments in a basin where sedimentation and subsidence were approximately balanced. After 440 ka, freshwater planktic diatoms dominate, implying that the Owens basin became deeper, perhaps as a result of increased rates of tectonic subsidence. The stratigraphic distribution of saline benthic and planktic diatoms record comparatively short intervals when the lake was shallow and saline. Nevertheless, periodic overflow during these times prevented deposition of evaporites.

According to a chronology based on sediment mass-accumulation rates, the alternation of saline and freshwater diatom assemblages approximately tracks the progression of oxygen isotope stages recorded in marine deposits. Even-numbered isotope stages representing glacial conditions are matched by episodes where freshwater planktic diatoms dominate, indicating abundant precipitation in the Sierra Nevada in response to a southward shift of storm tracks originating in the North Pacific around the Aleutian Low.

INTRODUCTION

The large closed-basin lakes and playas in the Great Basin just east of the Sierra Nevada, the Lahontan and Owens paleolake systems, figure prominently in studies of climate history of the Great Basin (Smith and Street-Perrott, 1983; Benson et al., 1990). Both lake systems are fed by rivers flowing east from the Sierra Nevada crest and both fill or desiccate in response to the difference between precipitation in the mountains and evaporation over the basins. The Owens system, with its progressively linked downstream basins (Owens, China, Searles, Panamint, and Death Valley), forms a comparatively simple hydrologic series in which the upstream lakes must fill and overflow before the lakes lower in the chain can receive significant water (Smith, 1976). Mono Lake, about 200 km north-northwest of Owens Lake, may have joined the chain of lakes during the high stand correlated to the penultimate glaciation (Benson et al., 1990). Long Valley Lake, formed by the caldera responsible for the middle Pleistocene Bishop ash, drained into the Owens River about 150 km northwest of Owens Lake until sill erosion sometime after 160 ka eliminated this basin (Sarna-Wojcicki et al., this volume).

Smith (1979) documented paleohydrologic and climate changes at Searles Lake, the third lake in the chain, by strati-

Bradbury, J. P., 1997, A diatom-based paleohydrologic record of climate change for the past 800 k.y. from Owens Lake, California, *in* Smith, G. I., and Bischoff, J. L., eds., An 800,000-Year Paleoclimatic Record from Core OL-92, Owens Lake, Southeast California: Boulder, Colorado, Geological Society of America Special Paper 317.

graphic analysis of alternating deposits of more or less pure evaporites (salts) and organic, calcareous muds. Salt units in Searles Lake required moderate discharge from Owens Lake under climates of high evaporative stress, whereas the mud units are assumed to have represented comparatively fresher and deeper lakes and greater discharge through the Owens River system to supply water and detrital clastic sediment to the basin. With the exception of pollen, rare fish fossils, and *Artemia* (brine shrimp) chitin in the muds, other indicators of lake and climate change, such as ostracodes and diatoms, are absent. Searles Lake may have been too saline to support or preserve such microfossils. Nevertheless, salt and mud units in Searles Lake reflect paleohydrologic flow in the Owens River system as a corresponding succession of dry, shallow, deep, and occasionally overflowing lakes in the Searles basin.

Diatoms and ostracodes are conspicuous in the sediments deposited in Owens Lake (Smith and Pratt, 1957) and their value in reconstructing paleohydrology and climate of Great Basin lakes (Bradbury et al., 1989) prompted studies of these fossils in Owens Lake drill hole OL-92. Because Owens Lake is the first principal lake in the chain of lakes on the Owens River, variations in the size and salinity of downstream lakes directly relate to the flow of water through Owens Lake given current climatic conditions that govern precipitation and the flow of the Owens River. The alternation of freshwater and saline diatoms in the Owens Lake core documents this hydrologic variability. For example, the presence of saline diatoms in the Owens stratigraphic record must indicate periods of extreme aridity in downstream lakes, with only sporadic deposition (and possibly erosion) of evaporites or other sediments.

The freshwater diatom assemblages, in addition to recording episodes of Owens Lake overflow, reveal much about seasonal changes in water flux, nutrient conditions, turbidity and stability of littoral habitats along the lake margin. This paper focuses on the variations in numbers and types of diatoms in Owens Lake that provide paleohydrologic and paleoclimatic data for the Sierra Nevada and the Great Basin for the past 800 k.y. The Owens Lake paleolimnology based on diatoms can also help clarify past limnological fluctuations in downstream lakes and evaluate the chronology of such changes.

METHODS

Coring

Three cores were raised from the south-central part of Owens Lake playa, Inyo County, California, in the spring of 1992 with a truck-mounted, rotary coring rig. Core OL-92-1 (5.5–61.4 m), OL-92-2 (61.3–323 m), and OL-92-3 (0–7.2 m) collectively recovered ~80% of the section (Smith and Bischoff, this volume). The three cores are combined to produce a largely continuous record of Owens Lake sedimentation. Stratigraphic overlap between the cores is minor or nonexistent according to the distribution of distinctive diatoms (Bradbury, 1993).

Diatom sampling, preparation, and analysis

Samples for diatom analysis (154 from core OL-92-1; 129 from core OL92-2; and 11 from core OL-92-3) representing stratigraphic thicknesses of ~1 cm each were removed from the split core with a clean spatula and stored in airtight plastic bags. The sediments were processed for diatoms by hot acid digestion and the cleaned residue was settled on coverslips and mounted in Hyrax (Bradbury, 1993; Battarbee, 1973).

Diatoms were enumerated at 1,000× magnification along transects measured by stage micrometer until at least 30 mm of transect were examined or until at least 300 diatoms were counted. All slides were made with the same volume of residue, and therefore the diatom valves encountered per millimeter of transect in each sample gives a semiquantitative approximation of diatom concentration in the sample (Bradbury, 1993). Because the diatom concentration of the Owens Lake record is highly variable, relative proportions of diatom taxa were calculated using a standard sample count of 300 diatom valves in order to graphically illustrate relative changes weighted by low and high concentrations.

Typically, diatom preservation is variable and often quite poor in Owens Lake sediments because of breakage and corrosion. Counts represent whole valves and (or) large (>40%) fragments of valves.

Age control

The paleomagnetic record of core OL-92-2 from tephra layers and diatom-rich silts and clays between 323 m and 311 m depth shows several abrupt swings between normal and reversed polarity that represent the terminal Matuyama Polarity Chron (783 ka) (Glen and Coe, this volume). The last significant episode of reversed polarity (311.3 m) is considered to represent the Matuyama/Brunhes magnetostratigraphic boundary by Sarna-Wojcicki et al. (this volume) and the base of the core is estimated to date at ca. 800 ka. Crystal-rich deposits of the Bishop ash (760 ka) that represent air fall tephra from the eruption of Long Valley caldera (Sarna-Wojcicki et al., this volume) occur at a depth of 309.15 m. Radiocarbon dates provide age control for the upper 23 m of the composite core, and an age-depth model based on mass-accumulation rates and provisionally tied to paleomagnetic excursions has been established for the cored record (Bischoff et al., this volume, Chapter 8).

Although the age-depth model of core OL-92 is tentative, in fact it appears reasonable and is adopted for this study. Comparison of global climatic events and trends of both terrestrial and marine records must, of course, be cautiously made considering the uncertain nature of pre-radiocarbon chronologies.

SETTING

Climate and hydrology

The Owens hydrologic system receives virtually all of its water from melting Sierra Nevada snows (Smith and Street-

Perrott, 1983) although streams in the White and Inyo Mountains to the east feed the Owens River and recharge ground water to the basin respectively (Hollett et al., 1991). Winter precipitation ultimately relates to the strength and persistence of the Aleutian Low at corresponding latitudes during that season. However, secondary precipitation maxima occur in the Owens drainage in April and August (Pyke, 1972). Although April precipitation probably cannot be easily differentiated from the hydrologic flux of winter snow melt to Owens Lake, increases in August or fall seasonal precipitation are likely to be limnologically visible in terms of the productivity and kinds of diatoms living in the lake. Significant late summer and fall storms typically represent tropical cyclones that bring moisture to the Great Basin from the Gulf of California or the equatorial Pacific Ocean (Hansen et al., 1981; Kay, 1982). Although the frequency and magnitude of such storms is variable, they can account for substantial precipitation during the season when evaporative stress on lakes is highest. In addition, such storms supply water to a significant (~80%) part of the Owens Lake drainage basin that lies outside the Sierra Nevada. Consequently, precipitation on and east of Owens Lake could have a significant impact on Owens limnology during past climates.

The Owens basin has a mean evaporation rate of about 1.65 m/yr (Smith, 1976). The volume of evaporites resulting from desiccation since 1913 suggests several thousand years of salt accumulation assuming modern Owens River salinity and no subsurface discharge. This study of OL-92 has shown that the last significant overflow of Owens Lake was during early Holocene (Smith et al., this volume). Periodic overflow of Owens Lake before and after this overflow probably accounts for the lack of evaporite deposits in its sediment record.

DIATOM ECOLOGY

The diatoms in the Owens Lake sediments may be grouped into taxa with ecological preferences for fresh (<3‰ tds) and saline (>5‰ but typically <50‰ tds) waters. Many species found in the cores overlap in their salinity preferences and tolerances and these guidelines cannot be taken too literally. For example, although most of the taxa considered "fresh" have optima well below 1‰ tds, they may inhabit waters 2–3 times more saline. "Saline" taxa, on the other hand, often tolerate (or even prosper in) a broad range of salinities above 5‰ tds, but may be present below 5‰ tds. Today, the Owens River has a salinity of about 0.2‰ tds (Friedman et al., 1976), and this level of salinity or less may have been typical of Owens Lake when it was overflowing.

In addition to this broad salinity classification of diatoms present in the Owens Lake cores, the species occupy different physical habitats within the photic zone of the lake: planktic = floating or suspended by turbulence in open water; benthic semi-attached = loosely attached to substrates such as submerged vascular plants; benthic adnate = firmly attached to substrates by stalks or pads of mucilage; and benthic motile = species that are capable of moving in and through the mud within illuminated areas of the lake bottom.

As with salinity, there are overlaps in the habitats diatoms occupy, especially on a seasonal basis. Semi- or loosely attached species may become planktic with sufficient turbulence, as can some motile species of *Campylodiscus, Nitzschia,* and *Surirella.* Planktic species often settle to the lake bottom until favorable conditions return to support them in the open water photic zone. By using these criteria and combining several habitat options, five ecological groups that have more or less coherent stratigraphic distributions have been established (Fig. 1).

1. Freshwater planktic diatoms: all species of *Stephanodiscus, Asterionella formosa, Fragilaria crotonensis, Cyclotella ocellata, C. bodanica, Cyclostephanos* species, and all *Aulacoseira* species.

2. Freshwater benthic, non-motile diatoms: *"Fragilaria"* species: *Pseudostaurosira brevistriata, Staurosira construens, Staurosirella leptostauron,* and *S. pinnata.* These taxa were once grouped together under the genus *Fragilaria* (Williams and Round, 1987).

3. Other freshwater (mostly benthic) taxa: Tightly attached species = *Achnanthes, Amphora ovalis, A. perpusilla, Cocconeis placentula, Cymbella, Epithemia, Gomphonema, Rhoicosphenia curvata, Rhopalodia gibba.* Loosely attached species = *Fragilaria vaucheriae, Fragilaria capucina, Melosira varians, Synedra mazamaensis, S. rumpens, S. ulna, S. acus.* Motile species = *Caloneis, Gyrosigma, Hantzschia, Pinnularia, Navicula* (excluding *N. subinflatoides* and *N. pygmaea*), *Nitzschia.*

4. Saline planktic diatoms: *Chaetoceros muelleri* spores, *Cyclotella caspia, C. quillensis, C. meneghiniana.*

5. Saline benthic attached and motile diatoms: *Amphora coffaeiformis, Anomoeoneis costata, Campylodiscus clypeus, Navicula subinflatoides, N. pygmaea, Nitzschia frustulum, N. monoensis, N. pusilla, Rhopalodia constricta, R. gibberula, Surirella hoefleri, S. ovalis, S. striatula.*

Because of the great variety and abundance of freshwater planktic diatoms in the Owens Lake core, some comments about their ecology and paleolimnologic-paleoclimatic significance are appropriate.

Stephanodiscus species and *Asterionella formosa* generally dominate phytoplankton blooms in the spring or fall (Bradbury, 1988a; Kilham et al., 1986). In the Owens Lake system, they are interpreted to indicate major hydrologic input in the spring, probably associated with melting of the Sierra Nevada snowpack.

Aulacoseira species, such as *A. subarctica* and *A. islandica,* bloom during the cold spring or fall seasons in north temperate and boreal lakes; sometimes under the ice when light levels first begin to increase in the spring. In Owens Lake, they should reflect cold-season, low-light, freshwater limnologic conditions and may imply seasonal ice cover.

On the other hand, *Aulacoseira granulata* and *A. ambigua* are warm-season freshwater phytoplankton that require high light levels coincident with abundant nutrients. These species are found in temperate and warm temperate shallow freshwater lakes

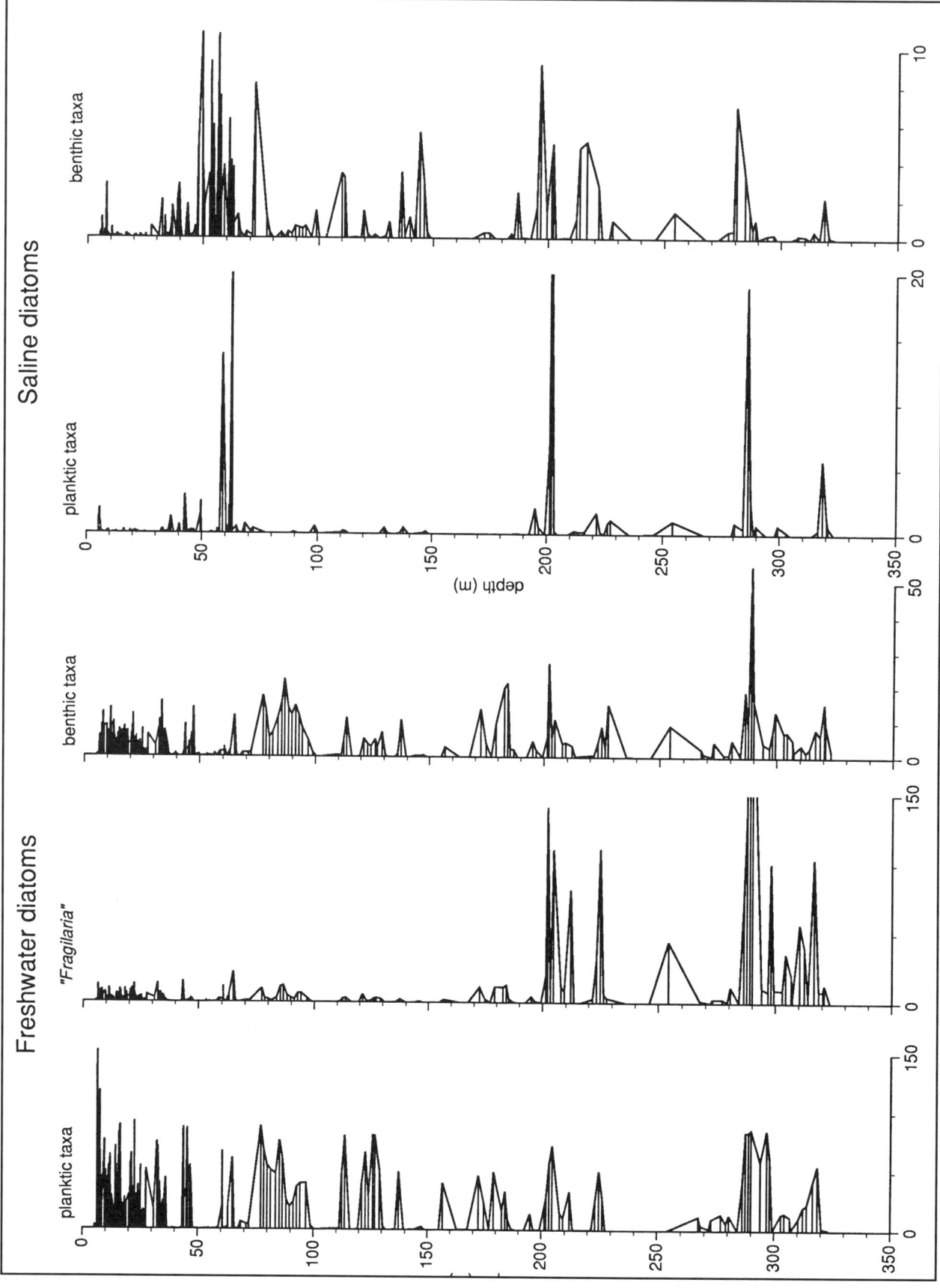

Figure 1. Concentration (diatoms/mm microscope transect) of principal diatom habitat groups versus depth in the Owens Lake core. Horizontal bars in the silhouettes represent stratigraphic position of samples.

that have sufficient turbulence to keep the heavy cells suspended in the photic zone (Kilham et al., 1986). In Owens Lake they imply hydrologic input of freshwater during the summer or fall. *Aulacoseira solida* now only survives in monomictic Lake Biwa, Japan, where it blooms as lake stratification begins to break down in the fall and circulation brings nutrients to the photic zone (Negoro, 1960). The dominance of this species in Owens Lake may suggest comparatively moist climates with winters warm enough for monomictic circulation (Bradbury, 1991).

Fragilaria crotonensis opportunistically blooms when phosphorus supply ratios are low, typically in the late spring, summer, or early fall (Kilham and Kilham, 1978). In Owens Lake, its presence suggests stable, low-conductivity limnologic conditions throughout the open-water season, and nutrient supplies that are governed by phytoplankton succession to favor development of this species.

Cyclotella bodanica is a planktic diatom of alpine, temperate, and north temperate dimictic lakes. Its nutrient requirements and succession dynamics are not known, but it certainly indicates cold and reasonably moist climatic conditions in Owens Lake. Similar conditions probably apply to the presence of *Cyclotella ocellata,* which occupies oligotrophic parts of the Great Lakes (Stoermer and Yang, 1970).

Very little is known about the ecology and requirements of *Cyclostephanos* species, although some appear to prosper in highly eutrophic freshwater systems (Håkansson and Kling, 1990).

RESULTS AND INTERPRETATION

General considerations

Concentrations of diatom ecological groups in the composite, 323-m stratigraphic sequence (Fig. 1) fluctuate widely over short stratigraphic intervals, particularly in zones where saline diatoms are abundant. Overall stratigraphic continuity of freshwater diatoms indicates a longer and more persistent limnologic record of a fresh and overflowing lake system than of a shallow, saline system. The concentration of freshwater diatoms is generally an order of magnitude greater than concentrations of saline diatoms, partly reflecting poor diatom preservation in saline systems. Alkaline corrosion of biogenic silica, slow deposition rates, and exposure of lake sediments to shallow, high energy environments and intermittent desiccation probably account for much of the diatom destruction in such habitats. Therefore, rare and short-lived episodes of high concentrations of saline planktic diatoms, implying large and possibly deep saline lakes, may be under-represented in the Owens Lake record and contrast with the paleolimnology of endorheic Great Basin lakes such as Walker Lake (Bradbury et al., 1989).

Throughout its 800-k.y. history, Owens Lake was extraordinarily sensitive to hydrologic and climatic change, sometimes oscillating between a large, through-flowing, freshwater lake, a shallow freshwater marsh, and a saline lake or playa over short periods of time (Fig. 2). Peaks of freshwater planktic diatoms logically correlate with increased flow of the Owens River and reflect moist climates, possibly associated with expansion of Sierran glaciers. Saline planktic and benthic diatoms correlate with arid, interglacial hydro-climatic environments and reduced flow of the Owens River.

Species of *"Fragilaria"* and other freshwater benthic diatoms have a stratigraphic distribution similar to planktic freshwater species, although benthic *"Fragilaria"* species dominate below 200 m (Fig. 1). Large concentrations of *"Fragilaria"* species indicate periods of time when Owens Lake was shallow, but fresh; probably a through-flowing marsh system. It is reasonable to suppose that tectonic rejuvenation of the basin or a reduction of clastic sedimentation (Smith, this volume) was required to reestablish a deeper lake suitable for planktic diatoms. Perhaps the stratigraphic distribution of *"Fragilaria"* species documents periods of tectonic quiescence that allowed the Owens basin to fill with sediment and develop persistent shallow but fresh marsh environments.

Paleolimnologic history of Owens Lake

From 800–ca. 735 ka (323–298 m). Core OL-92-2 bottomed slightly below tephra-rich sediments possibly related to Glass Mountain eruptions of tephra (Sarna-Wojcicki et al., this volume). Rare attached and motile freshwater diatoms characterize the basal, ash-rich sample of the Owens Lake core (323.28 m; Fig. 1) implying that the ash fell or was washed into a freshwater marsh. A short interval of saline benthic diatoms and *Cyclotella meneghiniana* (Fig. 2), a planktic diatom tolerant of elevated salinity, characterizes the 318-m level of the core and may indicate influx of chloride, sulfate, and bicarbonate ions associated with the eruption and aqueous reworking of deposited ash.

Between 790 ka and 735 ka, clayey silts contain abundant freshwater diatoms characterized by *"Fragilaria"* and planktic assemblages (Fig. 2). *Aulacoseira solida* and other warm-season *Aulacoseira* species are abundant (790–765 ka) (Fig. 3) and indicate moderate lake depths and either warm-season precipitation or at least absence of extreme evaporative stress during the summer. Although these freshwater diatom assemblages may have accumulated between 790 ka and 765 ka, their paleoclimatic and paleolimnologic interpretation makes it unlikely that they directly relate to nearby deposits of the Sherwin glaciation that preceded the eruption of the Bishop ash.

The primary air-fall of the Bishop ash (309.15 m) resulting from an eruption at Long Valley Caldera (760 ka) is preceded by a related tephra layer, possibly a Bishop ash precursor, at 311.4 m. Reworked Bishop ash dominates the core lithology to a depth of 298 m (Sarna-Wojcicki et al., this volume). Diatom concentrations are generally low, perhaps diluted by ash, and are dominated by cyclic pulses of *"Fragilaria"* and benthic freshwater taxa (Figs. 1, 2) that indicate shallow, marsh-like environments. Significant redeposition of ash may have occurred for several centuries, and it is likely that paleolimnological conditions at Owens Lake relate as much to the sedimentological and

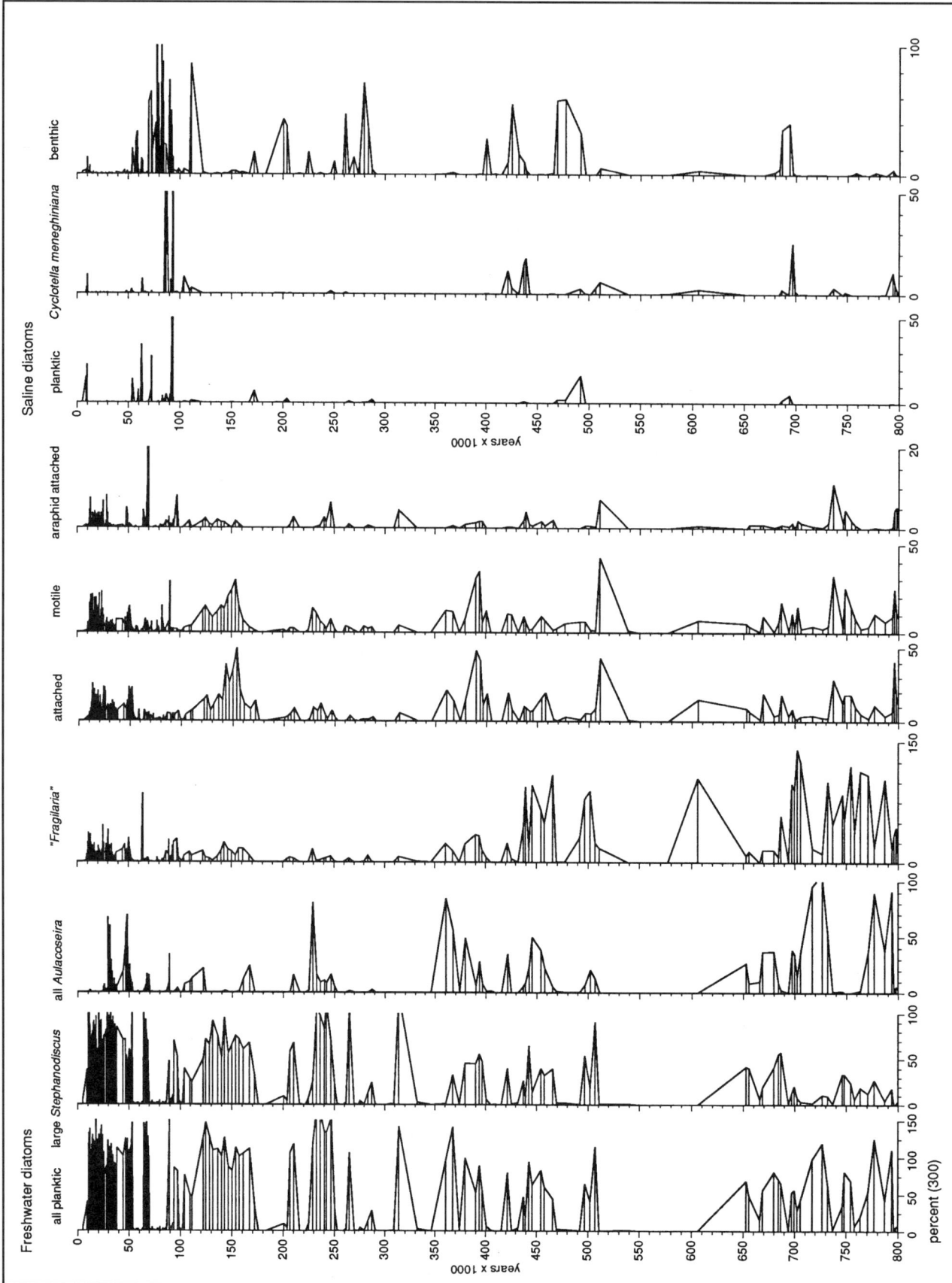

Figure 2. Weighted percentages of diatom habitat groups versus age in the Owens Lake core.

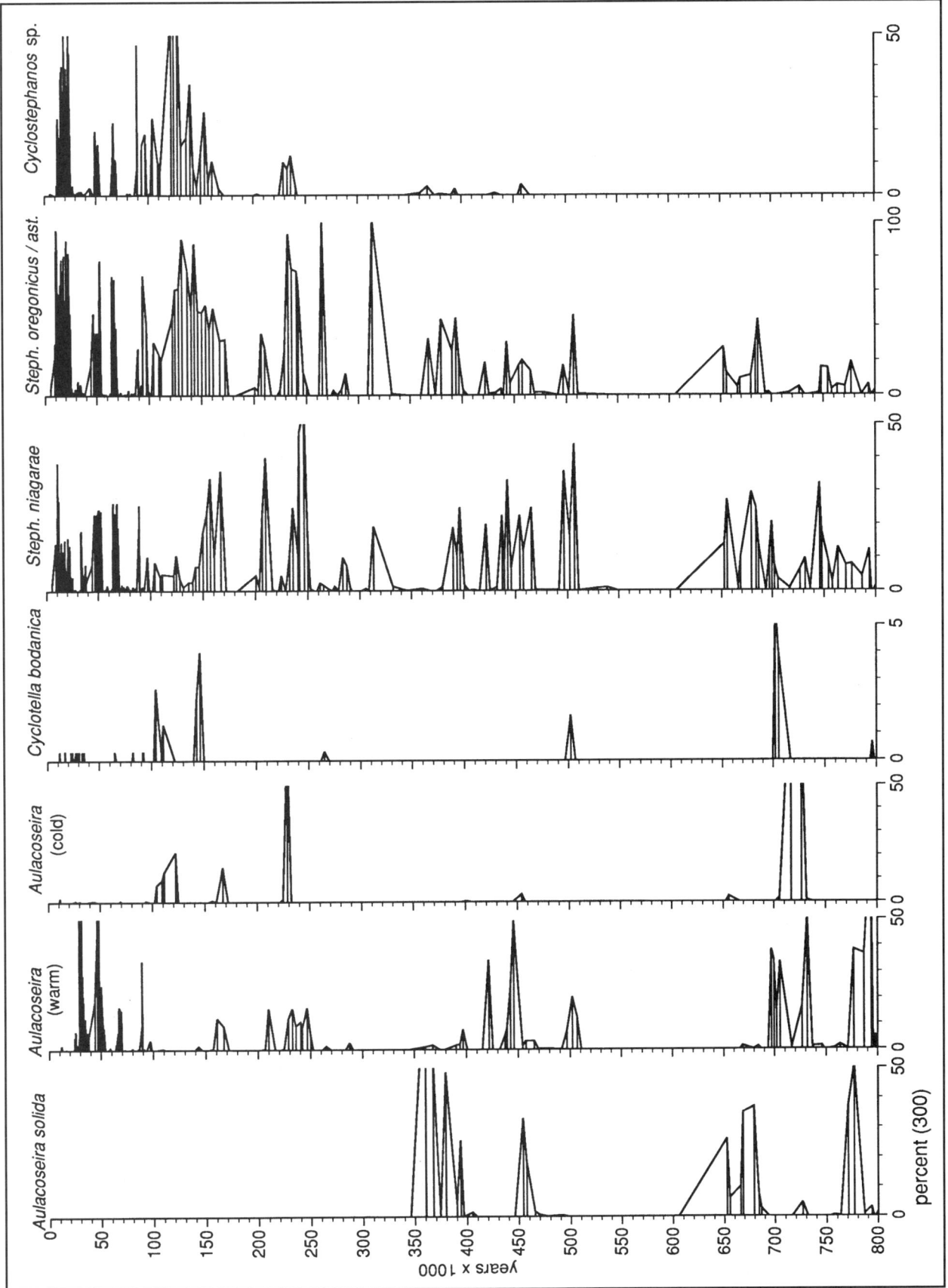

Figure 3. Weighted percentages of principal planktic diatom taxa versus age in the Owens Lake core.

geochemical effects of ash redeposition as to climatic conditions at that time.

From ca. 735–650 ka (298–267 m). The direct limnological effect of redeposited Bishop ash disappears above 298 m, a horizon arbitrarily dated to ca. 735 ka, but possibly less than a millennium after the eruption (Sarna-Wojcicki et al., this volume). Owens Lake became a shallow, marsh-fringed, open-water lake with large numbers of *"Fragilaria"* and freshwater planktic diatoms and lesser numbers of benthic freshwater diatoms (Figs. 1, 2). The planktic diatoms were characterized by fluctuations of warm- and cold-season *Aulacoseira* species (Fig. 2) that probably document overall increased effective moisture. The prominent spike of cold-season *Aulacoseira* species (in this case, *A. subarctica*) between 735 ka and 710 ka (Fig. 3) may document glacial environments related to oxygen isotope stage 18 (the first glacial stage after the Matuyama/Brunhes boundary) although more samples are needed to fully characterize this event. However, the presence of very cold water fish, trout and whitefish, in samples from ca. 725–700 ka support this interpretation (Firby et al., this volume). High values of magnetic susceptibility in this interval (Glen and Coe, this volume) may correlate to glacial conditions, such as suggested by ice-rafted moraine material on an island in Long Valley Lake that formed after the eruption of the Bishop ash (Smith et al., 1983). If the peak of *Aulacoseira subarctica* indicates glacial conditions, then the previous and following peaks of warm-season *Aulacoseira* species imply interglacial environments at least partly characterized by significant summer precipitation.

Between 700 ka and 682 ka (288–280 m) saline planktic (*Cyclotella meneghiniana*) and benthic diatoms dominate the record (Fig. 2). Freshwater benthic and attached diatoms are present, but not especially common. This assemblage indicates a shallowing of Owens Lake and evaporative concentration of its water and implies an arid climate phase. The Owens Lake chronology suggests it may directly precede isotope stage 16, but its small magnitude does not clearly fix it to published possibilities (Williams et al., 1988).

From 682 ka to 650 ka (280–267 m), increased moisture in the Owens Lake drainage supported the planktic freshwater diatoms *Stephanodiscus niagarae, S. oregonicus,* and *Aulacoseira solida* (Fig. 3). Diatom concentration throughout this interval (Fig. 1) is low, however, possibly because of dilution by clastic detritus eroding from the Sierra Nevada as recorded by high magnetic susceptibility at these levels (Glen and Coe, this volume). Freshwater planktic diatoms in combination with increased deposition of clastic detritus suggests glacial conditions at that time.

From 650–520 ka (267–230 m). Fine to medium sand and intervals of poor core recovery characterize this section of the Owens Lake core (Smith, 1993). Of the five samples available for analysis in this interval, only one (254 m), dominated by *"Fragilaria"* species and occasional fresh and saline benthic diatoms, indicates shallow, freshwater marsh environments at an interpolated age of 605 ka. Otherwise, this section of the core apparently represents fluvial or shallow lake environments in which deposition and preservation of diatoms was not favored.

From 520–435 ka (230–200 m). *"Fragilaria"* species accompanied by freshwater planktic warm-season *Aulacoseira* and large *Stephanodiscus* species (Figs. 2, 3) dominate in this interval indicating generally shallow, open-water conditions with extensive fringing marsh habitats. Two pulses of freshwater assemblages (520–490 ka and 470–435 ka) are separated by abundant saline benthic diatoms (Fig. 2) that represent a 20-k.y. interval of hydrologic deficit and only intermittent flushing of Owens Lake. These strong fluctuations in available moisture may relate to glacial and interglacial climates expressed by the marine oxygen-isotope record of stages 14, 13, and 12, although as dated, they neither closely coincide nor do they span sufficient amounts of time. Of the two freshwater pulses, the one dated between 470 ka and 435 ka coincides with a strong increase of magnetic susceptibility (Glenn et al., 1993) that suggests it may be a candidate for correlation to contemporaneous Sierran glaciation, possibly coeval in part with isotope stage 12. Nevertheless, the prominence of warm-season *Aulacoseira* species and of *A. solida* imply climates lacking continental extremes of temperature.

From 435–175 ka (200–100 m). Above 200 m in the core, *"Fragilaria"* species never regain dominance. Instead, freshwater planktic diatoms more completely characterize the Owens Lake record (Fig. 1), and their prevalence suggests that Owens Lake became larger and deeper than before as a result of tectonic and (or) sedimentologic controls on basin size and depth. Irrespective, the stratigraphic distribution of freshwater planktic diatoms occurs as short, quasi-regular, pulses within this 260 k.y.- interval that are separated by modest percentages of saline benthic diatoms or barren intervals that record intervening periods of dry climate.

Multiple peaks of freshwater planktic diatoms, first consisting of large *Stephanodiscus* species (Fig. 2) and later of *Aulacoseira solida* (Fig. 3) and small *Stephanodiscus* species (Fig. 4), occur between 435 ka and 350 ka. The distinctive interval dominated by *Aulacoseira solida* approximately correlates with isotope stage 10 and suggests repetition of climate characteristics suggested for stage 12. Modest percentages of *"Fragilaria,"* attached, and motile freshwater benthic diatoms indicate proximity of shallow-water habitats (Fig. 2). After 350 ka, *Aulacoseira solida* disappears from the record, perhaps implying a major change at that time to more extreme continental climates. At about 350 ka, sediment samples are barren of diatoms possibly reflecting excessive salinities or desiccation of the lake. Very high percentages of *Stephanodiscus niagarae* and *S. oregonicus* appear at ca. 312 ka (Fig. 3), but additional samples are required to document the beginning of this event. Levels between ca. 312 ka and 290 ka are generally barren of diatoms, although a few samples have saline benthic or freshwater planktic species in low numbers. A dry or rapidly fluctuating fresh to saline lake might explain these data.

Saline benthic diatoms fluctuate between high and low percentages between 290 ka and 250 ka, alternating between

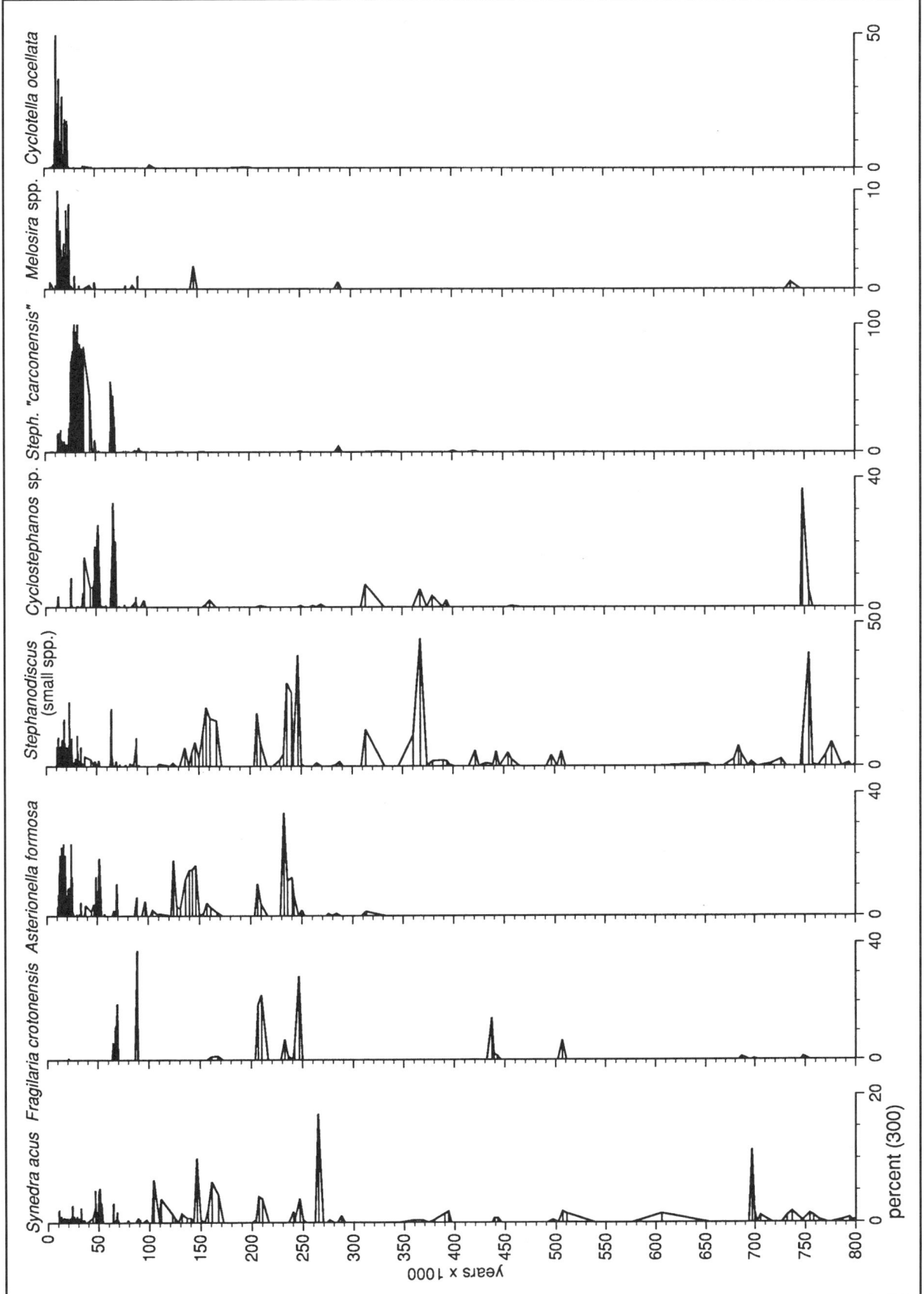

Figure 4. Weighted percentages of secondary planktic diatom taxa versus age in the Owens Lake core.

smaller peaks of large *Stephanodiscus* species (Fig. 2). This suggestion of prevalent dry climates between 290 ka and 250 ka conflicts with the oxygen-isotope record of glacial conditions (stage 8) at that time (see also Bischoff et al., 1985). The chronology of this part of the Owens Lake record requires confirmation to evaluate this discrepancy.

Wet conditions are implied by high percentages of large *Stephanodiscus* and *Aulacoseira* species between 250 ka and 220 ka and again at 210 ka (Fig. 2). *Asterionella formosa, Synedra acus,* small *Stephanodiscus* species, and *Fragilaria crotonensis* are also common during this period, particularly at the beginning and end of it (Fig. 4). These common temperate-lake diatoms suggest seasonally stable water levels and nutrient cycling in response to seasonal circulation. By extension, it seems that evaporative stress during the summer months was offset by an extended Owens River runoff period, midsummer to fall precipitation east of the Sierra Nevada, or both. The symmetrical distribution of this assemblage below and above levels dominated by large *Stephanodiscus* species suggests in this case that more equable, moist climates preceded and followed intensely cold climates documented at 230 ka by cold-season *Aulacoseira* species (Fig. 3). High values of magnetic susceptibility between 260 ka and 210 ka (Glen et al., 1993; Fig. 5) may reflect erosion of Sierra Nevada granite by glacial activity and transportation of silt-rich glacial flour to Owens Lake. Correlation to isotope stage 8, however, is not substantiated by the Owens Lake chronology.

Between 220 ka and 211 ka, samples are barren and the sediments are composed of fine silt-size particles. Conceivably the barren, silt-rich samples represent inhospitable conditions for diatom growth as a result of a large influx of glacial flour and consequent excessive turbidity. Nevertheless, after 210 ka saline benthic diatoms dominate to about 200 ka. The sediment between 200 ka and 180 ka (~110–100 m) was either unrecovered or barren.

From 175–0 ka (100–0 m). Close sampling intervals and more consistent diatom representation provide a high-resolution paleolimnologic history of this interval. Between 175 ka and 120 ka, freshwater planktic diatoms, mostly large *Stephanodiscus* species and *Cyclostephanos* (Figs. 2, 3) are consistently abundant and record a long interval of positive hydrologic balances. *Asterionella formosa* and small *Stephanodiscus* species also occur at this time (Fig. 4). This long interval probably relates to oxygen-isotope stage 6 and correlates to glaciation in the Sierra Nevada documented by "older" Tahoe-age moraines with cosmogenic chlorine-36 ages between 218 ka and 189 ka (Phillips et al., 1990). Very high values of magnetic susceptibility (Fig. 5) also probably reflect glaciation in the Owens River drainage.

The end of this probable glacial episode is characterized by sharp, short-term fluctuations of freshwater planktic and benthic diatoms (Fig. 2) alternating with samples with low diatom concentrations and a mix of saline and freshwater assemblages. Benthic and planktic saline diatoms and *Cyclotella meneghiniana*, a species tolerant of moderate to high salinities, dominate

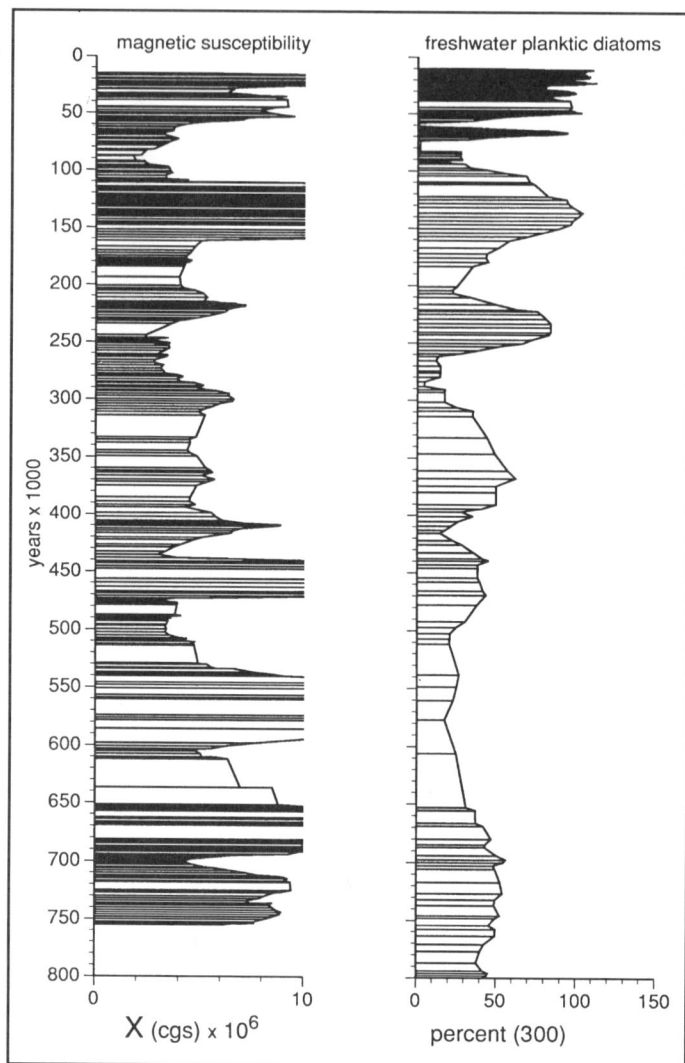

Figure 5. Magnetic susceptibility (Glen et al., 1993) and weighted percent of freshwater planktic diatoms versus age in the Owens Lake core. Both data sets are smoothed by 10-sample averaging.

from 95 ka to about 70 ka (Fig. 2) indicating a comparatively long period of negative hydrologic balance and probably arid climates. The general stratigraphy of the Owens Lake core would suggest correlation of this interval to isotope stage 5, although the age model (Bischoff, 1993; Bischoff et al., this volume, Chapter 8) does not suggest exact equivalency.

A prominent spike of freshwater planktic diatoms such as *Stephanodiscus niagarae, S. oregonicus* type, *Cyclostephanos* spp., and warm-season *Aulacoseira* spp., (Fig. 3) occurs between 72 ka and 65 ka, interrupting the declining trend of saline benthic diatoms (Fig. 2). After 65 ka, saline diatoms return in decreasing percentages until about 50 ka. The diatom record of Tule Lake, California, shows a similar, isolated interval of freshwater planktic diatoms between 75 ka and 70 ka (Bradbury, 1991) and a possibly coeval—suggested by tephra correlations (Sarna-Wojcicki et al., 1993)—spike of freshwater planktic diatoms occurs in a similar stratigraphic relationship with saline diatoms at Walker

Lake, Nevada. These stratigraphically abrupt and ecologically distinctive occurrences of diatoms indicating increased effective moisture and probably reduced summer temperatures have an intriguing potential relationship to climatic changes after stage 5 associated with the gigantic Toba eruption at ca. 74 ka in Sumatra (Rampino and Self, 1993).

Short intervals dominated by freshwater planktic diatoms between 65 ka and 45 ka and between 40 ka and 30 ka (Figs. 3, 4) are separated by very brief intervals in which planktic and benthic saline diatoms flourished (Fig. 2). Collectively, these rapid and short-term fluctuations in the hydrologic balance of Owens Lake appear to correlate with the variable climates of isotope stage 3. Conceivably, the major freshwater intervals between 72 ka and 65 ka and between 55 ka and 45 ka are correlative with "younger" Tahoe glacial moraines dated between 70 ka and 50 ka (Phillips et al., 1990). As noted earlier, the presence of warm-season *Aulacoseira* species suggests favorable hydrologic and limnologic conditions persisting into the summer and fall, perhaps reflecting greater persistence of past precipitation patterns such as fall cyclonic storms that today are rare and anomalous (Pyke, 1972; Hansen et al., 1981).

The dominance of *Stephanodiscus* "carconensis" between 72 ka and 65 ka and in the interval preceding the last full glacial (50–25 ka) (Fig. 4) presumably reflects distinctive limnologic conditions associated with seasonality of runoff, turbulence, and turbidity unique to those times. The species (presently undescribed and not equal to *S. carconensis* Grunow) apparently became extinct in Owens Lake by the Holocene, and it has not been recorded as an extant diatom elsewhere. It was common, however, in the late Pleistocene (isotope stage 6?) sediments of Clear Lake, Lake County, California (Bradbury, 1988b). This occurrence, in the generally more equable, Mediterranean climatic setting of California west of the Sierra Nevada, and the fact that warm-season *Aulacoseira* species precede and accompany large percentages of *Stephanodiscus* "carconensis" (Figs. 3, 4) suggest that it does not relate to the coldest and most intense phases of glacial environments in Owens Lake. Its stratigraphic distribution at levels correlated to isotope stage 3 implies intermediate climates with moisture throughout much of the year.

Nevertheless, sediment magnetic susceptibility progressively increased, starting about 80 ka (Fig. 5), and by 50–30 ka, high values suggest input of glacial detritus via the Owens River. Still higher values of magnetic susceptibility occur between 25 ka and 15 ka (Glen et al., 1993; Glen and Coe, this volume), according to the time scale developed by Bischoff et al. (this volume, Chapter 8), that probably correlate with Tioga moraines of the last full-glacial episode. *Stephanodiscus* "carconensis" had fallen to much lower levels by the time of the full glacial, and *S. niagarae, S. oregonicus, Cyclostephanos* sp., *Melosira* spp., and *Cyclotella ocellata* collectively characterize climatic and hydrologic conditions at 25–10 ka (Figs. 3, 4). The additional presence of *Asterionella formosa* and small *Stephanodiscus* species may imply cold, dimictic lake circulation and possibly a reduction in the length of the runoff and open-water season.

Glacial conditions as documented by the diatoms and magnetic susceptibility precipitously decrease at 15 ka and reach very low values by 10 ka. *Cyclotella ocellata,* a freshwater planktic diatom that can tolerate high alkalinities (Gasse and Tekaia, 1983), remains with high percentages until 10 ka and persists into the early Holocene. Saline planktic and benthic diatoms attain significant percentages in the early Holocene, but above 5 m (ca. 8 ka) no diatoms are preserved presumably because the lake became too shallow and saline.

Correlation of planktic diatoms with measures of precipitation and glaciation in Owens Valley

The correspondence between high values of magnetic susceptibility and large percentages of freshwater planktic diatoms in the Owens Lake core (Fig. 5) has been invoked in the previous section to suggest a relationship between glaciation and increased effective moisture filling the lake. Glaciers in Sierra Nevada drainages are presumed to have supplied eroded granitic bedrock in the form of glacial flour to tributaries of the Owens River. The high magnetic susceptibility of this material may relate to the concentration of iron-rich minerals in the bedrock (Glen et al., 1993). In the comparison of magnetic susceptibility and freshwater planktic diatoms (Fig. 5), an approximate correspondence exists for the past 250 k.y. Typically, the increasing trend in percentages of planktic diatoms precedes the more abrupt increase in magnetic susceptibility by 10 k.y. or more, implying that diatoms in Owens Lake recorded increased effective moisture that ultimately allowed glaciers to expand and enter canyon tributaries of Owens Lake.

Before 250 ka, the relationship between magnetic susceptibility and freshwater planktic diatoms is less clear, partly because of coarser sampling intervals and partly because tectonic and volcanic impacts in the Owens basin may have been responsible for the mineral character of sediments entering the lake. For example, the high susceptibility values between 600 ka and 550 ka have no corresponding high percentages of planktic diatoms and probably relate to the tectonic and geomorphic events that account for the abundant sand in this part of the core. Nevertheless, the susceptibility peak between 470 ka and 440 ka does coincide to planktic diatom percentages which began to increase at that time (Figs. 2, 5). Between 400 ka and 300 ka magnetic susceptibility shows only a subdued relationship to planktic diatom percentages (Fig. 5) possibly because the short-lived intervals of high planktic diatom percentages did not record climate changes of sufficient scale or duration to generate significant glaciation in the Sierra Nevada.

Percentages of Cupressaceae "juniper" pollen (TCT = Taxodiaceae, Cupressaceae, Taxaceae) analyzed by Litwin et al. (this volume) also show a generally positive relationship to percentages of planktic diatoms (Fig. 6). Increased effective moisture recorded by the freshwater diatoms may have been, in some cases, responsible for increasing percentages of "juniper" pollen.

Like increased magnetic susceptibility, the most convincing correspondences occur in the past 250 k.y. High percentages of this pollen type in the late Pleistocene are probably referable to juniper. *Juniperus osteosperma* forms a significant component of late Wisconsin packrat middens (31.5–9.5 ka) in the Alabama Hills (elevation 1,460 m) 9 km northwest of Owens Lake (Koehler and Anderson, 1995) as well as in middens north-northeast of Owens Lake between 1,100 m and 1,155 m elevation. At levels dated between 17.7 ka and 16.1 ka by radiocarbon, the 1,155 m midden also includes *Juniperus scopulorum,* a mesic juniper species of the Rocky Mountains and Pacific Northwest (Koehler and Anderson, 1994). Packrat midden records of these full and late glacial juniper species imply increased soil moisture, less summer moisture stress, and cooler summer temperatures. *Juniperus osteosperma* grows today in the Sierra Nevada and White Mountains west and east of Owens Lake respectively at elevations of about 2,000 m (Lloyd and Mitchell, 1973).

Cooler temperatures are also suggested by the distribution of *Artemisia* pollen that has a loose positive correlation with increased percentages of freshwater planktic diatoms (Fig. 6). *Artemisia* is an associate of pinyon juniper woodland plant communities on the lower slopes of the Sierra Nevada and White-Inyo ranges bordering Owens Valley (Woolfenden,

1993) and is one of the characteristic pollen types found in late Pleistocene lake deposits in the Great Basin (VanDevender et al., 1987). Although *Artemisia* is characteristic of cold, dry desert environments, it often correlates positively with evidence of high lake stands and increased effective moisture in Quaternary lake deposits. Both lakes and *Artemisia* respond positively to increased precipitation in the winter season when soils are more likely to be saturated and losses to evapo-transpiration are low (Prentice et al., 1992).

The provisional nature of the age-depth curve for Owens Lake qualifies comparisons of that record to the abundantly dated record of oxygen and carbon isotopes at Devils Hole, Nevada (Coplen et al., 1994; Winograd et al., 1992) and to the generalized (SPECMAP) marine oxygen-isotope record (Imbrie et al., 1984). At first approximation, periods of freshwater planktic diatoms in Owens Lake seem to fit more closely to SPECMAP ages of even-numbered isotope stages than to the similar fluctuations in Devils Hole (Fig. 7). Nevertheless, the SPECMAP chronology is based on an assumption of orbital forcing of Pleistocene climate changes that remains controversial, at least as it may apply to terrestrial records. There is certainly the possibility that some continuous, well-dated terrestrial records will reflect climate changes also recorded by oxygen isotopes of marine foraminifera, whereas others may relate more closely to terrestrial climates that both lead and lag orbital changes of global insolation. As the chronologies and interpretations of all these records become better understood, so will our understanding of the dynamics of climate change in western North America.

CONCLUSIONS

The 800-k.y. diatom record of Owens Lake is characterized overall by freshwater taxa. Shallow but freshwater diatoms were more abundant between 760 ka and 435 ka than open-water planktic taxa, and they probably reflect marsh-like environments in the Owens basin under a climatic regime of greater-than-modern effective moisture. Nevertheless, expansion of open-water habitats and increased percentages of planktic freshwater taxa occurred between 730 ka and 710 ka, 687–650 ka, 510–490 ka, and 470–435 ka. After 435 ka, the Owens basin deepened, probably under tectonic influence, and freshwater planktic taxa became more common. Intervals of significantly higher effective moisture occurred at 420 ka, 400–350 ka, 250–220 ka, ca. 210 ka, 180–120 ka, 72–65 ka, and generally after 50 ka until the early Holocene. Shallow-water saline diatoms, and occasionally saline planktic species, alternate with some of the intervals characterized by freshwater planktic species and suggest comparatively short periods of negative hydrologic balance. Peaks of freshwater diatoms plausibly correlate with glacial epochs defined by the oxygen isotope record for the past 250 k.y. Correlation of freshwater planktic diatom peaks to Sierra Nevada glaciations is very probable for stage 6 ("older" Tahoe), stage 4? ("younger" Tahoe), and stage 2 (Tioga). It is likely that the diatom record of Owens

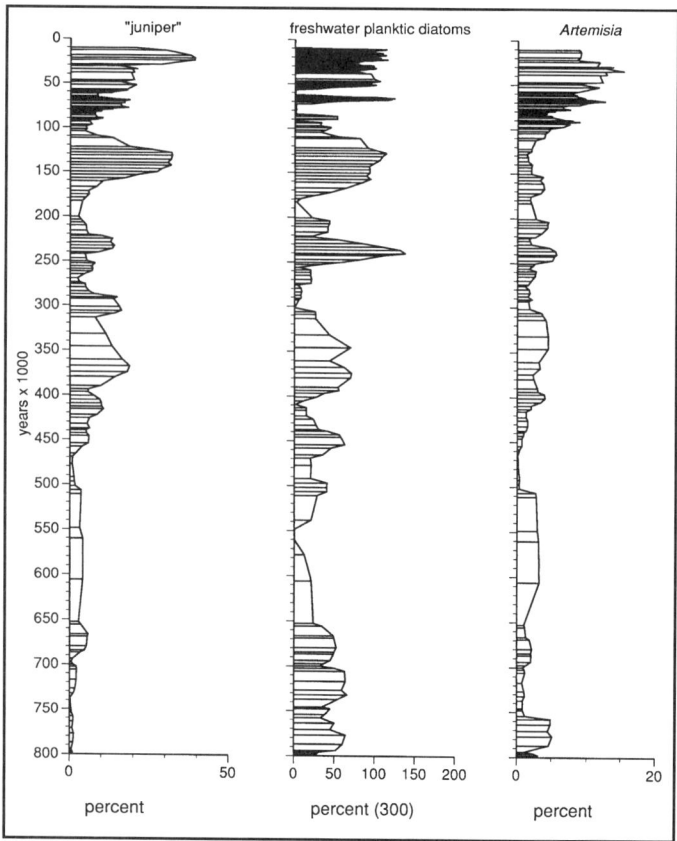

Figure 6. Percent "juniper" pollen, weighted percent freshwater planktic diatoms, and percent *Artemisia* pollen in the Owens Lake core versus age. Pollen data from Litwin et al. (1993) and Woolfenden (1993). All data smoothed by 5-sample averaging.

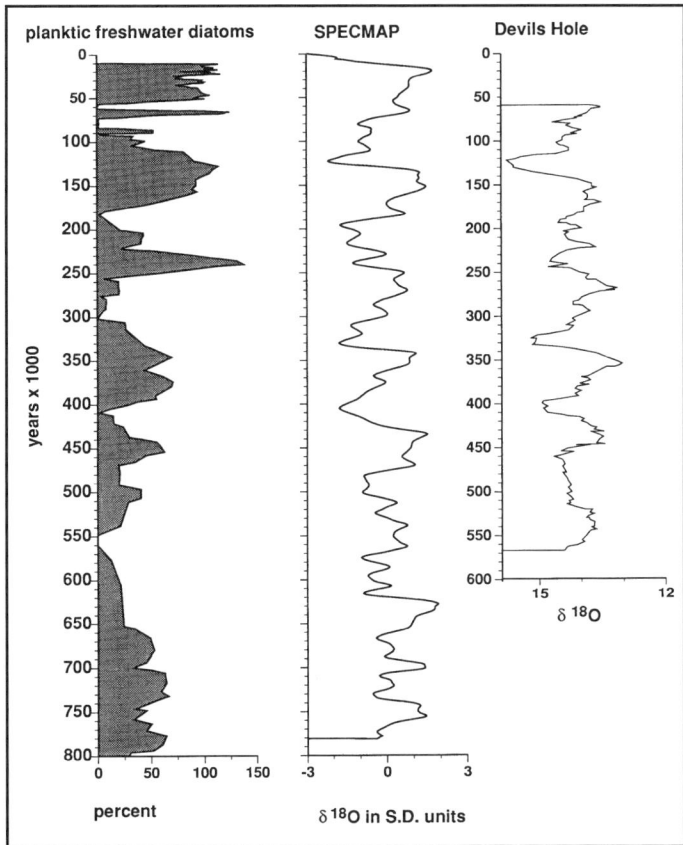

Figure 7. Chronostratigraphic comparison of freshwater planktic diatoms in the Owens Lake record with the stacked SPECMAP oxygen-isotope record for the past 800 ka and with the Devils Hole oxygen-isotope record of the past 567 k.y. SPECMAP data from Imbrie et al. (1984); Devils Hole data from Coplen et al. (1994). Diatom data smoothed by 5-sample averaging. S.D. = standard deviation.

Lake is a consistent and accurate measure of glacial-type climates in the Sierra Nevada, especially when augmented by the record of lake sediment chemistry changes (Bischoff et al., this volume, Chapter 4) and magnetic susceptibility that partly reflects outwash of glacial flour into the Owens River (Glen and Coe, this volume). The correspondence of pollen taxa, suggesting cooler and more mesic environments and peaks of freshwater planktic diatoms, indicates that at least for the past 250 k.y., Owens Lake responded chiefly to climate change rather than tectonic or volcanic perturbations of its watershed.

Although comparison of the Owens Lake record to Searles Lake downstream should be straightforward, major problems may exist with the Searles chronology before the last glacial that frustrate linking these two records. The abundance of salt in the Searles Lake record, however, suggests a dominance of arid climatic environments even during cooler and moister epochs of the Quaternary. The mud units presumably represent high-water, flow-through conditions, and it is important to establish to what extent they can be linked to influx of water from Sierran glaciations as opposed to local sources of terrestrial detritus and aquatic organic production.

The abundance of warm-season *Aulacoseira* taxa at certain stages of Owens Lake history implies significant climatic changes that maintained hydrologic input to the lake during the summer and fall, possibly even against strong evaporation gradients. There is a possibility that warm-season precipitation suggested by *Aulacoseira* may have characterized transitions between interglacial and glacial periods at Owens Lake. Understanding the climate dynamics associated with seasonality of precipitation in terrestrial environments will require Pacific Ocean records closely correlated (if not closely dated) to terrestrial records.

ACKNOWLEDGMENTS

I gratefully acknowledge the help of Vera Markgraf, R. S. Thompson, Andrei Sarna-Wojcicki, and David Herbst for reviewing and providing additional insights into the character of the Owens Lake record. David P. Adam supplied carefully curated samples from the core, and Kathy Dieterich-Rurup prepared them for diatom analysis.

REFERENCES CITED

Battarbee, R. W., 1973, A new method for the estimation of absolute microfossil numbers, with reference especially to diatoms: Limnology and Oceanography, v. 18, p. 647–653.

Benson, L. V., Currey, D. R., Dorn, R. I., Lajoie, K. R., Oviatt, C. G., Robinson, S. W., Smith, G. I., and Stine, S., 1990, Chronology of expansion and contraction of four Great Basin lake systems during the past 35,000 years: Palaeogeography, Palaeoclimatology, Palaeoecology, v. 78, p. 241–286.

Bischoff, J. L., 1993, Age-depth relations for the sediment column at Owens Lake, California: OL-92 drill hole, in Smith, G. I., and Bischoff, J. L., eds., Core OL-92 from Owens Lake, southeast California: U.S. Geological Survey Open File Report, OFR 93-683, p. 251–260.

Bischoff, J. L., Rosenbauer, R. J., and Smith, G. I., 1985, Uranium-series dating of sediments from Searles Lake: differences between continental and marine climate records: Science, v. 227, p. 1222–1224.

Bradbury, J. P., 1988a, A climatic- limnologic model of diatom succession for paleolimnological interpretation of varved sediments at Elk Lake, Minnesota: Journal of Paleolimnology, v. 1, p. 115–131.

Bradbury, J. P., 1988b, Diatom biostratigraphy and the paleolimnology of Clear Lake, Lake County, California, in Sims, J. D., ed., Late Quaternary climate, tectonism, and sedimentation in Clear Lake, Northern California Coast Ranges: Boulder, Colorado, Geological Society of America Special Paper 214, p. 97–129.

Bradbury, J. P., 1991, The late Cenozoic diatom stratigraphy of Tule Lake, Siskiyou County, California: Journal of Paleolimnology, v. 6, p. 205–255.

Bradbury, J. P., 1993, Diatoms in sediments, in Smith, G. I., and Bischoff, J. L., eds., Core OL-92 from Owens Lake, southeast California: U.S. Geological Survey Open File Report, OFR 93-683, p. 261–301.

Bradbury, J. P., Forester, R. M., and Thompson, R. S., 1989, Late Quaternary paleolimnology of Walker Lake, Nevada: Journal of Paleolimnology, v. 1, p. 249–267.

Coplen, T. B., Winograd, I. J., Landwehr, J. M., and Riggs, A. C., 1994, 500,000 year stable carbon isotope record from Devils Hole, Nevada: Science, v. 263, p. 361–365.

Friedman, I., Smith, G. I., and Hardcastle, K. G., 1976, Studies of Quaternary saline lakes—II. Isotopic and compositional changes during desiccation of the brines in Owens Lake, California, 1969-1971: Geochimica et

Cosmochimica Acta, v. 40, p. 501–511.

Gasse, F., and Tekaia, F., 1983, Transfer functions for estimating paleoecological conditions (pH) from East African diatoms: Hydrobiologia, v. 103, p. 85–90.

Glen, J. M., Coe, R. S., Menking, K., Boughn, S. S., and Altschul, I., 1993, Rock and paleo-magnetic results from core OL-92, Owens Lake, California, in Smith, G. I., and Bischoff, J. L., eds., Core OL-92 from Owens Lake, southeast California: U.S. Geological Survey Open File Report OFR 93-683, p. 127–183.

Håkansson, H., and Kling, H., 1990, The current status of some very small freshwater diatoms of the genera *Stephanodiscus* and *Cyclostephanos*: Diatom Research, v. 5, p. 273–287.

Hansen, E. M., Schwarz, F. K., and Riedel, J. P., 1981, Meteorology of important rainstorms in the Colorado River and Great Basin drainages: Hydrometeorological Report 50, National Oceanic and Atmospheric Administration, Silver Spring, Maryland, 167 p.

Hollett, K. J., Danskin, W. R., McCaffrey, W. F., and Walti, C. L., 1991, Geology and water resources of Owens Valley, California: U.S. Geological Survey Water-Supply Paper 2370, Chapter B, 77 p.

Houghton, J. G., 1969, Characteristics of rainfall in the Great Basin: Reno, University of Nevada, Desert Research Institute, 205 p.

Imbrie, J., Hays, J. D., Martinson, D. G., McIntyre, A., Mix, A. C., Morley, J. J., Pisias, N. G., Prell, W. L., and Shackleton, N. J., 1984, The orbital theory of Pleistocene climate: Support from a revised chronology of the marine $\delta^{18}O$ record, in Milankovich and climate, Berger, A., Imbrie, J., Hays, J., Kukla, G., and Saltaman, B., eds.: Dardrecht, Boston, and Lancaster, D. Reidel, p. 269–305.

Kay, P. A., 1982, A perspective on Great Basin paleoclimates in Madsen, D. B., and O'Connell, J. F., eds., Man and the environment in the Great Basin: Society of American Archaeology Papers no. 2, p. 76–81.

Kilham, S. S., and Kilham, P., 1978, Natural community bioassays: Predictions of results based on nutrient physiology and competition: Verhandlungen des Internationalen Vereins für Limnologie, v. 20, p. 68–74.

Kilham, P., Kilham, S. S., and Hecky, R. E., 1986, Hypothesized resource relationships among African planktonic diatoms: Limnology and Oceanography, v. 31, p. 1169–1181.

Koehler, P. A., and Anderson, R. S., 1994, Full-glacial shoreline vegetation during the maximum highstand at Owens Lake, California: Great Basin Naturalist, v. 54, p. 142–149.

Koehler, P. A., and Anderson, R. S., 1995, Thirty thousand years of vegetation changes in the Alabama Hills, Owens Valley, California: Quaternary Research, v. 43, p. 238–248.

Litwin, R. J., Frederiksen, M. D., Adam, D. P., Anderle, V. A. S., and Sheehan, T. P., 1993, Continental-marine correlation of Late Pleistocene climate change: Census of palynomorphs from core OL-92, Owens Lake, California: U.S. Gelogical Survey Open-file Report OFR 93-683, p. 333–391.

Lloyd, R. M., and Mitchell, R. S., 1973, A flora of the White Mountains, California and Nevada: Los Angeles, University of California Press, 202 p.

Negoro, K., 1960, Studies on the diatom vegetation of Lake Biwa-ko (First report): The Japanese Journal of Limnology, v. 21, nos. 3-4, p. 200–220.

Phillips, F. M., Zreda, M. G., Smith, S. S., Elmore, D., Kubik, P. W., and Sharma, P., 1990, Cosmogenic chlorine-36 chronology for glacial deposits at Bloody Canyon, eastern Sierra Nevada: Science, v. 248, p. 1529–1532.

Prentice, I. C., Guiot, J., and Harrison, S. P., 1992, Mediterranean vegetation, lake levels, and paleoclimate at the last glacial maximum: Nature, v. 360, p. 658–660.

Pyke, C. B., 1972, Some meteorological aspects of the seasonal distribution of precipitation in the western United States and Baja California: University of California Water Resources Center, Contribution 139, 205 p.

Rampino, M. R., and Self, S., 1993, Climate-volcanism feedback and the Toba eruption of ~ 74,000 years ago: Quaternary Research, v. 40, p. 269–280.

Sarna-Wojcicki, A. M., Meyer, C. E., Wan, E., and Soles, S., 1993, Age and correlation of tephra layers in Ownes Lake drill core OL92-1 and -2: U.S. Geological Survey Open-file Report OFR 93-683, p. 184–245.

Smith, G. I., 1976, Paleoclimatic record in the upper Quaternary sediments of Searles Lake, California, U.S.A., in Horie, S., ed., Paleolimnology of Lake Biwa and the Japanese Pleistocene, v. 4: Kyoto, Japan, Kyoto University, p. 577–604.

Smith, G. I., 1979, Subsurface stratigraphy and geochemistry of late Quaternary evaporites, Searles Lake, California: U.S. Geological Survey Professional Paper 1043, 130 p.

Smith, G. I., 1993, Field log of core OL-92, in Smith, G. I., and Bischoff, J. L., eds., Core OL-92 from Owens Lake, southeast California: U.S. Geological Survey Open File Report OFR 93-683, p. 1–57.

Smith, G. I., and Pratt, W. P., 1957, Core logs from Owens, China, Searles, and Panamint basins, California: U.S. Geological Survey Bulletin 1045-A, 62 p.

Smith, G. I., and Street-Perrott, F. A., 1983, Pluvial lakes of the western United States, Chapter 10, in Wright, H. E., ed., Late Quaternary environments of the United States: Minneapolis, University of Minnesota Press, p. 190–212.

Smith, G. I., Barczak, V. J., Moulton, G. F., and Liddicoat, J. C., 1983, Core KM3 a surface-to-bedrock record of late Cenozoic sedimentation in Searles Valley, California: U.S. Geological Survey Professional Paper 1256, 24 p.

Stoermer, E. F., and Yang, J. J., 1970, Distribution and relative abundance of dominant plankton diatoms in Lake Michigan: Ann Arbor, Great Lakes Research Division, Publication 16, University of Michigan, 64 p.

VanDevender, T. R., Thompson, R. S., and Betancourt, J. L., 1987, Vegetation history of the deserts of southwestern North America; the nature and timing of the Late Wisconsin–Holocene transition, in Ruddiman, W. F., and Wright, H. E., eds., North America and adjacent oceans during the last deglaciation: Boulder, Colorado, Geological Society of America, The Geology of North America, v. K-3, p. 323–352.

Williams, D. F., Thunell, R. C., Tappa, E., Rio, D., and Raffi, I., 1988, Chronology of the oxygen isotope record: 0–1.88 m.y., B.P.: Palaeogeography, Palaeoclimatology, Palaeoecology, v. 64, p. 221–240.

Williams, D. M., and Round, F. E., 1987, Revision of the genus *Fragilaria*: Diatom Research, v. 2, p. 267–288.

Winograd, I. J., Coplen, T. B., Landwehr, J. M., Riggs, A. C., Ludwig, K. R., Szabo, B. J., Kolesav, P. J., Revesz, K. M., 1992, Continuous 500,000-year climate record from vein calcite in Devils Hole, Nevada: Science, v. 258, p. 255–260.

Woolfenden, W. B., 1993, Pollen present in cores OL-92-2, and -3, in Smith, G. I., and Bischoff, J. L., eds., Core OL-92 from Owens Lake, southeast California: U.S. Geological Survey Open File Report, OFR 93-683, p. 313–332.

Manuscript Accepted by the Society June 17, 1996

Geological Society of America
Special Paper 317
1997

Ostracodes in Owens Lake core OL-92: Alternation of saline and freshwater forms through time

Claire Carter
U.S. Geological Survey, 345 Middlefield Road, MS 975, Menlo Park, California 94025

ABSTRACT

Ostracode species' geographic distributions are limited by parameters such as water temperature, salinity, and dissolved-ion composition. Because these parameters are, in part, determined by climate, ostracode biogeographic distributions serve as proxies for past climates. Therefore, the ostracodes in Core OL-92 from Owens Lake, southeast California, reveal climatic oscillations during the past 800,000 yr. The climatic history of the Owens Lake area, as indicated by the fossil ostracode record, reflects a number of high-latitude glacial and interglacial episodes in which glacial-period terminations fall at approximately 120 ka (Termination II), 225 ka (Termination III), 340 ka (Termination IV), and 438 ka (Termination V). A plot of saline versus freshwater ostracodes over time agrees quite well with a number of other geochemical and biological climatic indicators from the Owens Lake core.

INTRODUCTION

Ostracodes are microscopic aquatic crustaceans with a bivalved, calcitic carapace. Continental ostracodes are found in a wide range of environments, such as lakes, springs, and wetlands. Within each of those environmental categories, particular species of ostracodes inhabit various ranges of water temperatures, chemistries, and other parameters. In other words, each environment (stream-fed lake, ground-water-fed lake, aquifer, spring, marsh, and so forth) has characteristic species of ostracodes, and each of those species has its own preferred range of water temperatures, major-dissolved-ion compositions, and salinities (total dissolved solids, TDS). Therefore, where found in the fossil record, they can be used as indicators of past conditions. For example, if the occurrences in a core of ostracodes known to thrive in saline water are plotted against those known to prefer fresher water, the fluctuations in salinity through time are revealed.

MATERIALS AND METHODS

The Owens Lake core was sampled for ostracodes at approximately 1-m intervals, except in the absence of suitable lithologies—sandy beds are less likely to yield ostracodes than are finer grained sediments. The samples ranged in weight from 11.4–37.2 g, with the majority weighing about 15–25 g (Carter, 1993). Each raw sample was frozen, thawed, and placed in a beaker to which approximately 300 ml of hot distilled water and a rounded teaspoon of baking soda were added. After cooling, about two tablespoons of Calgon (sodium hexamataphosphate, sodium carbonate and bicarbonate, and soap) were added, and the sample was allowed to sit for 24 hr. It was then wet sieved under a shower-type flow over a 100-mesh sieve, and the resulting residue was allowed to air dry. Each sample was dry sieved and the resulting size fractions were examined for adult carapaces and valves which were picked onto standard paleontological slides. The results are presented in tabular form in Carter (1993).

OSTRACODES

The Owens Lake ostracode fauna (listed in part in Table 1) consists mainly of *Limnocythere ceriotuberosa* Delorme, *L. sappaensis* Staplin, *L. friabilis* Benson and MacDonald, *L.* cf. *L. bradburyi* Forester, *Candona* aff. *C. caudata* Kaufmann, and *Cytherissa lacustris* (Sars). A number of other species, including *L. itasca* Cole, *L.* aff. *L. paraornata* Delorme, *L. platyforma* Delorme, and *Cytheromorpha* sp. are also present in a few scat-

Carter, C., 1997, Ostracodes in Owens Lake core OL-92: Alternation of saline and freshwater forms through time, in Smith, G. I., and Bischoff, J. L., eds., An 800,000-Year Paleoclimatic Record from Core OL-92, Owens Lake, Southeast California: Boulder, Colorado, Geological Society of America Special Paper 317.

TABLE 1. OSTRACODE FAUNAS (IN PART) FROM THE OWENS LAKE CORE*

	Depth	Age (ka)	L. sappa	L. cerio	L. brad	C. cauda	L. friab	C. lacust
Flow through	5.59	9.8						
	7.59	13.6						
	9.83	17.1						
	17.84	26.3		26		53		211
	21.09	29.6						
Flow through	22.57	31.2						
	25.26	34.7						
	27.56	38.4		54		23		8
Flow through	32.31	46.7						
Variable, seasonal	33.68	49.1		7		7		
Ground water, alkaline	38.82	57.4	371					
	40.96	60.6	273					
	43.75	64.6						
Flow through	45.38	66.9						
	46.50	68.4						
	49.60	72.8	136	63				
	53.35	78.0						
Ground water, alkaline	54.80	80.2	162					
	56.50	82.8	91					
	57.76	84.8	91	61				
	58.95	86.7		83				
	60.63	89.5	7	119				
Variable, seasonal	61.73	91.4	5	22		j		
	62.96	93.6		248				
	64.94	97.2		20		j		
	65.94	99.1	j					
Drier, warmer, some seasonaality	67.49	102.1						
	68.59	104.3		j		4		
	71.06	109.4	6			j		
	72.26	111.8	j	26	16			
	77.16	122.2						
	78.39	124.9		j		4		
	79.90	128.3				4	41	
	81.42	131.6				21	58	11
Heavy runoff into large, cold, nonvariable, very freshwater lake	83.55	136.4				18	135	18
	85.17	140.0		j		30	135	35
	86.55	143.1		j		6	11	17
	88.06	146.6		j		4	j	9
	89.65	150.2				25	j	10
	91.17	153.8				11	11	40
	92.60	157.2		46		46	8	77
	94.12	160.9				101		84
Variable, seasonal	96.66	167.2		13		25		j
	98.64	172.1		434				
	100.14	175.8	36					
	101.59	179.4	4	j				
Ground water, alkaline	103.07	183.2	12	j		4		
	109.96	200.8	j	j				
	111.24	204.1	43					
Cooler, wetter	112.33	206.9				j		
	113.64	210.2				j		
	116.04	216.2	4	107				
Ground water, alkaline	117.41	219.6	23					
	118.48	222.2	21					
	119.65	225.0	j			j		
	121.32	228.9						
	122.95	232.8				?		
Cooler, wetter, some flow through	124.44	236.2						
	126.31	240.4				?		

TABLE 1. OSTRACODE FAUNAS (IN PART) FROM THE OWENS LAKE CORE* (page 2)

	Depth	Age (ka)	L. sappa	L. cerio	L. brad	C. cauda	L. friab	C. lacust
	127.17	242.4				?		
	129.18	247.0		j		15		
	130.62	250.2	6					
	131.74	252.7	24					
Ground water, alkaline	133.29	256.2	j	j		?		
	134.61	259.0	229					
Variable, seasonal	135.85	261.7		10		j		
	137.43	265.2		32		51		
Ground water, alkaline	139.54	269.9	105					
	141.42	274.1		j				
	142.29	276.1	j					
	143.91	279.8	13	17				
Ground water, alkaline	145.69	283.9	8	j		j		
	147.24	287.7	18					
	148.43	290.9						
	149.19	292.9	3	13				
	152.48	301.7				j		
Warmer,	154.06	306.1		j		j		
variable, seasonal	155.11	309.0		j		j		
	156.71	313.7		j		21		
	162.89	332.1	121	995		66		
	167.49	346.1						
	172.43	360.8						
	174.76	367.7						
Cooler, wetter,	177.05	374.2				j		
some flow through	179.00	379.6						
	182.79	390.2				j		
	184.18	393.8				39		
	185.34	396.8						
	186.92	400.9	29			32		
	188.63	405.3	9	j				
	190.15	409.2	18					
	191.53	412.9	3	14	22	19		
Ground water, alkaline	192.61	415.7						
	195.08	421.9	4			43		
	196.54	425.6	j			j		
	199.29	432.4	74					
	201.30	437.4	j		5	24		
	202.02	439.3				27		
	203.40	442.9						
Cooler, wetter,	204.60	445.9						
some flow through	207.89	454.3		j		j		
	209.26	457.9		j		j	19	
	212.41	466.3		j		7		
	213.49	469.2	j					
Ground water, alkaline	216.52	477.5	33					
	221.48	491.9	3	j	j			
	223.17	496.9			j	j		
	226.26	506.8				20		
	235.61	538.3						
	238.50	548.6						
	241.70	560.3						
	246.20	576.8						
	254.10	605.9						
?	267.53	652.8						
	268.43	655.6						
	271.98	666.3						
	273.00	669.0						

TABLE 1. OSTRACODE FAUNAS (IN PART) FROM THE OWENS LAKE CORE* (page 3)

	Depth	Age (ka)	L. sappa	L. cerio	L. brad	C. cauda	L. friab	C. lacust
Ground water, alkaline, warmer water	277.15	679.7		j	191	j		
	279.30	684.4	14	75	822			
Variable, seasonal	280.59	686.8	j	104				
	284.34	693.8	j	50				
	286.25	697.4						
	287.50	699.9						
	288.74	702.4						
	290.00	705.7						
	293.90	717.0						
?	296.60	726.7						
	297.93	732.0						
	299.17	736.9						
	302.82	752.9						
	304.41	759.9						
	306.64							
	307.24							
	310.37							
	312.33			52				
Variable, seasonal	313.97			28		41		
	316.45					12		
	318.36		j	j				

*Numbers represent whole adult carpaces (or equivalent in single valves) per gram of sample, multiplied by 100. For complete list of species from entire core, see Carter, 1993. Depth is in meters, age in thousands of years before present. Abbreviations as follows: "L. sappa" = Limnocythere sappaensis; "L. cerio" = L. ceriotuberosa; "L. brad" = L. cf. L. bradburyi; "C. cauda" = Candona aff. C. caudata; "L. friab" = L. friabilis (= Limnocythere sp. A in Carter, 1993); "C. lacust" = Cytherissa lacustris; "j" = juveniles (uncounted). Column one: "Variable, seasonal" = lake is variable and largely maintained by seasonal runoff, with evaporation exceeding precipitation during part of the year. "Flow through" = water undersaturated in $CaCO_3$ is flowing into and out of lake. "Ground water, alkaline" = warmer climate that reduces seasonal runoff and causes lake to become less variable, more ground-water supported. "Cooler, wetter" = transition period with more precipitation and cooler temperatures than preceding warm, dry period.

tered horizons, but they mostly represent estuarine and spring environments on the margins of the lake and therefore are excluded from the analysis.

ENVIRONMENTAL PREFERENCES

Limnocythere ceriotuberosa is a very common species today in lakes of western North America from California to the Canadian prairies (Smith and Forester, 1994). It is euryhaline and tolerates a seasonal TDS (total dissolved solids) change within the range of about 100 mg/L to about 5,000 mg/L (Delorme, 1989). Its favored habitat is a fresh or saline lake that undergoes the seasonal variability in temperature and water chemistry common to semi-arid and arid regions (Forester, 1991). *L. ceriotuberosa* is often present in lakes that receive a fair amount of seasonal (spring) runoff, and it is more common in alkaline-saline lakes north of the frostline, because it probably requires, or at least can tolerate, cold water (Forester, 1991).

Limnocythere sappaensis has a modern distribution pattern the same as *L. ceriotuberosa*, but its range also extends south into central Mexico (Forester, 1985). It is a halobiont and requires alkaline-enriched, calcium-depleted saline waters, becoming less common as the TDS and chemical composition approach fresh water (Forester, 1991). It is able to thrive in a wide range of temperatures (~10 °C to greater than 30 °C) but does not seem to tolerate or is rare in lakes having seasonal fresh water input that greatly lowers TDS. Therefore, its presence indicates a lake that receives limited freshwater runoff and is supported more by ground water (R. M. Forester, oral communication, 1988).

Cytherissa lacustris lives only in cold, very freshwater, stenotopic (single habitat), boreal forest lakes (Delorme, 1989; Forester, 1991). Its very long life cycle (greater than one year) limits it to permanent waters having a year-round oxygenated sediment/water interface. It prefers water temperatures below 20 °C and perhaps below 15 °C (Delorme, 1989; Forester, 1991).

Limnocythere friabilis has been found in a few lakes in the interior plains of Canada (Delorme, 1971), a few freshwater (dilute) lakes in Oregon, and in Lake Michigan (Forester et al., 1994), where it may be an indicator of shallow-water, nearshore environments. Very little is known about its ecological requirements (Coleman et al., 1990), but it apparently prefers cold, fresh water and may live near or within aquifers (R. M. Forester, oral communication, 1994).

Limnocythere bradburyi is known from Quaternary sediments in the Great Basin in the United States and in Mexico, and it is commonly associated with *L. sappaensis* and *L. ceriotuberosa*. However, its modern geographic range is greatly reduced. It is currently found only in a few shallow, turbid, fresh to slightly saline lakes in the central Mexican Plateau, an area characterized by mild winters and warm summers (Forester, 1985), and in a few similar environments in southwestern New Mexico (Smith and Forester, 1994). The more northerly occurrences of *L. bradburyi* appear to be in larger lakes than in Mexico. These larger lakes probably buffer the species from subfreezing conditions in winter, but circulate freely in summer to bottom temperatures near or above 20 °C (R. M. Forester, written communication, 1994).

ENVIRONMENTAL INTERPRETATIONS

Table 1 shows the depth, age, and ostracode faunas (in part) from samples of the Owens Lake core. The oldest assemblages (318.36 m to 312.33 m, ca. 795 ka to ca. 778 ka) indicate a lake supported by ground water and stream discharge, which varied seasonally.

Ostracodes were absent from about 310 m to 286 m (ca. 773 ka to ca. 695 ka). This and other such intervals of nonoccurrence in the core (e.g., 273 m to 235.61 m, ca. 664 ka to ca. 535 ka; and 46.5 m to 43.75 m, ca. 67 ka to ca. 64 ka; Table 1) may indicate an environment where ostracodes could not live, such as anoxic bottom waters, or it may simply indicate depositional or postdepositional solution of their calcitic carapaces during a period of time when the lake was overflowing with water undersaturated in $CaCO_3$.

In the next interval (284.34 m to 277.15 m, ca. 678 ka to ca. 670 ka), *Limnocythere ceriotuberosa* appears and is later joined by *L. sappaensis* and *L.* cf. *L. bradburyi*. *L. ceriotuberosa* tolerates variable TDS and is indicative of warm, evaporative summers, colder winters, and seasonal influxes of more dilute runoff. The appearance of *L.* cf. *L. bradburyi* indicates that climatic seasonality has shifted from a strong winter regime to a weak winter, stronger summer climate regime (R. M. Forester, written communication, 1994). *L. sappaensis* also indicates less inflow from seasonal runoff, more inflow from ground water, and more loss of lake water to the atmosphere.

After another interval barren of ostracodes (273 m to 235.61 m, ca. 663 ka to 532 ka), the appearance of *Candona* aff. *C. caudata* signals a colder, fresher-water, less variable lake in which outflow is greater than evaporation (low residence time). Over the next 10 m of core (~30 k.y.), *L.* cf. *L. bradburyi* (juveniles only) appears, followed by *L. sappaensis*, while *Candona* disappears; this sequence indicates a trend toward warmer summers, more variability in TDS, more support from ground water, and more loss of water through evaporation.

The return of *Candona* at 212.41 m (ca. 465 ka), followed by *L. friabilis* at 209.26 m (ca. 456 ka) (Table 1), probably is evidence of cooler water, lower TDS, and less environmental variability in the lake. At 202.02 m (ca. 440 ka), *Candona* reappears after a short absence since 204.6 m (ca. 446 ka). Over the next 19 m, the ostracode assemblage swings back and forth between domination by *C.* aff. *C. caudata* and then by *L. sappaensis*, so the lake probably oscillated between lower TDS, colder water, less environmental variability, and low residence time of the water to higher TDS, warmer water, high to moderate variability, and high residence time.

Following an interval mostly without preserved ostracodes (182.79 m to 167.49 m, ca. 390 ka to ca. 343 ka), the lake returned to conditions very favorable to ostracodes and their preservation in the sediments. *L. ceriotuberosa* strongly dominates the assemblage, and because it was so much more productive than the other species found with it (*L. sappaensis, C.* aff. *C. caudata*), the environment must have been very favorable, i.e., warm, evaporative summers, colder winters, seasonal influx of dilute water, high variability, and seasonally variable residence time.

The next four samples (156.71 m to 152.48 m, ca. 305 ka to ca. 300 ka) have mainly juveniles of *Candona* and *L. ceriotuberosa*. At 149.19 m (ca. 286 ka), *L. sappaensis* appears and remains a significant, if not dominant, portion of the assemblage together with *L. ceriotuberosa* up to about 142 m (ca. 276 ka). This would seem to indicate a somewhat fluctuating trend toward higher residence times with somewhat more evaporation and more support from ground water. At 139.54 m (ca. 270 ka), *L. sappaensis* is the only species found, thus the lake apparently became more favorable for it in terms of more alkaline-enriched, calcium-depleted, saline water.

At 137.43 m (ca. 265 ka), *L. sappaensis* has disappeared and *L. ceriotuberosa* and *Candona* have reappeared, indicating a shift toward lower TDS, colder water, less variability, and low residence time with loss of lake water mostly to outflow. This suggests a cooler climate with more precipitation. At 134.61 m (ca. 259 ka), *L. sappaensis* returns strongly, and is the only species found (except for some juveniles of *L. ceriotuberosa*) for the next 4 m. Another shift toward lower TDS, colder water, less variability, and more outflow is indicated by the return of *C.* aff. *C. caudata* at 129.18 m (ca. 248 ka). These fluctuations between conditions favorable to *L. sappaensis* and those favorable to *L. ceriotuberosa* and *C.* aff. *C. caudata* continued up the core to 98.64 m (ca. 173 ka).

Adult carapaces of *Cytherissa lacustris* first appear at 94.12 m (ca. 161 ka) and, together with *Limnocythere friabilis* and *Candona* aff. *C. caudata*, indicate a climate change that supported a large, cold, nonvariable, very freshwater (i.e., glacial or near glacial) lake. These presumably glacial conditions lasted until sometime between 122 ka and 112 ka, when the appearance of *Limnocythere ceriotuberosa, L.* cf. *L. bradburyi*, and juveniles of *L. sappaensis*, signal a change in the environment. The latter two taxa indicate that the lake became smaller and more saline but was still deep enough to support *L.* cf. *L. bradburyi* at this location. This drastic change in ostracode faunas is interpreted to be coincident with Termination II, the approximate midpoint of the transition from full glacial to peak interglacial climates.

About 84 ka, *L. ceriotuberosa* and *L.* cf. *L. bradburyi* largely disappeared and *L. sappaensis* became dominant, implying a lake supported by less surface-water discharge and with higher TDS. During this *L. sappaensis*-dominated interval, a significant portion of Owens Lake's output probably went to the atmosphere rather than to outflow. The absence of *L.* cf. *L. bradburyi* could be the result of elevated TDS or of a lake so shallow as not to buffer that taxon from cold winter temperatures. Sometime ca. 38 ka there was an influx of more-dilute water into the lake and at about 26 ka heavy runoff again produced a large, cold, stable, very freshwater lake.

In modern environments *C. lacustris* and *L. ceriotuberosa* are not known to live together, but they are often found together in the fossil record (Forester, 1991) and, indeed, do occur together in core OL-92 at depths of 92.6 m (ca. 158 ka), 27.56 m (ca. 38 ka), and 17.84 m (ca. 25 ka). According to R. M. Forester (written communication, 1994), the co-occurrence of these species means that the lake was likely undergoing a rapid transition from a lake with a large outflow/evaporation ratio to one with a much lower ratio. For example, the barren interval at 25.26 m (ca. 36 ka) to 21.09 m (ca. 31 ka) may represent a very freshwater lake, with conditions favorable for *C. lacustris*, but so low in calcium carbonate that the ostracode valves dissolved after deposition. The evaporative episode at 17.84 m (ca. 26.3 ka) would have changed the water chemistry enough to enable preservation of the ostracode valves, while eventually killing off *C. lacustris*. Co-occurrence of *L. ceriotuberosa* and *C. lacustris* may also result from either slow deposition or various forms of reworking. The robust, heavily calcified valves of *C. lacustris* are particularly prone to working up section (R. M. Forester, written communication, 1994). The absence of ostracodes in an interval bracketed by *C. lacustris* (34.7 m to 29.6 m, ca. 51 ka to ca. 41 ka) may be caused by lake waters that were unsaturated with respect to calcite.

SALINITY INDEX

Assuming that large numbers of adults of any particular ostracode species represent favorable environmental conditions for that species and small numbers or complete absence represent unfavorable conditions, and assuming that each species' environmental requirements are known, then a continuous record of fossil ostracode assemblages through time (such as a lacustrine core) also should be a record of the corresponding environmental conditions. Of course, this record is an imprecise approximation because it is a sediment-averaged record that will contain mainly the common species that lived near or above the core site, and taxa that were rare in the living assemblage will rarely be found in the core (Forester et al., 1994). Figure 1 represents fluctuations in the salinity of Owens Lake water over time. A numerical "salinity factor" was assigned to each ostracode species according to its preferred water salinity

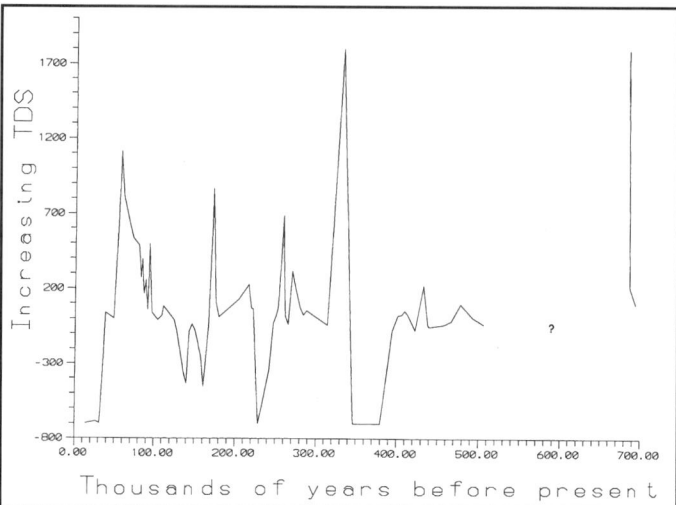

Figure 1. Salinity index of ostracodes from the Owens Lake core, using the formula:

$$Index = 3a + 2b - 2c - 3d$$

Where a = number of adult *L. sappaensis* carapaces, b = number of *L.* cf. *L. bradburyi* + *L. ceriotuberosa* carapaces, c = number of *C.* aff. *C. caudata* + *L. friabilis* carapaces, and d = number of *C. lacustris* carapaces. All numbers are per gram of sample. The algebraic sum of this calculation is plotted against the age of the sample (after Forester et al., 1994).

(the higher the TDS tolerance, the more positive the arbitrary salinity factor):

Species	Salinity factor
L. sappaensis	3
L. cf. *L. bradburyi*	2
L. ceriotuberosa	2
Candona aff. *C. caudata*	−2
L. friabilis	−2
C. lacustris	−3

The number of adult carapaces of each species per gram of sample was multiplied by the appropriate salinity factor and the resulting numbers were combined into a single number for each sample. The resulting numbers are plotted (Fig. 1) against the ages of the samples, which are based on calculations made by Bischoff (1993; Bischoff et al., this volume, chapter 8). For this plot, samples barren of ostracode valves were assumed to represent periods of time when the lake was overflowing with water undersaturated in $CaCO_3$; a value of −4 was assigned to them.

Comparing this ostracode graph (Fig. 1) with the graph for CO_3 values (Bischoff et al., this volume, chapter 4) reveals a good correlation that supports the validity of the ostracode data. The ostracode graph also correlates well with the smectite graph (Menking, this volume), at least as far back as about 400 ka. Saline ostracodes generally show peaks on the graph when freshwater diatoms and TCT pollen show troughs on their respective graphs and vice versa (Bradbury, this volume; Litwin et al., this volume). However, the match with the saline diatom graph (Bradbury, this volume) is not as good; some peaks match (40–120 ka, 250–280 ka, 400–440 ka, and ca. 690 ka), but others do not.

According to Bischoff (this volume, chapter 8), the Owens Lake section is continuous and no significant gaps occur in the record. Assuming that high salinity indicates a warmer climate with less precipitation and low salinity indicates a cooler climate and more precipitation, the ostracode salinity index (Figs. 1, 2) shows that an abrupt warming event in Owens Lake occurred at about 120 ka, although it could be argued that it is closer to the 128 ka event known as Termination II in the SPECMAP (Spectral Analysis Mapping Project) record of changes in foraminiferal ^{18}O (Imbrie et al., 1989; Winograd et al., 1992) (Fig. 2). However, the $CaCO_3$ record at Owens Lake shows an abrupt increase at 117 ka, which would mean that the rapid change from glacial to interglacial conditions at Owens Lake occurred more than 10,000 yr later than the collapse of the polar ice sheets (Bischoff, 1993; Smith et al., this volume). On the basis of ostracodes, another warming event coincides with the Owens Lake $CaCO_3$ record at about 225 ka, an age which is considerably later than the SPECMAP date of about 244 ka for Termination III. The ostracodes indicate the occurrence of another period of warming at about 340 ka, which nicely coincides with the SPECMAP date for Termination IV (ca. 338 ka), but the other Owens Lake parameters put it closer to 355 ka. A final warming episode apparently occurred earlier at Owens Lake (ca. 438 ka) than did Termination V in the SPECMAP record (ca. 420 ka).

CONCLUSIONS

The ostracode fossil assemblages found in samples from the core taken from Owens Lake identify numerous changes in the lake's limnology during the past 800,000 yr. Many of these changes were likely related to global climate change from glacial conditions to interglacial conditions and back, with various intermediate conditions as well. This ostracode record of climate variation is in fairly close agreement with those of the other parameters (mineralogical, geochemical, and biological) that have been obtained and analyzed from the Owens Lake core. A graphic representation of the alternation through time of fresh-water fossil ostracode assemblages with more-saline fossil ostracode assemblages reveals the timing of glacial cycles in the area of Owens Lake and supports the local "Termination II" date of 117 ka.

ACKNOWLEDGMENTS

Many thanks to Rick Forester for confirming ostracode identifications, suggesting the salinity index formula, and making numerous helpful comments on the manuscript. Thanks to Jim Bischoff for pointing out pertinent literature.

REFERENCES CITED

Bischoff, J. L., 1993, Age-depth relations for the sediment column at Owens Lake, California: OL-92 drill hole, *in* Smith, G. I., and Bischoff, J. L., eds., Core OL-92 from Owens Lake, southeast California: U.S. Geological Survey Open-File Report 93-683, p. 251–260.

Carter, C., 1993, Ostracodes present in sediments, *in* Smith, G. I., and Bischoff, J. L., eds., Core OL-92 from Owens Lake, southeast California: U.S. Geological Survey Open-File Report 93-683, p. 303–306.

Coleman, S. M., Jones, G. A., Forester, R. M., and Foster, D. S., 1990, Holocene paleoclimatic evidence and sedimentation rates from a core in southwestern Lake Michigan: Journal of Paleolimnology, v. 4, p. 269–284.

Delorme, L. D., 1971, Freshwater ostracodes of Canada. Part V. Families Limnocytheridae, Loxoconchidae: Canadian Journal of Zoology, v. 49, p. 43–64, pls. 1–19.

Delorme, L. D., 1989, Methods in Quaternary ecology 7. Fresh-water ostracodes: Geoscience Canada, v. 16, p. 85–90.

Forester, R. M., 1985, *Limnocythere bradburyi* n. sp.: A modern ostracode from central Mexico and a possible Quaternary paleoclimatic indicator: Journal of Paleontology, v. 59, p. 8–20.

Forester, R. M., 1991, Pliocene-climate history of the western United States derived from lacustrine ostracodes: Quaternary Science Reviews, v. 10, p. 133–146.

Forester, R. M., Coleman, S. M., Reynolds, R. L., and Keigwin, L. D., 1994, Lake Michigan's late Quaternary limnological and climate history from ostracode, oxygen isotope, and magnetic susceptibility: Journal of Great Lakes Research, v. 20, no. 1, p. 93–107.

Imbrie, J., McIntyre, A., Mix, A., 1989, Oceanic response to orbital forcing in the late Quaternary: Observational and experimental strategies, *in* Berger, A., Schneider, S., Duplessy, J. C., eds., Climate and geosciences: Boston, Kluwer, p. 121–164.

Smith, A. J, and Forester, R. M., 1994, Estimating past precipitation and temperature from fossil ostracodes, *in* Proceedings, Fifth Annual International High-Level Radioactive Waste Management Conference and Exposition, May 22–26, 1994, Las Vegas, Nevada: La Grange Park, Illinois, American Nuclear Society, and New York, American Society of Civil Engineers, p. 2545–2552.

Winograd, I. J., Coplen, T. B., Landwehr, J. M., Riggs, A. C., Ludwig, K. R., Szabo, B. J., Kolesar, P. T., and Revesz, K. M., 1992, Continuous 500,000-year climate record from vein calcite in Devils Hole, Nevada: Science, v. 258, p. 255–260.

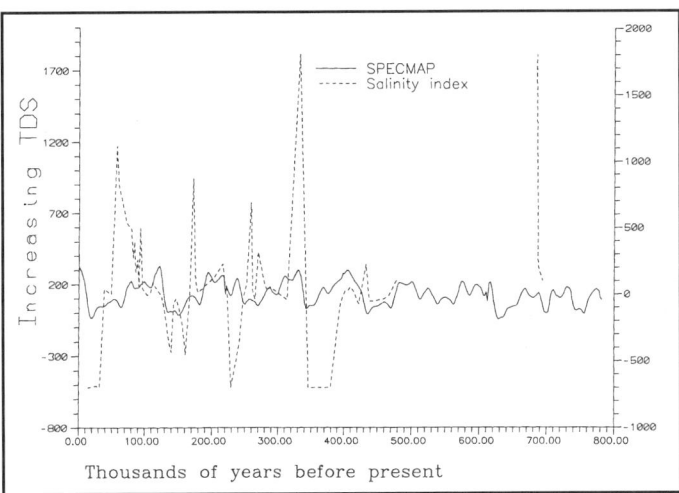

Figure 2. Salinity index of ostracodes from the Owens Lake core (Fig. 1) superimposed on the SPECMAP (Spectral Analysis Mapping Project) record of changes in foraminiferal ^{18}O (Imbrie et al., 1989).

Manuscript Accepted by the Society June 17, 1996

Geological Society of America
Special Paper 317
1997

Paleobiotic and isotopic analysis of mollusks, fish, and plants from core OL-92: Indicators for an open or closed lake system

James R. Firby
Department of Geological Sciences, Mackay School of Mines, University of Nevada, Reno, Nevada 89557
Saxon E. Sharpe
Desert Research Institute, 7010 Dandini Boulevard, Reno, Nevada 89512
Joseph F. Whelan
U.S. Geological Survey, Federal Center, Box 25046, MS 963, Denver, Colorado 80225
Gerald R. Smith
Museum of Paleontology, University of Michigan, Ann Arbor, Michigan 48109
W. Geoffrey Spaulding
Dames & Moore, 4220 S. Maryland Parkway, Suite 108, Las Vegas, Nevada 89119

ABSTRACT

Intervals of open versus closed lake systems for Pleistocene Owens Lake in California are suggested by a comparison of paleobiotic and isotopic evidence recovered from core samples of OL-92. Mollusks and fish were identified from 67 core samples, and their ecological requirements were noted. Carbon dioxide extractions for stable isotopes of ^{13}C and ^{18}O from aragonite of the molluscan shell material were obtained by standard procedures, and isotopic compositions were measured using a mass spectrometer. The values of ^{13}C and ^{18}O from sediment samples taken previously (Benson and Bischoff, 1993) were compared with ^{13}C and ^{18}O values from mollusk shells recovered from the core. These data indicate at least two times of open lake, very low salinity (or "fresh" water) episodes, which are in agreement with the interpretation of fish and mollusk paleoecology supporting open systems between 207 m and 208 m (ca. 450 ka to ca. 453 ka) and between 309 m and 313 m (ca. 765 ka to ca. 775 ka).

INTRODUCTION

Fossil mollusca and fish were recovered from 67 samples from core OL-92. Additional unidentifiable organic material, primarily referable to plants, was recovered from an additional 9 samples. The stratigraphic positions of all types of fossil material are given in Table 1.

Isotopic data (Table 1) from the aragonite of the molluscan shells for ^{18}O and ^{13}C were compared to the values obtained by Benson and Bischoff (1993) for concentrations of these isotopes in inorganic carbonate fractions of sediment samples. There was general agreement of values between these two sources indicating periods of open lake systems. However, no definitive data indicating a closed lake system could be identified using the molluscan or fish remains. It is possible that these data simply indicate that fish, mollusks, and sufficient plant material to form compressed organic and specifically nonidentifiable horizons are together indicative of an open system, and that their absence may or may not be coincident with corresponding closed lake systems.

DISCUSSION OF RESULTS

The two types of analysis conducted on the paleobiotic remains indicate periods of fresher water: the ecological interpretation of mollusks and fishes and the isotopic values for ^{13}C and ^{18}O of bivalve shell material. The other organic material, chiefly plants, is nondiagnostic because of poor preservation.

Firby, J. R., Sharpe, S. E., Whelan, J. F., Smith, G. R., and Spaulding, W. G., 1997, Paleobiotic and isotopic analysis of mollusks, fish, and plants from core OL-92: Indicators for an open or closed lake system, *in* Smith, G. I., and Bischoff, J. L., eds., An 800,000-Year Paleoclimatic Record from Core OL-92, Owens Lake, Southeast California: Boulder, Colorado, Geological Society of America Special Paper 317.

TABLE 1. PRESENCE OF MOLLUSKS, FISH, AND PLANT REMAINS BY DEPTH

Approx. Age (ka)	Depth (m)	Mollusks*	13C (PDB)	18O (PDB)	Fish Remains†	Plant Remains§	Approx. Age (ka)	Depth (m)	Mollusks*	13C (PDB)	18O (PDB)	Fish Remains†	Plant Remains§
	62.96				X			221.48				X	
ca. 100	64.94				X			225.73–225.96	X				
	92.60				X			226.26				X	
ca. 160	94.12				X			226.41–226.48	X				
	95.39				X			237.00					X
	96.66				X		ca. 550	237.57					X
ca. 180	100.90	X						240.82					X
	112.33				X			280.59				X	
	121.32				X			284.24	X	0.04	-5.15	X	
	127.17				X			284.64	X				
ca. 250	129.18				X			285.74				X	
	156.71				X			286.56				X	
	184.39				X			286.81				X	
	184-68				X			287.19				X	
ca. 400	186.71	X	0.86	-5.46	X		ca. 700	287.68				X	
	195.08				X			288.14				X	
	195.14	X						289.37				X	
	195.70–195.72	X	0.04	-2.57				293.90				X	
	199.80				X			295.50				X	
	200.96				X			296.23				X	
	201.66				X			296.86				X	
	203.10				X			297.26				X	
	204.00				X		ca. 750	302.67				X	
	204.27				X			302.90				X	
ca. 450	206.64				X			304.41				X	
	206.60–206.78	X						308.99				X	
	207.49–207.53	X						309.49	X	-0.04	-8.25	X	
	207.91–207.96	X	-5.81	-7.49				310.25	X				
	207.99					X		310.40	X	-3.59	-9.27		
	208.29	X	-4.05	-7.75				310.73					X
	208.39	X						312.15					X
	210.13	X	-3.15	-7.27				312.92	X	-4.54	-5.32	X	
	210.15	X						313.97	X				
	212.41				X			314.75	X	-3.37	-6.31		
	213.54	X						315.62				X	
	214.05	X	-1.69	-6.24				318.36	X			X	
	216.93–217.01	X	0.80	-6.78			ca. 800	320.34					X

*Description of this material is given by Firby, 1993.
†Description of this material is given by Smith, 1993.
§Description of this material is given by Spaulding, 1993.

Ecology

Seven taxa of mollusks were recorded from core OL-92, consisting of five gastropods and two pelecypods. Of the gastropods, three (*Tryonia*, *Amnicola*, and *Valvata*) are gilled species and two (*Helisoma* and *Lanx*) are pulmonates. The pelecypods, *Anodonta* and *Pisidium*, are widely distributed throughout the present sample. A discussion of the general ecologic requirements and geographic distribution of these taxa follows.

Four species of fish were recovered from the Owens Lake core: *Catostomus* cf. *fumeiventris* (sucker), *Gila* (*Siphateles*) *bicolor* (tui chub), *Prosopium* sp. (whitefish), and *Oncorhynchus* cf. *clarki* (trout). Two of these, the sucker and chub, represent lineages surviving in the Owens Valley today. Two other species (a small minnow and pupfish) are presently found in the Owens system (Hubbs and Miller, 1948; Smith, 1978), but were not recovered in the core. None is stratigraphically informative; the trout and whitefish, however, are ecologically informative as they have not been recorded from these elevations in the Owens Basin.

The tui chub is the most common fish represented in the core. It is found in all levels and most samples. Teeth are recognized by their oval or round cross section and hooked tips. Scales, usually 2–4 mm in diameter, are recognized by their asymmetrical shield shape, with strong radii in only one field. Vertebrae are long and round in cross section. Most specimens in the sample represent small individuals. This species is tolerant of a wide range of temperatures and water conditions (La Rivers, 1962). Examination of growth rings relative to body size indicates growth rates similar to those in modern Pyramid Lake, Nevada.

The Owens sucker is found in 13 samples from all parts of the core. It is recognized by distinctive skull bones, flattened teeth, and slightly flattened vertebrae. Most specimens in the sample represent juvenile individuals. Suckers also have broad ecological tolerance.

As compared to the rest of the core, three depths showed an abundance of fish remains: 200–204 m, 285–288 m, and 295–296 m. The approximate ages 435–443 ka, 690–700 ka, and 730 ka, respectively (Bischoff, 1993; Bischoff et al., this volume, Chapter 8), may indicate an increased number of individuals during these time periods, or fatal catastrophic events such as an overturn of the lake. Isotopic values and freshwater diatom abundance (Bradbury, 1993, this volume) indicate that the abundance of fish remains may correlate more closely with a fresher water environment during these three intervals rather than a fatal catastrophic episode.

Trout and whitefish are found in three samples: 286.81 m and 287.29 m (ca. 695 ka), and 296.23 m (ca. 730 ka). Trout and whitefish are recognized by their round scales with no radii. They are cold-water inhabitants and their presence indicates that sediments at core depths of 286–296 m (ca. 695 ka to ca. 730 ka) were deposited in a colder lake than sediments deposited at other times. Trout and whitefish are generally limited to areas where summer air temperatures do not exceed 32 °C for more than 60–90 days per year (U.S. Commerce Department, 1968).

Mollusks were not recovered from the core above 100.90 m (ca. 177 ka), although Holocene to late Pleistocene *Helisoma (Carinifex) newberryi* occurs in surficial deposits of Owens Lake (Firby, 1972, personal observation). The general absence of mollusks from the upper third of the core could be explained one of three ways. (1) The mollusks were present but were not preserved. Since ostracodes do occur in the upper part of the core, and since their composition is similar to the mollusks (calcitic versus aragonitic), it seems unlikely that the absence of molluscan fossils is a function of differential solution. (2) The location of the core in relationship to the shallow zones of the lake during the deposition of this interval was such that the mollusks did not inhabit the site of the core. (3) The environmental parameters during this period consistently exceeded the physical tolerance of the mollusks.

Following is a list of intervals where mollusca occur, in descending depth sequence, with a list of taxa present. We interpret each of these intervals as representative of freshwater regimes.

1. Interval 187–196 m. (ca. 400 ka to ca. 423 ka). Taxa present: *Lanx, Valvata, Pisidium, Helisoma*.
2. Interval 206–210 m. (ca. 450 ka to ca. 460 ka). Taxa present: *Anodonta, Pisidium*.
3. Interval 210–226 m. (ca. 460 ka to ca. 507 ka). Taxa present: *Helisoma, Amnicola, Valvata, Pisidium*.
4. Interval 284–296 m. (ca. 690 ka to ca. 730 ka). Taxa present: *Tryonia, Anodonta*.
5. Interval 309–318 m. (ca. 765 ka to ca. 790 ka). Taxa present: *Anodonta, Pisidium, Amnicola, Tryonia*.

A single specimen of *Lanx* sp., a fresh water limpet, was found at a core depth of 186.71 m (sample 802-B, ca. 400 ka). Since this genus is typically found in well-oxygenated streams, its presence in lake sediments probably represents post-mortem transport. The sub-central and elevated apex and concentric sculpture with no trace of radial striae mark this single specimen as *Lanx* s.s., several species of which are extant in California and Oregon.

Two occurrences of *Valvata* sp. cf *V. sincera* were recovered at 195.70–195.72 m (ca. 420 ka) and 226.41–226.48 m (ca. 507 ka). The general wide distribution of living *V. sincera* throughout much of Canada and the midwestern United States does not include Nevada or California, and if this specimen is assignable to that species, it represents a departure from the modern geographic distribution. Habitat range for the modern *V. sincera* includes lakes and streams, but it is more typical of lakes. *Valvata* is not usually tolerant of water with a pH of less than 7.0 (Pennak, 1953; Firby, 1966).

Several specimens of the small clam *Pisidium compressum* (Prime) were recovered from intervals throughout the core. It first occurs at a core depth of 186.71 m (ca. 400 ka). In some instances, both valves were still together, an indication of little if any post-mortem transport. *Pisidium compressum* has a preference for sandy substrate with vegetation, although it is not confined to that habitat, and has been recorded from water depths as great as 20 m (Herrington, 1962). Geographic distribution is very broad, encompassing most of Canada and the United States. Because of its wide distribution and general abundance within the core samples, this species was selected for isotopic analysis.

Helisoma (Carinifex) newberryi (Lea) occurs at several intervals in the upper part of the core, but is not observed below 217.01 m (ca. 480 ka). A single fragmentary specimen of this genus, not specifically identifiable but probably representing this species, was recovered from a core interval of 225.96 m (ca. 507 ka). This species is abundant in more modern sediments of Owens Lake, from sub-Holocene and Holocene. *Helisoma (Carinifex) newberryi* is more typical of shallow (<3 m) water and seems tolerant of a wide variety of bottom environments within that depth range. Geographic distribution of *H.(C.) newberryi* includes Idaho, Utah, Nevada, Oregon, and California (Burch, 1989).

Anodonta sp. is typically lacustrine, although some species have been recorded from streams. Most species of the genus exhibit a preference for muddy or fine-sand bottom environments, rarely occurring in coarser facies. This is consistent with the matrix of the samples from core OL-92. Largest of the mollusca from the core samples, it is the least likely to have undergone any significant degree of post-mortem transport, and is here presumed to represent its life habitat in all samples. The glochidia larval stage of almost all species of the genus attach themselves to the gills of various fish, which are largely responsible for their distribution. Association of *Anodonta* with fish is evident between 284 m and 296 m (ca. 690 ka to ca. 730 ka) and may represent an open lacustrine system.

All species of *Amnicola* prefer perennial water habitats with aquatic vegetation (Clarke, 1981). The wide geographic distribution of the genus in North America includes most of Canada, and extends to Florida, Texas, Utah, California, and Nevada, at least as fossil occurrences.

Because of the high degree of morphologic similarities between the shells of many species within the *Tryonia* group, assignment to species solely on shell characteristics is not attempted here. In the initial identification of mollusks from core OL-92 (Firby, 1993) this taxon was reported as *Hydrobia*, a commonly used generic designation, especially in older literature. However *Hydrobia* should probably be restricted to European species, following the usage of Burch (1989). *Tryonia* first occurs at 284.24 m depth (ca. 690 ka) and is more or less continuous to the bottom of the core. This restriction may be indicative of a generally more open system at those intervals. Distribution of the genus is restricted to the western United States, and includes Texas, Arizona, California, and Nevada (Burch, 1989).

Stable isotopes

Stable isotope analyses were performed on bivalves to determine if any correlation existed with other studies conducted on the Owens Lake core (Benson and Bischoff, 1993; Bischoff et al., this volume, Chapter 4). Fifteen samples were hand picked from twelve depths in the sediment cores. The X-ray diffraction patterns indicated that the small shell fragments were still aragonite, despite visual degradation. Carbon dioxide extractions for ^{13}C and ^{18}O were performed on the larger samples by routine procedures as described by McCrea (1950), and the stable C and O isotopic compositions measured on a Finnigan MAT 252 mass spectrometer. Two smaller samples were extracted by an automated Kiel carbonate device and their stable C and O isotopic compositions determined on a Finnigan MAT 251 mass spectrometer. Stable isotopic compositions are reported as the per mil deviations of the samples from the International standards (PDB) and are accurate to ±0.1‰. Isotope values were averaged when more than one sample was analyzed from a given depth.

The bivalves chosen for analysis, *Pisidium* and *Anodonta*, were the most widely distributed and the most abundant mollusks found throughout the samples. These genera suffered little post-mortem transport, as evidenced by the original size of *Anodonta* and the (generally) large size of shell fragments. The occurrence of *Pisidium*, frequently found with appressed valves, also indicates *in situ* deposition.

Although the data set is small, the more negative ("lighter") shell aragonite values of both ^{18}O (Fig. 1) and ^{13}C (Fig. 2) appear to be indicative of open lake systems. The ^{18}O curve of the bivalves roughly matches that of the ^{18}O of inorganic carbonates in the sediments (Benson and Bischoff, 1993). The bivalves show ^{18}O enrichment compared to the sediments; this may reflect cooler water temperatures at the bottom of the water column (increased fractionation) than in the carbonate found in sediments (which probably precipitated near the surface). Both ^{18}O records (mollusks and sediment) indicate open lake systems at ~208 m and ~310 m depth (ca. 450 ka and ca. 765 ka).

The ^{13}C record is not as clear. The bivalves record a much lighter isotopic signature than the sediments (Benson and Bischoff, 1993); the bivalves may indicate open systems at the 207 m (ca. 450 ka) and 313 m (ca. 775 ka) depths. The paucity of molluscan shell material makes further analysis impossible.

SUMMARY AND CONCLUSIONS

The ecology of the fish and mollusks indicate an open lacustrine system at intervals 187–196 m (ca. 400 ka to ca. 423 ka), 200–204 m (ca. 435 ka to ca. 443 ka), 210–226 m (ca. 460 ka to

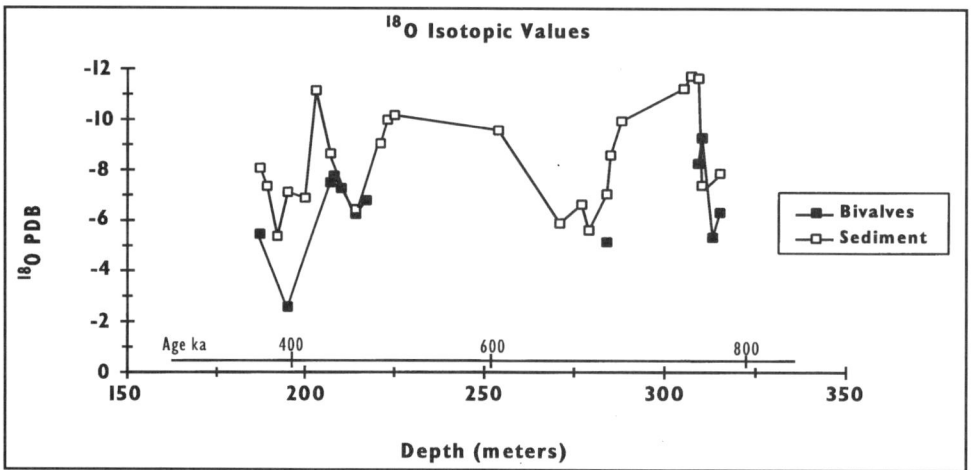

Figure 1. ^{18}O stable isotope analyses of bivalve (*Anodonta* and *Pisidium*) shell aragonite and sediment channel samples (Benson and Bischoff, 1993), by depth.

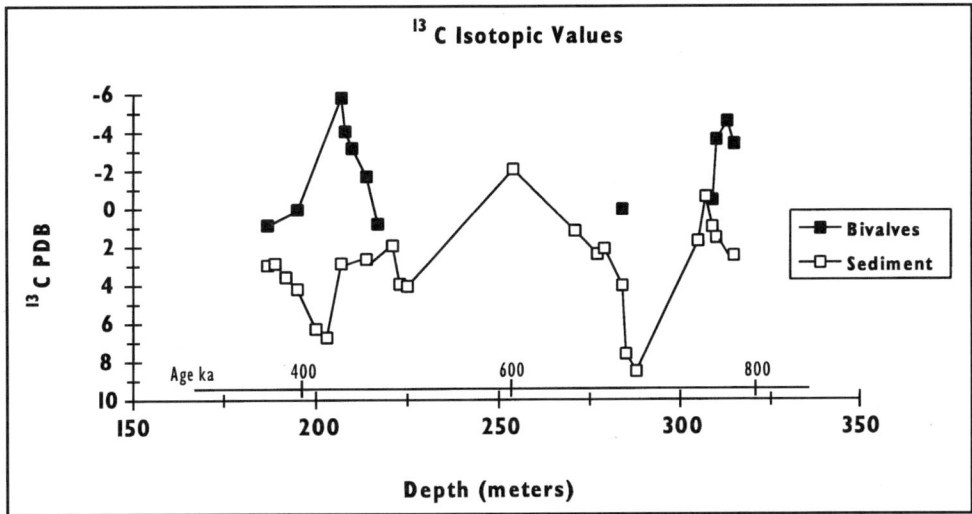

Figure 2. ^{13}C stable isotope analyses of bivalve (*Anodonta* and *Pisidium*) shell aragonite and sediment channel samples (Benson and Bischoff, 1993), by depth.

ca. 507 ka), 284–296 m (ca. 690 ka to ca. 730 ka), and 309–318 m (ca. 765 ka to ca. 775 ka). An absence of mollusks in the intervening depths may indicate less favorable conditions, such as increased salinity, or a less "fresh" aquatic environment. The occurrence of *Anodonta* and fish together is at least suggestive of an open system, as the distribution of *Anodonta* is dependent on spending a part of its larval (glochidia stage) development as a resident on the gills of host fish.

Isotopic values from both sediment samples (Benson and Bischoff, 1993) and molluscan shells indicate the most open systems occurred during deposition between 207 m and 208 m (ca. 450 ka) and again between 309 m and 313 m (ca. 765 ka to ca. 775 ka). The heavier isotopic values depicted in Figures 1 and 2 still indicate open systems, based on the occurrence of fish and mollusks, although the water is possibly not as fresh as times of more negative values. The presence of the pulmonate gastropod *Helisoma* at depth 187 m (ca. 400 ka), 195 m (ca. 420 ka), 210 m (ca. 460 ka), 213 m (ca. 468 ka), 217 m (ca. 480 ka), and 225 m (ca. 505 ka) is suggestive of a less open system when considered with the general lack of gill-breathing mollusks from the same samples. Some of these depths also correspond to the heavier ^{18}O values (Fig. 1), also indicating less open systems.

Comparison between isotopic and paleobiotic data indicate at least a degree of correlation between oxygen and carbon isotope values and the paleoecologic requirements of the mollusca and the fish. Isotopic data indicating an open system are in agreement with what would be expected from the evidence of the biota.

ACKNOWLEDGMENTS

We thank Kelly Conrad for preparing the mollusks for isotopic analysis and Emily Nelson for assistance with graphics. Thanks are also due Larry Benson for his suggestions.

REFERENCES CITED

Benson, L. and Bischoff, J. L., 1993, Isotope geochemistry of Owens Lake Drill Hole OL-92: U.S. Geological Survey Open-File Report 93-683, p. 106–109.

Bischoff, J. L., 1993, Age-depth relations for the sediment column at Owens Lake, California: OL-92: U.S. Geological Survey Open-File Report 93-683, p. 251–260.

Bradbury, J. P., 1993, Diatoms in sediment: U.S. Geological Survey Open-File Report 93-683, p. 261–302.

Burch, J. B., 1989, North American freshwater snails: Hamburg, Michigan, Malacological Publications, 365 p.

Clarke, A. H., 1981, The freshwater molluscs of Canada: Ottawa, Ontario, National Museum of Natural Sciences, 446 p.

Firby, J. R., 1966, New non-marine Mollusca from the Esmeralda Formation, Nevada: California Academy of Sciences, Proceedings, 4th series, v. 33, no. 14, p. 453–479.

Firby, J. R., 1993, Identification of Mollusca from Core OL-92, Owens Lake, California: U.S. Geological Survey Open-File Report 93-683, p. 307–309.

Herrington, H. B., 1962, A revision of the Sphaeriidae of North America (Mollusca: Pelecypoda): Ann Arbor, University of Michigan, Miscellaneous Publications, Museum of Zoology, no. 18, 74 p.

Hubbs, C. L., and Miller, R. R., 1948, The zoological evidence: Correlation between fish distribution and hydrographic history in the desert basins of western United States: University of Utah Bulletin, Biological Series, v. 10, p. 17–166.

La Rivers, I., 1962, Fishes and fisheries of Nevada: Carson City, Nevada State Fish and Game Commission.

McCrea, J. M., 1950, On the isotopic chemistry of carbonates and a paleotemperature scale: Journal of Chemical Physics, v. 18, p. 849–857.

Pennak, R. W., 1953, Fresh water invertebrates of the United States: New York, Ronald Press, 769 p.

Smith, G. R., 1978, Biogeography of intermountain fishes: Great Basin Naturalist, Memoir, v. 2, p. 17–42.

Smith, G. R., 1993, Owens Lake Core OL-92 fish remains: U.S. Geological Survey Open-File Report 93-683, p. 310–311.

Spaulding, W. G., 1993, Macroscopic organic material from Owens Lake Core: U.S. Geological Survey Open-File Report 93-683, p. 312–313.

U.S. Department of Commerce, 1968, Climatic atlas of the U.S.: Washington, D.C., U.S. Government Printing Office, 80 p.

Manuscript Accepted by the Society June 17, 1996

An 800,000-year pollen record from Owens Lake, California: Preliminary analyses

R. J. Litwin
U.S. Geological Survey, MS 955, National Center, Reston, Virginia 22092
D. P. Adam
U.S. Geological Survey, 345 Middlefield Road, MS 939, Menlo Park, California 94025
N. O. Frederiksen
U.S. Geological Survey, MS 970, National Center, Reston, Virginia 22092
W. B. Woolfenden*
University of Arizona, Tucson, Arizona 85721

ABSTRACT

A long sequence of fossil palynomorph assemblages from a 323-m-long core taken at Owens Lake has enabled us to evaluate the gross vegetational trends for the Owens Valley region of California over the past ~800,000 years. Shifts in vegetation composition and abundance in the study area during the Pleistocene were indicated in core sediments by marked fluctuations in the pollen frequencies of pines, junipers, and, to a lesser extent, of big sagebrush, composites, and chenopods/amaranths. The modern vegetation distribution and modern pollen rain on the eastern flank of the Sierra Nevada indicate that maximal abundances of these taxa generally characterize higher elevation subalpine and montane coniferous forests, lower elevation coniferous woodland, steppe, and desert scrub environments. Pollen frequencies in the upper part of core OL-92 corroborate vegetational trends documented previously from late Wisconsin and Holocene *Neotoma* middens in the Great Basin. These trends and evidence from this study suggest that woodland taxa expanded their range down the slope of the eastern flank of the Sierra Nevada and were established in (and immediately adjacent to) Owens Valley during moderated climates of the late Wisconsin, apparently in response to decreases in temperature and increases in precipitation, but retreated upslope toward their present position starting as long ago as ca. 20 ka.

More importantly, pollen evidence from core OL-92 documents that the southern Sierra Nevada has experienced nine major cool-to-warm vegetation shifts (in addition to the late Wisconsin-early Holocene warming) during the time interval spanning the middle Pleistocene to early Holocene (Brunhes Normal Polarity Chron). We believe that at least six consecutive cool-to-warm shifts (the most recent ones) represent transitions from full-glacial to full-interglacial conditions on the basis of the magnitude of vegetation change in this portion of the pollen record. These marked changes in the frequency curves of dominant palynomorph taxa enabled us to identify boundaries that define 19 (?20) pollen zones in OL-92. The excursions of the pollen frequency curves within and across zone boundaries approximate the nature, duration, and timing of the middle and late Pleistocene climatic trends documented by geochemical ($\delta^{18}O$) evidence from OL-92 and from Devils Hole (DH-11) in the Amargosa Desert of Nevada.

---

*Present address: U.S. Forest Service, Bishop, California 93514.

Litwin, R. J., Adam, D. P., Frederiksen, N. O., and Woolfenden, W. B., 1997, An 800,000-year pollen record from Owens Lake, California: Preliminary analyses, *in* Smith, G. I., and Bischoff, J. L., eds., An 800,000-Year Paleoclimatic Record from Core OL-92, Owens Lake, Southeast California: Boulder, Colorado, Geological Society of America Special Paper 317.

INTRODUCTION

This study is a preliminary assessment, based on fossil pollen, of the gross vegetational trends in the southern part of the Owens Valley of California and the southern part of the eastern flank of the Sierra Nevada (Fig. 1) during the past ~800,000 years. Even for the last full-glacial cycle, published terrestrial records are relatively rare in North America (e.g., Atwater et al., 1986; Adam, 1988; Adam et al., 1989; Heusser and Heusser, 1990). Palynological analyses from cores spanning shorter time intervals have been documented by Martin (1963a), Martin and Mehringer (1965), Anderson et al. (1985), Hevly (1985), and Madsen (1985), among others, for the latest Pleistocene and Holocene vegetation of the southwestern United States. In addition, midden studies from the southwestern United States have documented numerous short-interval, site-specific vegetational trends during the latest Pleistocene and Holocene; examples include Wells and Berger (1967), Van Devender (1977), Van Devender and Spaulding (1979), Thompson and Mead (1982), Spaulding (1985), Van Devender et al. (1987), Jennings and Elliot-Fisk (1993), Koehler and Anderson (1994, 1995) and others. The plant macrofossils preserved in middens routinely can be identified to species-level resolution. Although this resolution is not as easily accomplished for fossil pollen, long pollen records have different merits. Thompson (1990) has noted that many to most of the midden records from the Great Basin have radiocarbon dates clustering in the late Wisconsin or early Holocene (less than ca. 10–14 ka); the oldest dated middens of which we are aware are less than 50,000 years old (Spaulding, 1985). Although middens are an excellent tool for documenting geologically instantaneous local vegetational assemblages at specific elevations throughout the Great Basin, the length and continuity of the OL-92 pollen record enable us to filter out single-sample biases in the vegetation signal and to examine longer duration vegetational trends for a larger geographic area. Midden and core studies thus complement each other in increasing our understanding of the nature and timing of climatic change in the southwestern United States since the middle Pleistocene.

This initial report does not distinguish among the species of pines (per Martin, 1963b), propose a climatic mechanism for vegetation changes, or provide a detailed chronology of serial vegetation change. These aspects are well beyond the scope of this paper because of the reconnaissance-level sampling reported here and may be properly addressed only after we have sampled the entire core more thoroughly and have documented the vegetation changes noted in this study with greater temporal resolution. More detailed palynological (paleovegetation) analyses of the Owens Valley region are in progress, and their results will be presented elsewhere.

MATERIALS AND METHODS

Sample preparation procedures have been described by Litwin et al. (1993) and Woolfenden (1993). A "reconnaissance" pollen percentage for each taxon was calculated on minimum counts of 300 total specimens per sample, corrected to the terrestrial pollen component of each sample.

The sample set was calibrated to the age-depth model of Bischoff (1993; Bischoff et al., this volume, chapter 8), which was derived from the following control points: (1) ^{14}C-dated samples in the upper part of core OL-92, (2) the age and depth of the top of the Bishop ash bed, (3) the suggested position of the Matuyama-Brunhes paleomagnetic reversal near the base of the core (Glen et al., 1993; Glen and Coe, this volume), and (4) 10 paleomagnetic excursions detected in the cored sediments. The resulting age estimates for inflections in the OL-92 pollen curve differ somewhat from age values proposed for presumably equivalent trends in SPECMAP (Imbrie et al., 1984; Martinson et al., 1987) and in Devils Hole (DH-11) (Winograd et al., 1992). At present, the palynomorph frequency curves presented in this report were calibrated to absolute ages calculated exclusively from evidence in OL-92 in order to maintain consistency with other geophysical, mineralogical, and geochemical evidence derived from this core. The average time resolution of the 143 samples that comprise this reconnaissance study is less than 5,900 years. The net average sedimentation rate in Owens Lake is calculated to be 40.1 cm/k.y. (Bischoff, 1993; Bischoff et al., this volume, chapter 8); it was derived by dividing the depth to the top of the Bishop ash bed (303.9 m) (Smith, 1993; Sarna-Wojcicki et al., 1993) by its estimated age (758 ka) (Izett et al., 1988; Sarna-Wojcicki and Pringle, 1992; Sarna-Wojcicki et al., 1993).

MODERN VEGETATION DISTRIBUTION

The modern vegetation distribution and pollen rain were used as a present baseline reference for this study. Although fossil packrat midden evidence suggests that the compositions of past plant communities likely were different from those of communities present in the study area today, the environmental preferences for each taxon were presumed here to be similar to those found within each taxon's present area of distribution.

Küchler (1977) identified 10 modern vegetation zones in the study area, covering an altitudinal range of more than 3,300 m (Fig. 1B); Solomon and Silkworth (1986) later added an eleventh zone. These zones exhibit a strong correlation with altitude and therefore comprise an approximately parallel series of vegetation bands from Owens Lake to the crest of the Sierra Nevada. This vegetation pattern is typical for the Great Basin (Thompson, 1990). Currently, the distribution of three of these zones (the Joshua tree scrub, the Sierran montane forest, and the northern Jeffrey pine forest) are in large part outside the Owens Lake drainage. During the Pleistocene, however, biogeographic patterns and plant community compositions probably differed moderately to significantly from those of the present.

Seven main vegetation zones presently occur in close proximity to Owens Lake; the elevational ranges listed here are approximate. The desert saltbush-shadscale zone occupies the lake fringe and floor of Owens Valley (~1,200–1,300 m). This zone is characterized by *Atriplex confertifolia*, *A. polycarpa*

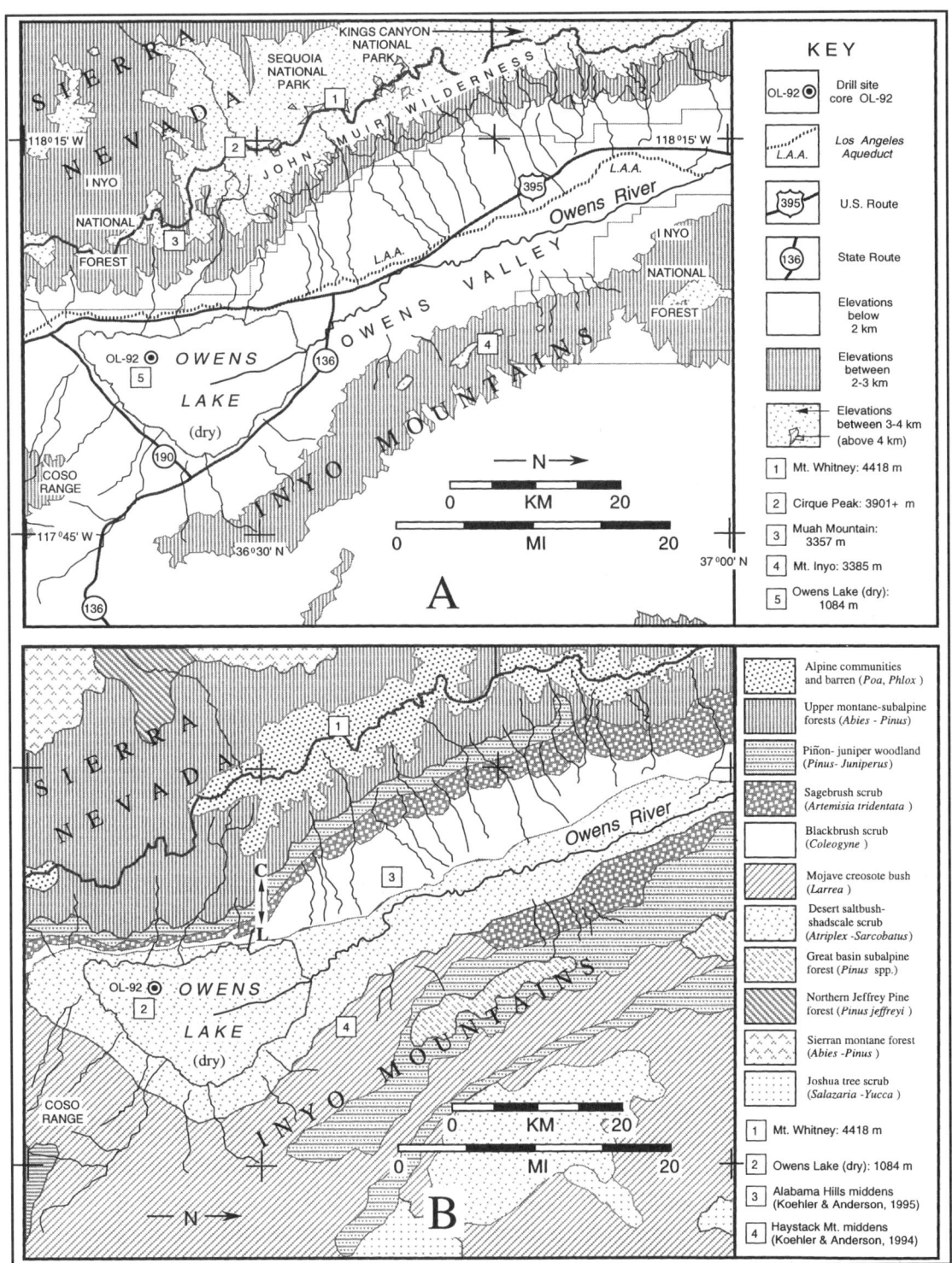

Figure 1. A. Site map for core OL-92 showing major physiographic features in the Owens Valley study area. Elevation zones are in kilometers. B. Modern vegetation map (generalized) for the Owens Valley, southern Sierra Nevada, and Inyo Mountain areas showing terrestrial vegetation zones discussed in this report (from Küchler, 1977; Solomon and Silkworth, 1986). Location of modern pollen rain transect of Solomon and Silkworth (1986) shown (northeast of Owens Lake, stations C–L).

(Chenopodiaceae), *Sarcobatus vermiculatus*, *Suaeda* spp., *Distichlis spicata*, *Artemisia spinescens*, *Psorothamnus fremontii*, and *Ephedra nevadensis*, among other taxa. Three higher zones divide Owens Valley latitudinally into cold desert (to the north), hot desert (to the south), and a transitional zone in between them along the western side of Owens Valley. The southern component, the Mojave creosote bush scrub, is characterized by *Larrea tridentata*, *Ambrosia dumosa* (Compositae), and *Opuntia* spp., among other taxa. It lies above the saltbush-shadscale zone (~1,300–1550 m) and occupies the hot desert south of Owens Lake and alluvial fans along the east side of Owens Valley; the study area comprises part of its northernmost range. On the upper bajadas of the Sierra Nevada is the sagebrush scrub zone, characterized by *Artemisia tridentata*, *Ephedra viridis*, and *Purshia tridentata*. It represents an extreme southern extension of a cooler sagebrush steppe community that becomes more dominant north of Bishop. *Artemisia* spp. also extend (albeit more sparsely) up into both forested zones on the eastern flank of the Sierra Nevada (Young et al., 1977; Küchler, 1977). The western side of Owens Valley hosts a transitional zone below the sagebrush scrub; this diverse blackbrush scrub community is dominated by *Coleogyne ramosissima*, interspersed with shrubs such as *Grayia spinosa*, *Tetradymia axillaris*, *Ephedra nevadensis*, *Ericamieria cooperi*, *Eriogonum fasciculatum*, and *Hymenoclea salsola* (Küchler, 1977; Solomon and Silkworth, 1986).

The lower forested zone, the piñon-juniper woodland, occupies the foothills of the eastern Sierran escarpment (~1,550–1,800 m) and Inyo Mountains and generally is characterized by *Pinus monophylla* (single leaf piñon pine) and *Juniperus osteosperma*. *Juniperus osteosperma* is absent from this zone along the eastern flank of the Sierra Nevada in the study area but present along the flanks of the Inyo Mountains. *Quercus* spp. also are found in this zone but presently are found only locally in the study area as small isolated stands along tributaries draining the eastern side of the Sierra Nevada. *Shepherdia argentea* (silver buffaloberry) is associated locally in the eastern Sierra Nevada.

The higher forested zone, the upper montane-subalpine forest, is characterized by *Abies magnifica* and *A. concolor* in the lower elevations (~1,800–2,750 m) and *P. contorta* var. *murrayana* (lodgepole pine), *P. albicaulis* (whitebark pine), *P. balfouriana* (foxtail pine), *P. monticola* (western white pine), and *P. flexilis* (limber pine) in the higher elevations (~2,750–3,600 m) (Rundel et al., 1977). This zone also includes *P. jeffreyi* (Jeffrey pine), *Juniperus occidentalis* var. *australis*, *Salix* spp., *Tsuga mertensiana* (north of Owens Lake), and *Poa* spp., among other taxa. It also includes local montane-subalpine meadows of grasses and sedges (Poaceae and Cyperaceae).

Above timberline (>3600 m) are the alpine communities, which are characterized by *Draba*, *Festuca*, *Poa rupicola*, *Carex* (sedges), *Phlox*, and *Epilobium* spp. (Major and Taylor, 1977). An eighth zone, which occupies the crest of the Inyo mountains east of Owens Valley (Vasek and Thorne, 1977), the Great Basin subalpine forest, is characterized by *Pinus longaeva* (bristlecone pine), *P. flexilis*, *Artemisia* spp., and *Poa*.

In summary, the modern vegetation of the southern part of Owens Valley and the eastern flank of the southern Sierra Nevada comprises low-elevation desert scrub communities that change at progressively higher elevations into sagebrush, coniferous woodland, montane/subalpine coniferous forest, and alpine plant communities.

MODERN POLLEN DISTRIBUTION

Solomon and Silkworth (1986) examined the spatial distribution of the present-day pollen rain on the eastern flank of the Sierra Nevada, the Owens Valley floor, and the western flank of the Inyo Mountains. The transect that they studied crossed the northern third of Owens Lake, at approximately 36°30′ N latitude (Fig. 1). They did not resolve the modern pollen rain to species level but selected five categories of pollen that they observed to characterize the upper montane, upper mountain slope, lower mountain slope, and valley floor environments (Fig. 2): pine (*Pinus* subgenera *Haploxylon* and *Diploxylon*, here combined), big sagebrush (*Artemisia tridentata*), Mormon tea or joint-fir (*Ephedra* spp.), ragweed or bursage (*Ambrosia* spp.), and chenopods (Chenopodiaceae-Amaranthaceae).

Solomon and Silkworth (1986) concluded that modern pollen rain on the eastern flank of the Sierra Nevada had two characteristics. First, maximum pollen-deposition rates occurred in source plant areas and diminished rapidly across the zonal boundaries that delimited each source plant. Second, most stations of their transect downwind from the source area exhibited approximately similar pollen-concentration values. Pollen concentrations of these five taxa typically were similar (and low) across the valley floor and generally minimal to low at lake-edge stations. An estimate of the relative frequency (per station) of these five taxa, from the Sierran crest to the edge of Owens Lake (calculated from data in Solomon and Silkworth, 1986, table 1, stations C–M″) is given in Figure 2B. Stations east of Owens Lake have been excluded here because wind-frequency evidence from Owens Valley (Bishop, California) suggests that east-to-west transport of pollen by winds is minimal at present (Solomon and Silkworth, 1986).

When the data are replotted as relative frequency values (Fig. 2B), their modern pollen rain samples exhibit several predictable trends. First, each of the five pollen groups identified in the Sierra Nevada exhibits maximum values at station intervals approximately coinciding with the elevational range of its maximum vegetational abundance. Along the transect from west to east (as well as from coldest to warmest mean average temperature) the pollen groups are pine and big sagebrush, pine and joint-fir, big sagebrush, ragweed (composites), and goosefoot (chenopods). Second, each of the pollen curves decreases in absolute value laterally away from these maxima. We conclude that the maximum pollen frequency of a taxon usually occurs within its zone of vegetation abundance and reflects the preferred environmental parameters for that species or group. On the eastern flank of the Sierra Nevada, maximum pine abundances indicate cooler conditions currently present at middle to high elevations along the mountain front, whereas high chenopod and composite abundances indicate significantly warmer conditions present along the valley floor and lower mountain slope. No modern pollen rain values were given by Solomon and Silkworth (1986) for fir, hemlock, juniper, or greasewood in the study area, but we expect that their greatest vegetational abundance and highest pollen frequencies would correlate with high elevations and cool mean annual temperature (MAT), middle elevations and moderate MAT, and low elevations and warm MAT, respectively.

As noted above, exceptions to these generalized relative abundance patterns occurred at stations on rocky outcrops and at the mouths of large canyons. The former showed aberrantly low pollen concentration levels, and the latter showed a disproportionate increase in pine pollen from higher elevations. We attribute the skewing of the pollen rain in both cases to airflow complexities

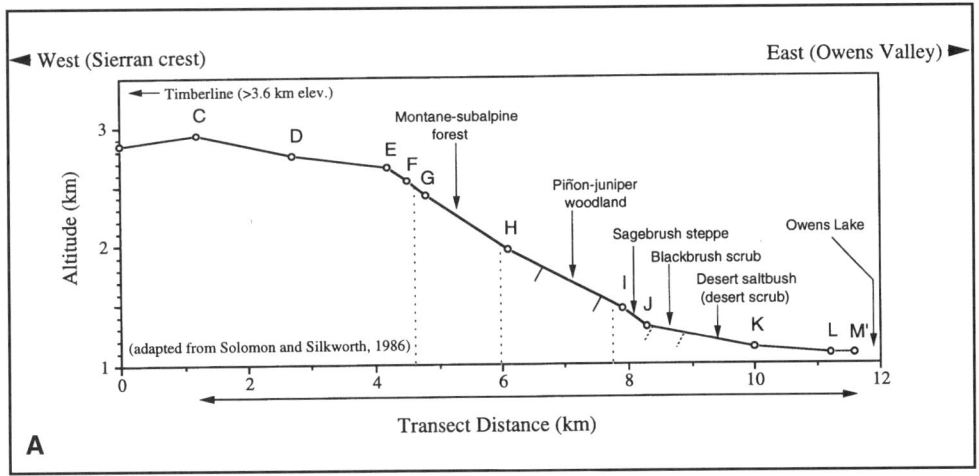

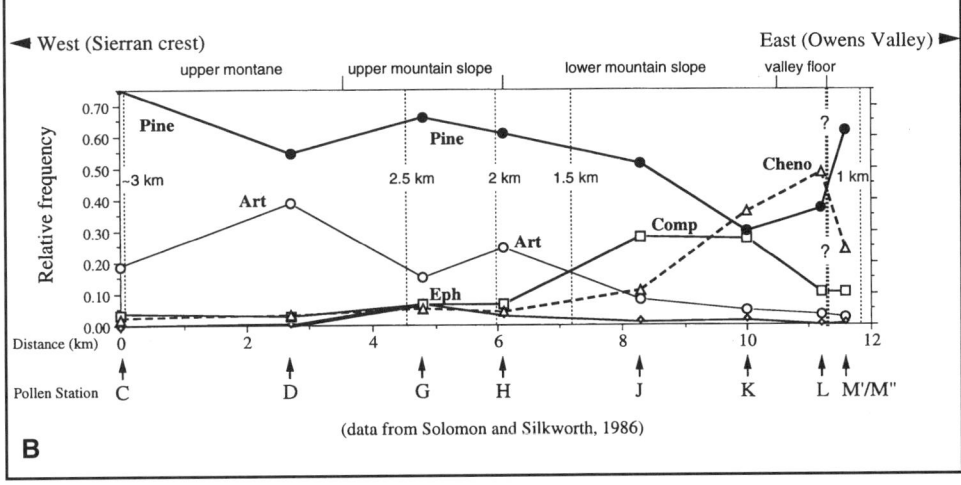

Figure 2. A. Vertical profile of the eastern flank of the Sierra Nevada showing modern vegetational zones (adapted from Solomon and Silkworth, 1986). Line below the x axis indicates the portion of the transect used to illustrate modern pollen rain in Figure 2B. Vertical dotted lines mark the intercepts of elevation contours at 2.5 km, 2.0 km, and 1.5 km elevation, respectively (from left to right). B. Relative frequency of modern pollen rain on the eastern flank of the Sierra Nevada, calculated from Solomon and Silkworth (1986). The most abundant component (pine) is denoted by solid circles; *Artemisia* ("Art") or big sagebrush is denoted by open circles; the Compositae ("Comp") are denoted by open squares; chenopods and amaranths ("Cheno") are denoted by open triangles; *Ephedra* ("Eph") is denoted by open diamonds. Three of the stations along the western half of the Solomon and Silkworth (1986) transect were omitted or queried here. The values at stations E and F were queried by Solomon and Silkworth (1986) because of aberrantly low pollen capture, attributed to placement of these stations on isolated, exposed rocky outcrop. Our replot of the pollen transect omits them. The third station of questionable accuracy is M (M'/M"). Stations M' and M" were placed to the south of the study transect, at the mouths of two major canyons on the eastern Sierran front. It appears that the pollen values from these two (combined and averaged) stations reflect atypical pine concentration owing to increased transport by canyon winds. Therefore, we have queried the pollen values at this station. Approximate intercepts for 2.5 km, 2.0 km, and 1.5 km elevations are noted by dotted lines.

induced by topography. For example, we ascribe the latter case to increased transport by canyon winds and believe it may be analogous to larger scale anomalies seen in pollen-distribution patterns of nearshore marine sediments at the mouths of major river systems (Muller, 1959).

PRELIMINARY PALYNOMORPH RESULTS FROM CORE OL-92

Figure 3 illustrates the first-order pollen profile of Owens Valley and the eastern flank of the southern Sierra Nevada for the past ~800,000 k.y. Percentages of each taxon in Figure 3 were calculated as a proportion of the terrestrial pollen component and represent more than 41,000 tabulated specimens. For purposes of discussion, we have divided the core into two time intervals (of unequal duration). The younger portion of the core, from ca. 200 ka to ca. 9 ka, comprises approximately 25% of the geologic time represented in core OL-92 and may be compared fairly directly (especially in its uppermost portion) to other fossil vegetation evidence from the study area, in particular to macrofossil evidence from isotopically dated *Neotoma* middens and to shorter (late Pleistocene and Holocene) serial records of fossil pollen assemblages in the Great Basin, the Sierra Nevada, and the southern Colorado Plateau. The older portion of core OL-92 spans a substantially longer interval (~75% of the time represented, from ca. 800 ka to ca. 200 ka) and is much older than most Quaternary vegetation records from the Great Basin. Pollen assemblages below ~35 m depth (>50 ka) in OL-92 are too old for direct comparison to other paleobotanical records from the region. We therefore infer vegetation responses for this older portion of OL-92 in part by comparison with the vegetation/climate interactions in the younger (upper) part of the core, which is relatively better understood.

We define 20 preliminary pollen zones in core OL-92, numbered from youngest (P1) to oldest (?P20) in Figures 4 and 5. (Pollen zone boundary placements suggested here supersede those suggested initially by Litwin et al. (1993).) Zone boundaries were selected visually at significant changes in pollen frequency; this procedure emphasized the most common pollen types (e.g., pines and junipers) in defining zones, but systematic changes in other taxa also figured in defining zone boundaries. We interpret odd-numbered zones to represent warmer (mostly interglacial) climatic conditions and even-numbered zones to represent cooler (mostly glacial) climatic conditions, on the basis of comparing the OL-92 pollen record to the modern vegetation, to the modern pollen rain of the study area, and to midden and pollen records of the late Wisconsin–Holocene transition. However, the pollen signal suggests that not all odd-numbered pollen zones were fully interglacial in intensity, nor were all even-numbered pollen zones fully glacial in intensity.

Some of the pollen zones in OL-92 appear to be divisible into several subzones (e.g., zones 5, 7, and 9), an indication that smaller-scale vegetation responses (trends) to climatic change were superimposed on larger scale vegetational trends. In addition, vegetation change between some of these zones appears to have been extreme (e.g., P6–P5 transition), whereas, in others it appears to have been much less so (e.g., P14–P13 transition); in these latter cases the cool-warm zonal boundaries are less easily established.

200–0 ka: Midden, pollen, and $\delta^{18}O$ evidence

The modern vegetation on the eastern flank of the southern Sierra Nevada shows altitudinal distribution (Küchler, 1977; Thompson, 1990), and the modern pollen rain reflects that zonation (Solomon and Silkworth, 1986; Fig. 2B). The pollen record of core OL-92 thus can be used to assess major vegetation trends in the southern part of Owens Valley and the eastern flank of the southern Sierra Nevada for the past ~800,000 years. Within the past 45,000 years, direct vegetation comparisons with Great Basin midden records and nearby dated pollen records are available. The isotopic record from Devils Hole (Winograd et al., 1992) spans this interval and a much longer one and is very well dated. However, the superbly dated bristlecone pine chronologies of the White Mountains (Ferguson, 1969; LaMarche, 1974) are too young to be useful for comparison here.

Comparison of midden macrofossil and OL-92 pollen trends. Numerous fossil packrat (*Neotoma*) sites have been found in Owens Valley and the surrounding areas, although not all of them are relevant to the OL-92 palynological record. Comparison of pollen and midden records is restricted by several factors. First, the entire suite of dated fossil *Neotoma* midden records from the Great Basin spans only approximately the past 45,000 years, whereas our pollen record is essentially continuous through (at least) the Bishop ash bed, which has been dated at ca. 758 ka. Additionally, the bulk of the middens from the Great Basin are less than ca. 20 ka (Wisconsin or younger), whereas our present youngest pollen sample has been estimated at ca. 9 ka; thus the number of middens of suitable age for comparison is limited. This subset is further reduced by geographic distance from the study area. Many middens are distant from the OL-92 study area; a review of general patterns in late glacial (ca. 14–10 ka) midden evidence across the Great Basin has been done by Thompson (1990).

The vegetation changes in midden records compared in this report mostly document the early Holocene warming and the climate transition since the last glacial maximum, although the oldest of these local midden records extend into the middle Wisconsin. The trends in these records form the basis for comparison with our OL-92 pollen record. The oldest local midden records were reported by Koehler and Anderson (1995), who analyzed 20 packrat middens from the Alabama Hills north and west of Owens Lake (Fig. 1B); radiocarbon analyses of these middens placed them as middle Wisconsin (ca. 31.5 ka) to late Holocene (ca. 2.8 ka) in age. These middens were recovered from an area that presently supports a saltbush-shadscale scrub community. Macrofossil evidence in their oldest middens (TG-E-4T and TG-E-1) suggests that Utah juniper (*Juniperus osteosperma*) and Joshua tree (*Yucca brevifolia*) inhabited low elevations in Owens Valley (in the Alabama

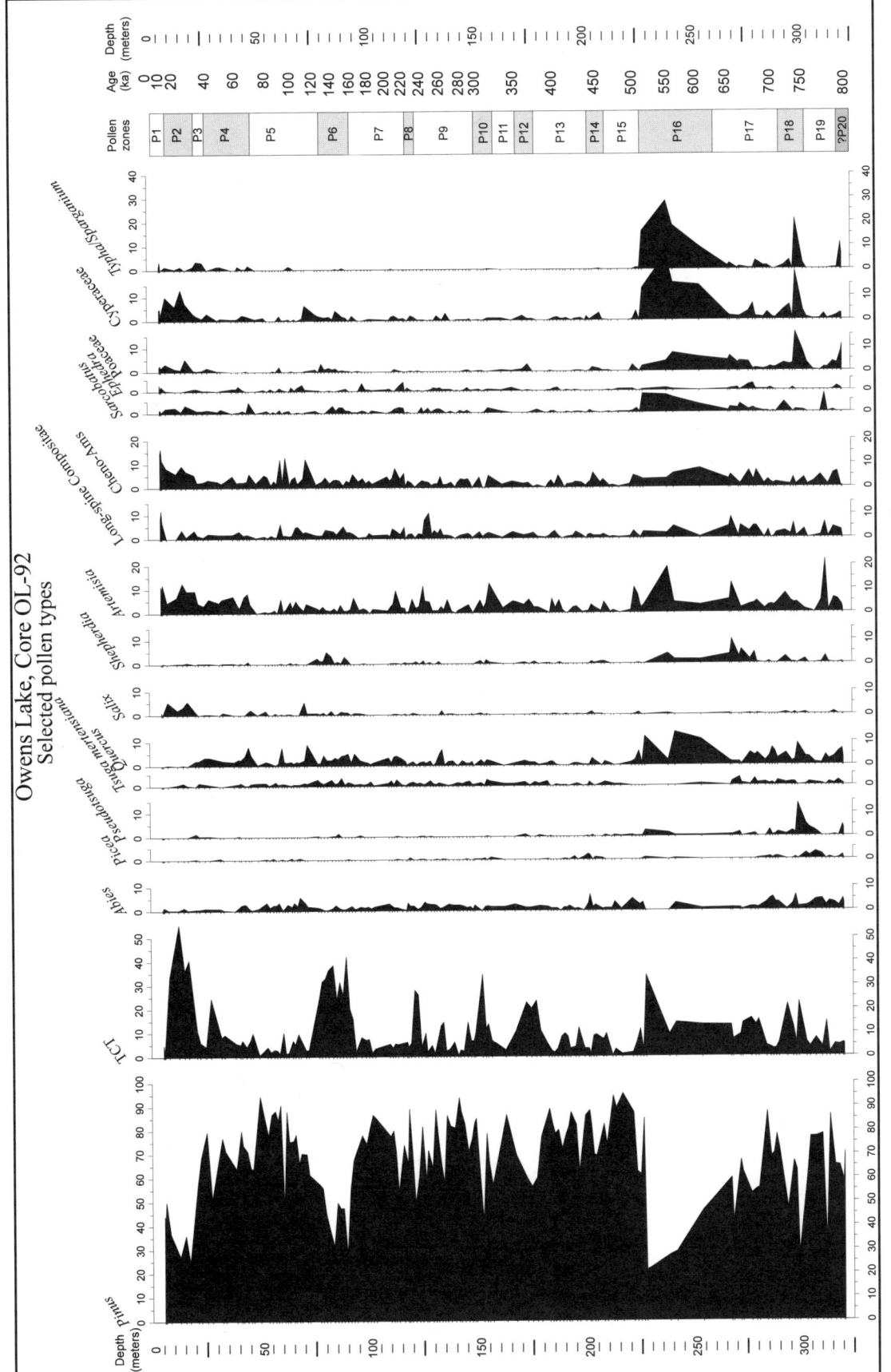

Figure 3. Selected preliminary vegetation response curves from core OL-92, plotted against depth. Pollen zones proposed in this report are numbered in order from youngest (P1) to oldest (?P20). The pollen group labeled TCT mostly represents undifferentiated *Juniperus* spp. (*J. osteosperma*, *J. scopulorum*, and *J. occidentalis*), although isolated grains of other taxa (*Sequoia*, *Calocedrus*, *Cupressus*, etc.) might be present as rare occurrences lower in the OL-92 record. Pollen frequencies are expressed as percentages for each taxon (x-axis).

Note added in proof: We have reexamined all *Picea* occurences shown in Figure 3 and have confirmed all of them below 119.65 m in depth (pollen interval P8). Occurences above this depth were not confirmed. We thank O. K. Davis for independent confirmation of our identification of this taxon.

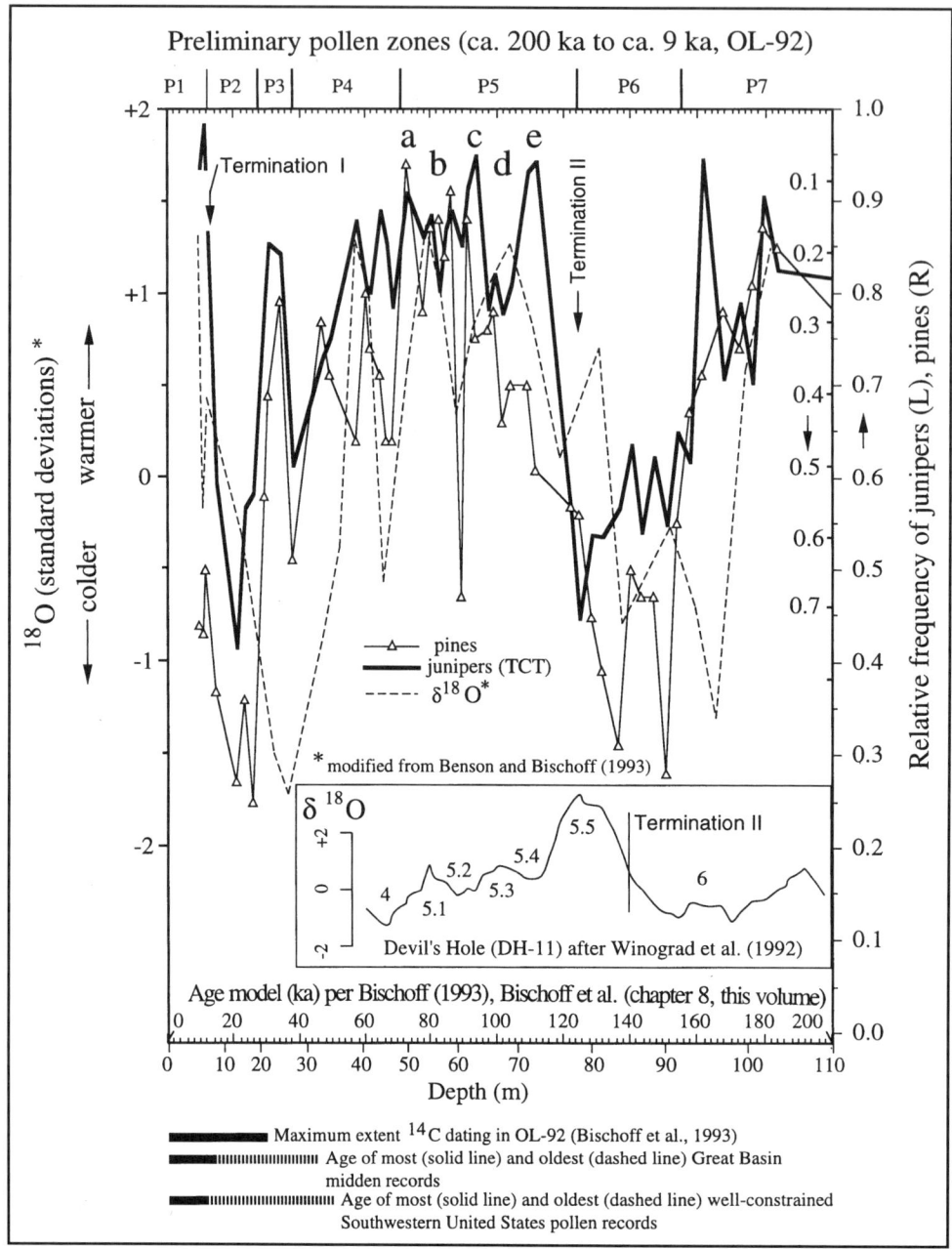

Figure 4. Comparison of δ¹⁸O curve (derived from vein calcite) of Devils Hole core DH-11 (inset) (Winograd et al., 1992) with the preliminary pine and juniper frequency curves from OL-92 (this study) and the δ¹⁸O curve from OL-92 (modified from Benson and Bischoff, 1993), from ca. 200 ka to ca. 9 ka. Ages for OL-92 pollen records are derived from current age model (Bischoff, 1993; Bischoff et al., this volume, chapter 8; J. L. Bischoff, 1994, oral communication); horizontal axis at the base of the diagram plots core depth and interpolated age (age scale applies to both OL-92 and DH-11 plots). Seven pollen zones have been identified in this interval. Even-numbered pollen zones represent predominantly cold vegetation intervals, and odd-numbered zones represent predominantly warm vegetation intervals, interpreted on the basis of modern vegetation distribution (Küchler, 1977; Solomon and Silkworth, 1986), modern pollen rain (Solomon and Silkworth, 1986), and continental δ¹⁸O trends (Winograd et al., 1992; Benson and Bischoff, 1993). Relative frequencies of pollen are plotted along the right vertical axis: pines (light line) are plotted along the right side of the axis (scale increases upward); junipers (heavy line) are plotted on the left side of the same axis, calculated as percentages of the terrestrial pollen component minus pines (scale increases downward). Limit of reliable AMS radiocarbon dating is plotted at bottom of figure, as are the common and maximum ages of Great Basin midden records (Thompson, 1990; Spaulding, 1985) and the common and maximum ages of well-constrained Southwestern United States pollen records (Hevly, 1985; Adam, 1967; Davis et al., 1985). Position of terminations I and II on the pollen records were established at the midpoints of the major changes in pollen response (the P1/P2 and P5/P6 boundaries, respectively) by using the nonaveraged dataset. Position of termination II in the DH-11 record is from Winograd et al. (1992); notation of substages on DH-11 record is from I. J. Winograd (1994, oral communication).

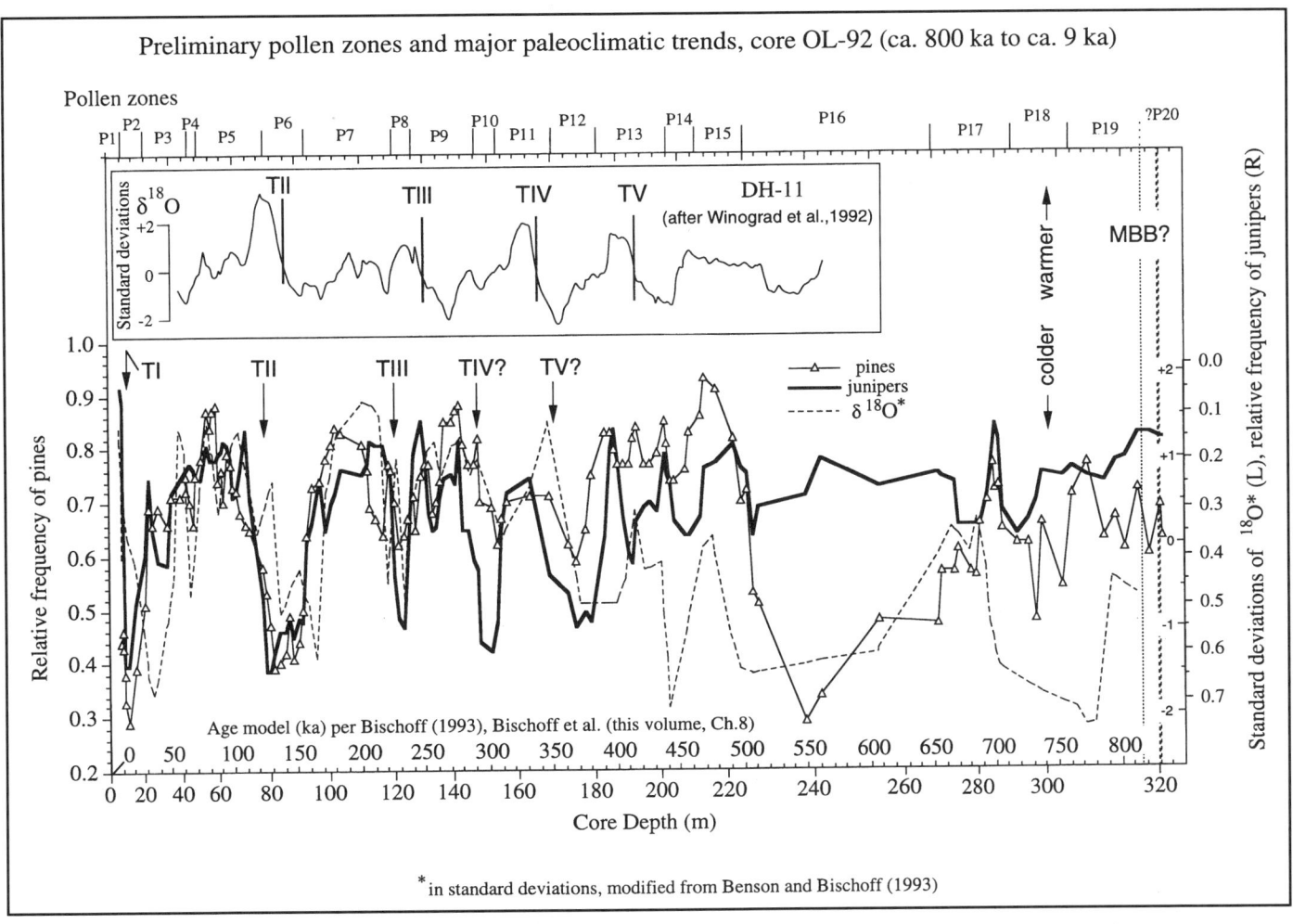

Figure 5. Summary diagram denoting cold and warm vegetation pulses (pollen zones P1–?P20) observed in core OL-92. Zones are based in part on inflections in the pine and juniper frequency curves; the $\delta^{18}O$ records of OL-92 (modified from Benson and Bischoff, 1993, to standard deviations) and core DH-11 from the Amargosa Desert (after Winograd et al., 1992) are plotted for comparison of climatic trends. Juniper frequencies were calculated as percentages of the terrestrial pollen component minus pine. Both pollen datasets were smoothed by running three-point averages to enhance visual comparison to the $\delta^{18}O$ record. Relative positions of glacial terminations I–V (TI–TV?) are noted with arrows on the OL-92 reconnaissance pollen record; correlative positions of terminations II–V (TII–TV) in DH-11 record after Winograd et al. (1992). The base of the lowest "warm" interval (pollen zone P19, light dotted line) compares closely in depth even with the lower placement of the Matuyama-Brunhes Chron boundary (MBB, dark hachured line), as determined from paleomagnetic evidence (Glen et al., 1993; Glen and Coe, this volume). This placement of the MBB approximates 783 ka, its age as estimated elsewhere (Baksi et al., 1992). Complications in defining the MBB are discussed by Glen and Coe (this volume).

Hills) from ca. 31.5 ka to ca. 19.1 ka (respectively), in association with Great Basin sagebrush (*Artemisia tridentata*) and bitterbush (*Purshia tridentata*). This middle Wisconsin interstadial plant association changed with the onset of maximum Wisconsin glaciation and resulted in a probable local disappearance of *Y. brevifolia* by ca. 19.1 ka.

The next oldest midden records near Owens Lake also were documented by Koehler and Anderson (1994), from Haystack Mountain on the eastern side of Owens Valley, in what is now a Mojave creosote bush scrub community (1,155 m elevation) (Fig. 1B). The full-glacial (Wisconsin) portion of these middens (ca. 22.9 ka to ca. 20.6 ka) is dominated by Utah juniper (*J. osteosperma*) and piñon pine (*Pinus monophylla*), associated with green joint-fir (*Ephedra viridis*) and Great Basin sagebrush (*A. tridentata*). The mesic Rocky Mountain juniper (*J. scopulorum*) is present in the late glacial portion of these middens, with its first appearance being at ca. 17.7 ka and its apparent last appearance being at ca. 14.9 ka, associated with Utah juniper (*J. osteosperma*), piñon pine (*P. monophylla*, ca. 17.7 ka), and green joint-fir (*E. viridis*, ca. 16 ka to ca. 14.9 ka). Middens TG-FA-2 (9.54 ka),

LC-SE (9.46 ka), and C-1 (8.7 ka) of Koehler and Anderson (1995) document the early Holocene warming, with Utah juniper disappearing from the southern part of the Alabama Hills (1,264 m elevation) by 9.46 ka and from the northern part (1,460 m elevation) by ca. 7.99 ka. Jennings (1988) and Jennings and Elliott-Fisk (1993) likewise documented that Utah juniper inhabited lower elevations (1,341 m elevation) in the northern part of Owens Valley as late as ca. 9.8 ka, on the basis of macrofossil evidence in two middens from the Volcanic Tablelands ~120 km north of Owens Lake. Their middens were radiocarbon dated at ca. 19.3 ka and ca. 9.8 ka (samples 3C and 2, respectively). The older, late Wisconsin midden (3C) was dominated by *J. osteosperma* and associated with an understory of *Purshia* and *Tetradymia*. The younger, early Holocene midden (sample 2) also was dominated by *J. osteosperma* but had a different understory component (*Artemisia, Chamaebatiaria, Prunus,* and *Ribes*). None of the taxa in midden 3C grow at that site today; both midden sites presently are located in modern desert scrub communities.

The OL-92 pollen record exhibits a pattern similar to that documented in these local midden records. Juniper frequency in OL-92 is low (<25% of the terrestrial component, exclusive of pine) during the middle Wisconsin (ca. 31 ka) but increases to a maximum (>60%) by ca. 24 ka and remains high as late as ca. 15 ka, after which time it decreases rapidly. (Ages suggested here for OL-92 pollen samples are constrained by the AMS radiocarbon dates of Bischoff et al. (1993).) By ca. 10 ka, juniper is dramatically decreased (<10%), and chenopods, composites, and big sagebrush are markedly more abundant (Figs. 3, 4), suggesting the presence of sagebrush steppe, sagebrush scrub, or desert scrub vegetation at the edge of Owens Lake at that time. The site of OL-92 thus had become significantly warmer by ca. 9.5 ka, and the lower edge of the juniper woodland most likely was located somewhere between the Owens Lake core site (1,084 m elevation) and the Volcanic Tableland 2 site (1,341 m elevation). Juniper occurrences in the northern part of the Alabama Hills already may have become isolated stands in the valley by this time (ca. 10 ka), as juniper pollen comprises less than 5% of the terrestrial component (exclusive of pine) in correlative OL-92 samples. Both the pollen evidence from OL-92 and the midden evidence from Owens Valley sites are consistent with evidence for late Pleistocene deglaciation (<ca. 15 ka) documented on the western flank of the Sierra Nevada by Smith and Anderson (1992). The OL-92 pine, juniper and fir records also are consistent with the suggestion of Phillips et al. (1990) that maximum Wisconsin glacial advance occurred prior to ca. 21 ka and perhaps as long ago as ca. 24 ka, dates that they determined on the basis of cosmogenic ^{36}Cl studies of Sierran glacial moraines.

Van Devender et al. (1987) and Jennings and Elliot-Fisk (1993) previously had suggested that piñon likely was not broadly associated with juniper woodland east of the Sierra Nevada during the late Pleistocene, owing possibly to rain-shadow effects along the lee slope of the mountain front. The presence of piñon and the mesophyte *J. scopulorum* in the Haystack Mountain middens of Koehler and Anderson (1994) at ca. 17.7 ka suggests, however, that pluvial highstands of Pleistocene Owens Lake may have had a strong moderating influence on the composition of local vegetation.

Owing to the rarity of valley floor midden sites, conflicting opinions exist on the nature of the vegetation occupying the valley floors of the Great Basin during glacial periods, specifically during the Wisconsin. Wells (1983) postulated that subalpine conifers probably formed a more or less continuous zone across the Great Basin during full glacial climate intervals. Thompson and Meade (1982) and Thompson (1990) disagreed and suggested rather that most subalpine conifer taxa likely had not colonized the relatively finer grained alluvial and lacustrine substrates of the Great Basin valley floors during the late Wisconsin but were restricted to the coarser, shallow rocky soils and bedrock outcrops along the lower mountain slope and to isolated bedrock outcrops along the valley floor. They further suggested that the valley floors may not even have been occupied by juniper woodland but were occupied by sagebrush and other shrubs. Pine and *Artemisia* pollen records from OL-92 provide only tentative evidence in support of their ideas. For example, we note in the OL-92 record that pine minima in the upper part of the core generally coincide with "cooler" pollen zones and subzones ("glacial" intervals, Fig. 4). We interpret this trend to be the result of (1) juniper providing a stronger, more local signal near the lake during these times, the result being a dilution of the pine component, and (2) displacement of montane-subalpine forest taxa from higher elevations and from canyon heads by alpine glaciers and moraines during glacial intervals. If pines had successfully colonized the fine-grained valley floors during glacials, their local abundance likely would have contributed a larger pollen component during these intervals than those we have observed. Furthermore, no macrofossil evidence of pines was noted in Wisconsin middens from the Alabama Hills bedrock outcrops by Koehler and Anderson (1995); pine also was greatly reduced in their pollen spectra from the Wisconsin portion of these middens. Despite the fact that *Artemisia* approached maximal values during parts of the Wisconsin (~20% of the terrrestrial component, exclusive of pine), its maxima do not consistently coincide with cooler intervals in core OL-92, as would be expected if sagebrush were the dominant valley floor vegetation during glacial periods. At present, it appears to us more likely that juniper frequently occupied significant portions of the lower slopes and upper valley floor of Owens Valley during glacial intervals.

Comparison of midden pollen and OL-92 pollen trends. Relatively few suitably long or well-dated pollen records exist through the ca. 200 ka to ca. 5 ka time interval for comparison with OL-92. As is the case for midden macrofossil records, pollen records from the western Great Basin and Sierra Nevada frequently are younger than ca. 12 ka or are geographically distant from the Owens Lake site. Some of the better documented pollen records from the southern Sierra Nevada (e.g., Adam, 1967; Davis et al., 1985; Anderson et al., 1985) noted increased aridity in the region during the late Pleistocene to early Holocene transition, but the bulk of these records mostly postdates the OL-92 record.

The pollen samples isolated by Koehler and Anderson (1995) from middens in the Alabama Hills provide perhaps the most suitable record for comparison with OL-92 (from ca. 32 ka to ca. 9 ka). The pollen record and macrofossil record of the Alabama Hills middens and the OL-92 pollen record show coincident trends in the major pollen components. Specifically, pine minima are noted at ca. 26 ka and ca. 21 ka (Wisconsin glacial), with more abundant pine in these records from ca. 11.5 ka to ca. 10 ka (late Pleistocene to early Holocene warming). Juniper increases markedly in both midden and core records from ca. 32 ka to ca. 21 ka (middle Wisconsin through late Wisconsin glaciation) and attains a maximum in both at ca. 21 ka (a date that provides additional evidence for maximum Wisconsin glaciation before ca. 20 ka). Likewise, juniper decreases in the two records from ca. 20 ka to ca. 10 ka (the sharpest decline starting ca. 15 ka), and thus indicates a relatively early transition to postglacial conditions. The OL-92 pollen record, although preliminary in nature, therefore indicates the same vegetation trends and the same chronology as those documented from the late Pleistocene and early Holocene midden records (and the pollen spectra derived from them).

Other pollen records also document vegetation changes that are concurrent with those of OL-92 and the Alabama Hills, although they are geographically distant from Owens Valley. One of the best-documented records for comparison is that of Adam (1988), from two closely sampled pollen records from Clear Lake in northern California. Clear Lake, located in the Coast Ranges of northern California and about 550 km northwest of Owens Lake, is presently surrounded by oak woodland. Accordingly, we believe that the frequency response of the *Quercus* spp. in core 4 from Clear Lake best reflects the climate signal from ca.120+ ka to ca. 6 ka. Adam designated warm and cold climatic intervals in core 4 as *thermomers* and *cryomers*, respectively, and named each climatic episode after Native American villages that once surrounded Clear Lake. Although the apparent absolute ages of the climatic features in the Clear Lake core 4 and OL-92 pollen records do not coincide completely, we believe that the climatic trends probably were coincident and that apparent age differences may be largely an effect of different dating methods and relative age models. In the younger part of core 4, the Tuleyome thermomer appears to correlate with our zone P1 (Fig. 4), the Cigom 2 cryomer appears to correspond with our P2, the Halika thermomers appear to correspond to our zone P3, and the Cigom 1 cryomer appears to correspond to our zone P4. In the lower half of Clear Lake core 4, the thermomers designated Boomli 4 and Boomli 3-1 appear to correspond to our zones P5a and P5c, respectively. The Konocti thermomer and Tsabal cryomer appear to correspond to zone P5e and P6 in OL-92.

Other long pollen records, several of which are geographically closer to Owens Valley, will need more detailed age control before similar comparisons can be made. For example, Atwater et al. (1986) documented the vegetation change in the late Pleistocene to Holocene Tulare Lake, which is in the San Joaquin Valley and approximately 150 km west of Owens Lake. In the Tulare Lake core, the juniper ("TCT" or Taxodiaceae-Cupressaceae-Taxaceae) component appears to parallel the response to climate change seen in the juniper component from OL-92. Similarly, Martin (1963a) documented an apparently longer Pleistocene pollen record from pluvial Lake Cochise in the Wilcox Playa of southeastern Arizona, but the respective age constraints of both of these records are not sufficiently established at present to support a strong comparison to OL-92.

Comparison of isotopic and OL-92 pollen trends. Beyond ca. 50 ka in core OL-92 opportunities for comparison of our pollen record to other paleobotanical records become more limited (except for the diatom record from OL-92) (Bradbury, this volume). In their absence, geochemical evidence can be used as a proxy baseline reference through this interval. The $\delta^{18}O$ record from Devils Hole (core DH-11) (Winograd et al., 1992) has a significant temporal overlap with both the OL-92 pollen record and the $\delta^{18}O$ record constructed for OL-92 by Benson and Bischoff (1993) and can be compared with each as a relatively well-constrained, baseline proxy climate record. We use the Devils Hole record of Winograd et al. (1992) instead of the SPECMAP record of Imbrie et al. (1984) as a basis for comparison for three reasons: (1) the Devils Hole record is geographically close (~130 km to the east), (2) it is a continental (nonmarine) isotopic record, and (3) it is internally calibrated by mass spectrometric uranium-series dating.

Figure 4 plots the younger portions of the DH-11 isotopic record and the pollen and isotopic records of OL-92, from ca. 200 ka to ca. 9 ka. The frequency curves of pines and juniper are plotted as proxy indicators of montane-subalpine coniferous forest and woodland zones, respectively, and the DH-11 isotopic record is plotted as a proxy indicator of climate (temperature) (Winograd et al., 1992). The OL-92 $\delta^{18}O$ record of Benson and Bischoff (1993) has been modified here; it has been scaled to the age model of Bischoff (1993) and Bischoff et al. (this volume, chapter 8), and converted to standard deviations. We focus on the long-term vegetational trends of these records by smoothing the pollen record with three-point sample averaging, to reduce single sample effects (short-term trends) and enhance the signal of long-term trends. Additionally, we calculate the juniper relative frequency curve as a percentage of the terrestrial pollen component, exclusive of pine.

In the youngest OL-92 samples (ca. 45 ka to ca. 9 ka), we were able to use the following criteria as our basis for interpretation: (1) the present distribution of pines along the eastern flank of the Sierra Nevada, (2) the presence and abundance of juniper macrofossils in glacial-age midden records from Owens Valley, and (3) the relative frequency of juniper pollen in glacial-age intervals of the midden and OL-92 pollen spectra. For samples older than 45 ka, we rely on a fourth criterion to support interpretation of the pollen record: the correlation of ^{18}O high relative abundance with warmer winter-spring temperatures and ^{18}O low relative abundance with cooler winter-spring temperatures for the DH-11 and OL-92 isotopic records (Winograd et al., 1992). On the basis of these criteria, we recognize seven pollen assemblage zones (P1–P7) through this upper geologic interval (ca. 200 ka to ca. 45 ka); even-numbered zones (P2, P4, P6) indicate major expansion of juniper woodland downslope into Owens Valley, a decline in montane pine forest at higher elevations along the eastern flank of the southern Sierra Nevada, and predominantly full-glacial conditions, and odd-numbered zones (P1, P3, P5, P7) indicate periods of relative decline in juniper in the vicinity of the OL-92 core site, an increase in montane pine forest at higher elevations, and predominantly full-interglacial conditions.

We identified terminations I and II in the OL-92 record as large-amplitude, relatively short duration increases in pine frequency and (conversely) large-amplitude decreases in juniper frequency (Fig. 4, passing from glacial to interglacial conditions). In general, the three chronologies (Owens Lake, Devils Hole, SPECMAP) appear to show much similarity, a notable exception being the (contested) absolute-age placement of termination II.

Given the reconnaissance-level resolution of our sample set and the current age model of Bischoff (1993) and Bischoff et al. (this volume, chapter 8), the relative age of termination II appears to be somewhat younger in the OL-92 record (ca. 125 ka) than it is in the isotope record from Devils Hole (ca. 140 ka). This age discrepancy may be partly real, because these records reflect responses to different environmental parameters (G. I. Smith, 1995, oral communication; I. J. Winograd, 1995, oral communication). Imbrie et al. (1984, 1993) favored placing termination II at ca.128 ka for the SPECMAP stacked isotopic record. Others such as Szabo et al. (1994) suggested that the age of termination II must exceed ca. 131 ka (an intermediate placement with respect to OL-92 and DH-11) by at least several thousand years, on the basis of ^{230}Th ages they derived from samples of now-emergent coral reefs that developed during the last sea-level high stand after this deglaciation and that now fringe Oahu, Hawaii. Crowley and Kim (1994) suggested a similar older age for this termination (ca. 134 ka) and noted the variance between termination ages proposed by Imbrie et al. (1984) and those proposed by Gallup et al. (1994) and Chen et al. (1991). The time differences for termination II in these records also must reflect differences in their respective age models/age control. Because the Devils Hole record and the Hawaiian corals were directly dated by thermal ionization mass spectrometry (TIMS) and the OL-92 ages are calculated on the basis of less directly dated criteria (Bischoff et al., this volume, chapter 8), we must consider present placement and absolute age assignment of glacial terminations in the OL-92 record to be preliminary. The OL-92, SPECMAP, and DH-11 records nonetheless remain similar in many trends; for example, pollen zone P5 of OL-92 and stage 5 of SPECMAP and DH-11 can be divided into three subintervals of decreasing warmth and two interspersed subintervals of cooling (pollen subzones P5a-P5e versus DH-11 substages 5.1-5.5). In OL-92, these zonal subdivisions are recorded best in the pine and juniper (TCT) frequency curves (Figs. 3, 4; see explanation of TCT in Fig. 3 caption). The present sample control of the OL-92 record currently precludes resolution of other important features, such as the duration of pollen subzone P5e. The duration of its counterpart, isotopic substage 5.5, initially was estimated at approximately 10 k.y. (Imbrie et al., 1984). Other more recent estimates (e.g., Lorius et al., 1985; Lambeck and Nakada, 1992; Winograd et al., 1992; Szabo et al., 1994) propose extending its duration significantly, to a minimum of ~17 k.y. Further pollen sampling in this interval is necessary before an accurate estimate of the duration of zone P5e can be made.

The current $\delta^{18}O$ isotope curve derived from OL-92 sediments also is plotted here with the pine and juniper frequency curves (Fig. 4). It is not as intensively sampled as the DH-11 $\delta^{18}O$ isotope curve; however, present evidence suggests a tentative position for termination II at ca. 130 ka. Although this placement appears to suggest a delay between the $\delta^{18}O$ and vegetation records, we consider the data to be preliminary and defer any conclusions until further analyses are completed. In general, major excursions in the OL-92 $\delta^{18}O$ isotope curve approximate those in the DH-11 record, and provide additional (internal) evidence for correlating increases in juniper pollen in the OL-92 record with cooler temperatures.

800–200 ka: Pollen and $\delta^{18}O$ evidence

Few paleobotanical records older than ca. 200 ka are available for comparison with the terrestrial plant record in OL-92, the exception (noted earlier) being diatoms (Bradbury, 1993, this volume). Because the OL-92 pollen record closely matched the pollen trends documented in the Alabama Hills midden record of Koehler and Anderson (1995) and because the $\delta^{18}O$ record from OL-92 corroborates these trends and compares closely with $\delta^{18}O$ trends in DH-11, we zoned the lower portion of the core (ca. 800 ka to ca. 200 ka) primarily on the basis of juniper and pine frequency and cross-checked the zonation against the $\delta^{18}O$ records of OL-92 and DH-11. As before, other taxa (e.g., composites, cheno-ams, fir) were used in establishing zone boundaries. As a result, we recognize 13 additional pollen zones (P8 to ?P20) in this lower interval (Fig. 5), for a total of 10 cold-to-warm climate shifts in the core during the Brunhes Normal Polarity Chron. Juniper relative frequency correlates with cold (mostly glacial) intervals moderately well as far back as the base of pollen zone P15. Below this level (~226 m depth), the pattern of the juniper response in OL-92 changes and the pine and composite frequency curves appear to define pollen assemblage zone boundaries more clearly. This change in the response of the juniper record in OL-92 approximately coincides with a change from dominantly finer grained clastics above to dominantly coarser grained clastics below ~220 m depth (Menking et al., 1993; Menking, this volume; Smith, this volume).

We have identified terminations III, IV, and V in the middle portion of OL-92 (Fig. 5), although comparison of the DH-11 isotopic record with the OL-92 isotopic and pollen records suggests different absolute ages for them. As before, these age discrepancies may be real or may be (partly or completely) an artifact of the respective age models for the two cores; further work is in progress.

Throughout the length of OL-92, there appears to be a close correlation between the $\delta^{18}O$ record and the pine-pollen frequency record. This similarity suggests that changes in the vegetation of the study area appear to have been related to the same climatic mechanism(s) responsible for isotopic changes recorded in cores OL-92 and DH-11. The evidence also preliminarily suggests that the vegetation shifts represented in the OL-92 pollen record may be closely related to the nature and duration of the marine isotope stages, as determined from the $\delta^{18}O$ variations preserved in the calcareous tests of benthic foraminifera from marine sediments. Differences in the ages of these trends may reflect relative age control and relative age models as well as differences introduced by iterative adjustments of the marine record to orbital cycles. Evidence for similarity in the marine isotopic and OL-92 pollen trends can be seen by comparing two alternative placements for the boundary between the Matuyama and Brunhes paleomagnetic chrons in core OL-92, determined separately on the basis of paleomagnetic and palynological evidence. Glen and Coe (this volume) tentatively place the Matuyama-Brunhes boundary (MBB) in OL-92 at either 320.2 m or 317.6 m, depending on paleomagnetic criteria used; Glen et al. (1993) previously placed it at 320.2 m depth. They assigned it an age of 783 ± 11 ka on the basis of work done on the chron boundary by Baksi et al. (1992) from other sites. This age correlates closely with that recently proposed by Shackleton et al. (1990), which was based on astronomical calibration. Although the age model of Bischoff (1993) appears to suggest an older fixed age for this depth (Fig. 5), he cautioned that depth-age relationships below the Bishop ash bed (ca. 760 ka) in the lower part of the core are much less well constrained than those in younger parts of the core, because deposition of the ash in Owens Lake, plus the ~5 m of post-airfall mudflows caused by it, may have been very rapid (J. L. Bischoff, 1995, oral communication). The MBB placement of Glen and Coe (this volume) falls within this interval of uncertain age.

As a test, we used the base of pollen zone P19 to approximate the chron boundary separately, by analogy to the marine record, for which the base of oxygen isotope stage 19 occurs at or just below the MBB. Our estimation of the MBB (and our placement of all assemblage zone boundaries) was made by examining pollen taxon percentages prior to three-point averaging. On this basis, our expected placement of the MBB (below ~316 m depth) reasonably approximates its placement (317.6 m or 320.2 m) on the basis of paleomagnetic evidence. Given the reconnaissance-level spacing of our pollen samples and problems in placing the MBB on the basis of the OL-92 paleomagnetic signal (Glen and Coe, this volume), we consider this first-order correlation of the OL-92 pollen record to the marine isotopic record to be reasonable.

Ideally, the relationship between the continental pollen record and the global $\delta^{18}O$ record should be established by using pairs of samples from a single section or core, but such parallel records are rare (e.g., Heusser and Shackleton, 1979; Heusser and Heusser, 1990). More often, long continental pollen records must be correlated with marine records by using combinations of (sometimes sparse) independent control points, such as magnetic reversals and excursions, tephra, and dated horizons as well as similarities in trends ("curve matching") among independent climate-response records (Adam, 1988; Hooghiemstra and Sarmiento, 1991). Because the mass accumulation rates in OL-92 greatly exceed the mass accumulation rates in marine sediments, it appears that our continuing work on OL-92 ultimately may attain or, in some intervals, exceed the level of temporal resolution present in the SPECMAP marine cores; it is hoped that core OL-92 will provide a level of resolution similar to that now established in the European continental pollen record (e.g., Woillard and Mook, 1982; Guiot et al., 1989).

CONCLUSIONS AND SUMMARY

Core OL-92 contains a long, nearly continuous terrestrial pollen record that documents the major vegetation responses to climate change during the Brunhes for the eastern flank of the southern Sierra Nevada and south-central Owens Valley. This paleovegetation record is constrained in age by AMS radiocarbon dates in its upper portion, $^{40}Ar/^{39}Ar$ dates on the Bishop ash bed in the lower part of the core, and the position of the MBB near the base of the core. Pollen spectra in OL-92 have been assigned here to 20(?) preliminary pollen assemblage zones on the basis of major shifts in taxon frequency and composition; these shifts perhaps are best illustrated by the frequency curves of pine, juniper, and the composites. These pollen curves appear to correspond to the same temperature shifts represented in the terrestrial isotopic record from Devils Hole and may also correspond to the oxygen isotope curves of the marine record. At least six of these zones (P2, P4, P6, P8, P10, P12) appear to represent glacial conditions; likewise, at least six other zones (P1, P3, P5, P7, P9, P11) appear to represent interglacial conditions.

The OL-92 pollen record indicates that the distribution and composition of the forest and woodland zones along the eastern flank of the Sierra Nevada have evolved in a dynamic fashion throughout the Brunhes. Because each modern plant species responds to climatic change uniquely, the overall compositions of the plant communities along the eastern flank of the Sierra Nevada undoubtedly have changed through this time interval. Fossil pollen evidence also suggests that warm periods and cold periods varied in both intensity and duration during the Brunhes; for example, the replacement of pine by juniper (TCT) is much better developed in zones P2 and P6 than it is elsewhere in the core, and the combination of high juniper (TCT) frequency and high *Quercus* frequency in zone P16 is likewise unusual. In general terms, however, vegetation belts probably were displaced to lower elevations during cooler intervals and to higher elevations during warmer intervals of the Brunhes. Cooler and (or) wetter intervals may have experienced significant reductions in the total vegetated area on the eastern flank of the Sierra Nevada as a result of expansions of both Owens Lake and Sierran alpine glaciers. However, the OL-92 pollen record suggests that not all cool periods necessarily produced glaciation in the southern Sierra Nevada.

Relatively few terrestrial pollen records in the Northern

Hemisphere span such a long interval of the Brunhes. Accordingly, OL-92 likely will become an important record for comparative studies of Quaternary terrestrial vegetation response in the western United States, along with records from Clear Lake (Adam et al., 1981; Adam, 1988) and Tulelake (Adam et al., 1989), as well as long terrestrial records from other continents: from Funza, Colombia (Hooghiemstra, 1989; Hooghiemstra and Sarmiento, 1991; Hooghiemstra and Melice, 1994); Grande Pile and Les Echets, France (Woillard, 1978; Woillard and Mook, 1982; de Beaulieu and Reille, 1984; Guiot et al., 1989; de Beaulieu and Reille, 1992; Guiot et al., 1992; Seret et al., 1992; Guiot et al., 1993); the Dead Sea, Israel (Horowitz, 1989); and Lake Biwa, Japan (Fuji and Horowitz, 1989). More detailed study of this exceptionally well-preserved long terrestrial pollen record is in progress.

ACKNOWLEDGMENTS

We acknowledge with sincere thanks the help of G. I. Smith, J. L. Bischoff, R. Z. Poore, P. E. Wigand, T. M. Cronin, I. J. Winograd, R. S. Thompson, J. P. Bradbury, O. K. Davis, and J. T. Barndt during various stages of this study and preparation of this manuscript. Additionally we thank V. A. Andrle, T. P. Sheehan, and N. J. Durika for processing samples examined for this report. This contribution is part of the U.S. Geological Survey Global Change and Climate History Program.

REFERENCES CITED

Adam, D. P., 1967, Late-Pleistocene and Recent palynology in the Central Sierra Nevada, California, *in*: Cushing, E. G., and Wright, H. E., Jr., eds., Quaternary paleoecology: New Haven, Connecticut, Yale University Press, p. 275–301.

Adam, D. P., 1988, Palynology of two Upper Quaternary Cores from Clear Lake, Lake County, California: U.S. Geological Survey Professional Paper 1363, 86 p.

Adam, D. P., Sims, J. D., and Throckmorton, C. K., 1981, 130,000-yr continuous pollen record from Clear Lake, Lake County, California: Geology, v. 9, p. 373–377.

Adam, D. P., Sarna-Wojcicki, A. M., Rieck, H. J., Bradbury, J. P., Dean, W. E., and Forester, R., 1989, Tulelake, California: The last 3 million years: Palaeogeography, Palaeoclimatology, Palaeoecology, v. 72, p. 89–103.

Anderson, R. S., Davis, O. K., and Fall, P. L., 1985, Late glacial and Holocene vegetation and climate in the Sierra Nevada of California, with particular reference to the Balsam Meadow site, *in* Jacobs, B. F., Fall, P. L., and Davis, O. K., eds., Late Quaternary vegetation and climates of the American Southwest: American Association of Stratigraphic Palynologists, Contribution Series, no. 16, p. 127–140.

Atwater, B. F., and 8 others, 1986, A fan dam for Tulare Lake, California, and implications for the Wisconsin glacial history of the Sierra Nevada: Geological Society of America Bulletin, v. 97, p. 97–109.

Baksi, A. K., Hsu, V., McWilliams, M. O., and Farrar, E., 1992, ^{40}Ar/^{39}Ar dating of the Brunhes-Matuyama geomagnetic field reversal: Science, v. 256, p. 356–357.

Benson, L., and Bischoff, J. L., 1993, Isotope geochemistry of Owens Lake core OL-92, *in* Smith, G. I., and Bischoff, J. L., eds., Core OL-92 from Owens Lake, southeast California: U.S. Geological Survey Open-File Report 93-683, p. 106–109.

Bischoff, J. L., 1993, Age-depth relations for the sediment column at Owens Lake, California, *in* Smith, G. I., and Bischoff, J. L., eds., Core OL-92 from Owens Lake, southeast California: U.S. Geological Survey Open-File Report 93-683, p. 251–260.

Bischoff, J. L., Stafford, T. W., Jr., and Rubin, M., 1993, AMS radiocarbon dates on sediments from Owens Lake Drill Hole OL-92, *in* Smith, G. I., and Bischoff, J. L., eds., Core OL-92 from Owens Lake, southeast California: U.S. Geological Survey Open-File Report 93-683, p. 246–250.

Bradbury, J. P., 1993, Diatoms in sediments, *in* Smith, G. I., and Bischoff, J. L., eds., Core OL-92 from Owens Lake, southeast California: U.S. Geological Survey Open-File Report 93-683, p. 261–302.

Chen, J. H., Curran, H. A., White, B., and Wasserburg, G. J., 1991, Precise chronology of the last interglacial period: ^{234}U-^{230}Th data from fossil coral reefs in the Bahamas: Geological Society of America Bulletin, v. 103, p. 82–97.

Crowley, T. J., and Kim, K.-Y., 1994, Milankovitch forcing of the last interglacial sea level: Science, v. 265, p. 1566–1568.

Davis, O. K., Anderson, R. S., Fall, P. L., O'Rourke, M. K., and Thompson, R. S., 1985, Palynological evidence for early Holocene aridity in the Southern Sierra Nevada, California: Quaternary Research, v. 24, p. 322–332.

de Beaulieu, J.-L., and Reille, M., 1984, A long Upper Pleistocene pollen record from Les Echets, near Lyon, France: Boreas, v. 13, p. 111–132.

de Beaulieu, J.-L., and Reille, M., 1992, The last climatic cycle at La Grande Pile (Vosges, France): A new pollen profile: Quaternary Science Reviews, v. 11, p. 431–438.

Ferguson, C. W., 1969, A 7104-year annual tree ring chronology for bristlecone pine, *Pinus aristata*, from the White Mountains, California: Tree-Ring Bulletin, v. 29, p. 1–29.

Fuji, N., and Horowitz, A., 1989, Brunhes epoch paleoclimates of Japan and Israel: Palaeogeography, Palaeoclimatology, Palaeoecology, v. 72, p. 79–88.

Gallup, C. D., Edwards, R. L., and Johnson, R. G., 1994, The timing of high sea levels over the past 200,000 years: Science, v. 263, p. 796–800.

Glen, J. M., Coe, R. S., Menking, K. M., Boughn, S. S., and Altschul, I., 1993, Rock- and paleo-magnetic results from Core OL-92, Owens Lake, California, *in* Smith, G. I., and Bischoff, J. L., eds., Core OL-92 from Owens Lake, southeast California: U.S. Geological Survey Open-File Report 93-683, p. 127–183.

Guiot, J., Pons, A., de Beaulieu, J.-L., Reille, M., 1989, A 140,000-year continental climate reconstruction from two European pollen records: Nature, v. 338, p. 309–313.

Guiot, J., Reille, M., de Beaulieu, J.-L., and Pons, A., 1992, Calibration of the climatic signal in a new pollen sequence from La Grande Pile: Climate Dynamics, v. 6, p. 259–264.

Guiot, J., Reille, M., de Beaulieu, J.-L., Cheddadi, R., David, F., Ponel, P., and Reille, M., 1993, The climate in western Europe during the last glacial/interglacial cycle derived from pollen and insect remains: Palaeogeography, Palaeoclimatology, Palaeoecology, v. 103, p. 73–93.

Heusser, C. J., and Heusser, L. E., 1990, Long continental pollen sequence from Washington State (U.S.A.): Correlation of upper levels with marine pollen-oxygen isotope stratigraphy through substage 5e: Palaeogeography, Palaeoclimatology, Palaeoecology, v. 79, p. 63–71.

Heusser, L. E., and Shackleton, N. J., 1979, Direct marine-continental correlation: 150,000-year oxygen isotope-pollen record from the North Pacific: Science, v. 204, p. 837–839.

Hevly, R. H., 1985, A 50,000 year record of Quaternary environments; Walker Lake, Coconino Co., Arizona, *in* Jacobs, B. F., Fall, P. L., and Davis, O. K., eds., Late Quaternary vegetation and climates of the American Southwest: American Association of Stratigraphic Palynologists, Contribution Series, no. 16, p. 141–154.

Hooghiemstra, H., 1989, The orbital-tuned marine oxygen isotope record applied to the Middle and Late Pleistocene pollen record of Funza (Columbian Andes): Palaeogeography, Palaeoclimatology, Palaeoecol-

ogy, v. 66, p. 9–17.
Hooghiemstra, H., and Melice, J. L., 1994, Pleistocene evolution of orbital periodicities in the high-resolution pollen record Funza I, Eastern Cordillera, Columbia, *in* de Boer, P. L., and Smith, D. G., eds., Orbital forcing and cyclic sequences: Special Publications of the International Association of Sedimentologists, v. 19, p. 117–126.
Hooghiemstra, H., and Sarmiento, G., 1991, Long continental pollen record from a tropical intermontane basin; late Pliocene and Pleistocene history from a 540-meter core: Episodes, v. 14, p. 107–115.
Horowitz, A., 1989, Continuous pollen diagram for the last 3.5 M.Y. from Israel: Vegetation, climate and correlation with the oxygen isotope record: Palaeogeography, Palaeoclimatology, Palaeoecology, v. 72, p. 63–78.
Imbrie, J., and 8 others, 1984, The orbital theory of Pleistocene climate: Support from a revised chronology of the marine ^{18}O record, *in* Berger, A., Imbrie, J., Hays, J., Kukla, G., and Saltzman, B., eds., Milankovitch and climate—Understanding the response to astronomical forcing: Boston, Massachusetts, I. Reidel Publishing Company, p. 269–305.
Imbrie, J., and 18 others, 1993, On the structure and origin of major glaciation cycles: 2. The 100,000-year cycle: Paleoceanography, v. 8, p. 699–735.
Izett, G. A., Obradovich, J. D., and Mehnert, H. H., 1988, The Bishop Ash bed (Middle Pleistocene) and some older (Pliocene and Pleistocene) chemically and mineralogically similar ash beds in California, Nevada, and Utah: U.S. Geological Survey Bulletin 1675, 37 p.
Jennings, S. A., 1988, Late Quaternary vegetation change in the White Mountain Region, *in* Hull, C. A., Jr., and Doyle-Jones, V., eds., Plant biology of eastern California. Natural history of the White-Inyo Range: Los Angeles, University of California, p. 139–147.
Jennings, S. A., and Elliot-Fisk, D. L., 1993, Packrat midden evidence of late Quaternary vegetation change in the White Mountains, California-Nevada: Quaternary Research, v. 39, p. 214–221.
Koehler, P. A., and Anderson, R. S., 1994, Full-glacial shoreline vegetation during the maximum highstand at Owens Lake, California: Great Basin Naturalist, v. 54, p. 142–149.
Koehler, P. A., and Anderson, R. S., 1995, Thirty thousand years of vegetation changes in the Alabama Hills, Owens Valley, California: Quaternary Research, v. 43, p. 238–248.
Küchler, A. W., 1977, The map of the natural vegetation of California, *in* Barbour, M. G., and Major, J., eds., Terrestrial vegetation of California: New York, John Wiley and Sons, p. 909–938.
LaMarche, V. C., Jr., 1974, Paleoclimatic inferences from long tree-ring records: Science, v. 183, p. 1043–1048.
Lambeck, K., and Nakada, M., 1992, Constraints on the age and duration of the last interglacial period and on sea-level variations: Nature, v. 357, p. 125–128.
Litwin, R. J., Frederiksen, N., Adam, D. P., Andrle, V. A. S., and Sheehan, T. P., 1993, Continental-marine correlation of Late Pleistocene climate change: Census of palynomorphs from core OL-92, Owens Lake, California, *in* Smith, G. I., and Bischoff, J. L., eds., Core OL-92 from Owens Lake, southeast California: U.S. Geological Survey Open-File Report 93-683, p. 333–391.
Lorius, C., Jouzel, J., Ritz, C., Merlivat, L., Barkov, N. I., Korotkevich, Y. S., and Kotlyakov, V. M., 1985, A 150,000-year climatic record from Antarctic ice: Nature, v. 316, p. 591–595.
Madsen, D. B., 1985, Two Holocene pollen records from the central Great Basin, *in* Jacobs, B. F., Fall, P. L., and Davis, O. K., eds., Late Quaternary vegetation and climates of the American Southwest: American Association of Stratigraphic Palynologists, Contribution Series, no. 16, p. 113–126.
Major, J., and Taylor, D. W., 1977, Alpine, *in* Barbour, M. G., and Major, J., eds., Terrestrial vegetation of California: New York, John Wiley and Sons, p. 601–675.
Martin, P. S., 1963a, Geochronology of pluvial Lake Cochise, southern Arizona. II. Pollen analysis of a 42-meter core: Ecology, v. 44, p. 436–444.
Martin, P. S., 1963b, The last 10,000 years: A fossil pollen record of the American Southwest: Tucson, University of Arizona Press, 78 p.
Martin, P. S., and Mehringer, P. J., 1965, Pleistocene pollen analysis and biogeography of the Southwest, *in* Wright, H. E., Jr., and Frey, D. G., eds., The Quaternary of the United States: Princeton, New Jersey, Princeton University Press, p. 433–451.
Martinson, D. G., Pisias, N. G., Hays, J. D., Imbrie, J., Moore, T. C., and Shackleton, N. J., 1987, Age dating and the orbital theory of the ice ages: Development of a high-resolution 0 to 300,000-year chronostratigraphy: Quaternary Research, v. 27, p. 1–30.
Menking, K. M., Musler, H. M., Fitts, J. P., Bischoff, J. L., and Anderson, R. S., 1993, Sediment size analyses of the Owens Lake core, *in* Smith, G. I., and Bischoff, J. L., eds., Core OL-92 from Owens Lake, southeast California: U.S. Geological Survey Open-File Report 93-683, p. 58–74.
Muller, J., 1959, Palynology of the Recent Orinoco Delta and shelf sediments: Micropaleontology, v. 5, p. 1–32.
Phillips, F. M., Zreda, M. G., Smith, S. S., Elmore, D., Kubik, P. W., and Sharma, P., 1990, Cosmogenic chlorine-36 chronology for glacial deposits at Bloody Canyon, eastern Sierra Nevada: Science, v. 248, p. 1529–1532.
Rundel, P. W., Parsons, D. J., and Gordon, D. T., 1977, Montane and subalpine vegetation of the Sierra Nevada and Cascade Ranges, *in* Barbour, M. G., and Major, J., eds., Terrestrial vegetation of California: New York, John Wiley and Sons, p. 559–599.
Sarna-Wojcicki, A. M., and Pringle, M. S., Jr., 1992, Laser-fusion ^{40}Ar/^{39}Ar ages of the tuff of Taylor Canyon and the Bishop Tuff, E. California–W. Nevada [abs.]: Eos (Transactions, American Geophysical Union), v. 73, p. 633.
Sarna-Wojcicki, A. M., Meyer, C. E., Wan, E., and Soles, S., 1993, Age and correlation of tephra layers in Owens Lake drill Core OL-92-1 and -2, *in* Smith, G. I., and Bischoff, J. L., eds., Core OL-92 from Owens Lake, southeast California: U.S. Geological Survey Open-File Report 93-683, p. 184–245.
Seret, G., Guiot, J., Wansard, G., de Beaulieu, J.-L., and Reille, M., 1992, Tentative paleoclimatic reconstruction linking pollen and sedimentology in La Grande Pile (Vosges, France): Quaternary Science Reviews, v. 11, p. 425–430.
Shackleton, N. J., Berger, A., and Peltier, W. R., 1990, An alternative astronomical calibration of the lower Pleistocene timescale based on ODP Site 677: Transactions of the Royal Society of Edinburgh: Earth Sciences, v. 81, p. 251–261.
Smith, G. I., 1993, Field log of Core OL-92, *in* Smith, G. I., and Bischoff, J. L., eds., Core OL-92 from Owens Lake, southeast California: U.S. Geological Survey Open-File Report 93-683, p. 1–53.
Smith, S. J., and Anderson, R. S., 1992, Late Wisconsin paleoecologic record from Swamp Lake, Yosemite National Park, California: Quaternary Research, v. 38, p. 91–102.
Solomon, A. M., and Silkworth, A. B., 1986, Spatial patterns of atmospheric pollen transport in a montane region: Quaternary Research, v. 25, p. 150–162.
Spaulding, W. G., 1985, Vegetation and climates of the last 45,000 years in the vicinity of the Nevada Test Site, south-central Nevada: U.S. Geological Survey Professional Paper 1329, p. 1–83.
Szabo, B. J., Ludwig, K. R., Muhs, D. R., and Simmons, K. R., 1994, Thorium-230 ages of corals and duration of the last interglacial sea-level high stand on Oahu, Hawaii: Science, v. 266, p. 93–96.
Thompson, R. S., 1990, Late Quaternary vegetation and climate in the Great Basin, *in* Betancourt, J. L., Van Devender, T. R., and Martin, P. S., eds., Packrat middens. The last 40,000 years of biotic change: Tucson, University of Arizona Press, p. 200–239.
Thompson, R. S., and Mead, J. I., 1982, Late Quaternary environments and biogeography in the Great Basin: Quaternary Research, v. 17, p. 39–55.
Van Devender, T. R., 1977, Holocene woodlands in the Southwestern deserts: Science, v. 198, p. 189–192.
Van Devender, T. R., and Spaulding, W. G., 1979, Development of vegetation and climate in the Southwestern United States: Science, v. 204,

p. 701–710.

Van Devender, T. R., Thompson, R. S., and Betancourt, J. L., 1987, Vegetation history of the deserts of southwestern North America; the nature and timing of the Late Wisconsin–Holocene transition, *in* Ruddiman, W. F., and Wright, H. E., Jr., eds., North America and adjacent oceans during the last deglaciation: Boulder, Colorado, Geological Society of America, The Geology of North America, v. K-3, p. 323–352.

Vasek, F. C., and Thorne, R. F., 1977, Transmontane coniferous vegetation, *in* Barbour, M. G., and Major, J., eds., Terrestrial vegetation of California: New York, John Wiley and Sons, p. 797–832.

Wells, P. V., 1983, Paleobiogeography of montane islands in the Great Basin since the last glaciopluvial: Ecological Monographs, v. 53, p. 341–382.

Wells, P. V., and Berger, R., 1967, Late Pleistocene history of coniferous woodland in the Mohave Desert: Science, v. 155, p. 1640–1647.

Winograd, I. J., Coplen, T. B., Landwehr, J. M., Riggs, A. C., Ludwig, K. R., Szabo, B. J., Kolesar, P. T., and Revesz, K. M., 1992, Continuous 500,000-year climate record from vein calcite in Devils Hole, Nevada: Science, v. 258, p. 255–260.

Woillard, G. M., 1978, Grande Pile peat bog: a continuous pollen record for the last 140,000 years. Quaternary Research, v. 9, p. 1–21.

Woillard, G. M., and Mook, W. G., 1982, Carbon-14 dates at Grande Pile: Correlation of land and sea chronologies: Science, v. 215, p. 159–161.

Woolfenden, W. B., 1993, Pollen present in cores OL-92-1, -2 and -3, *in* Smith, G. I., and Bischoff, J. L., eds., Core OL-92 from Owens Lake, southeast California: U.S. Geological Survey Open-File Report 93-683, p. 314–332.

Young, J. A., Evans, R. A., and Major, J., 1977, Sagebrush steppe, *in* Barbour, M. G., and Major, J., eds., Terrestrial vegetation of California: New York, John Wiley and Sons, p. 763–796.

MANUSCRIPT ACCEPTED BY THE SOCIETY JUNE 17, 1996

Synthesis of the paleoclimatic record from Owens Lake core OL-92

George I. Smith and James L. Bischoff
U.S. Geological Survey, 345 Middlefield Road, MS 902 and MS 910, Menlo Park, California 94025
J. Platt Bradbury
U.S. Geological Survey, Federal Center, Box 25046, MS 919, Denver, Colorado 80225

ABSTRACT

During much of the late Quaternary, Owens Lake overflowed into one or more of four successively lower-elevation basins. Most of the water came from the high, eastern slopes of the southern Sierra Nevada, and changes in the volumes of that water reflect a dominant climatic cycle of ~100 k.y.

Variations in the inflow to, and outflow from, Owens Lake since ca. 800 ka left biological, chemical, mineralogical, and geophysical evidence in the sediments of those changes. Biological evidence includes fossil ostracodes, diatoms, fish, and mollusks (and $\delta^{18}O$ data from their shells) which indicate fresh or brackish lake water on the basis of their modern habitats. Fossil pollens indicate ~20 regional vegetation cycles during the same period. Chemical evidence of high inflow and, commonly, outflow volumes is provided by the low inorganic- and organic-C content of some sediments, reflecting short lake-water residence times; long residence times produced higher and more variable quantities of these components. Mineralogical variations in illite/smectite ratios indicate changes in weathering processes and glacial comminution. High magnetic susceptibility correlates with other criteria that indicate high runoff.

Between 810 ka and 645 ka, Owens Lake was fresh, several meters deep, and depositing silt with a few beds of sand; it supported a flora and fauna now found in fresh, sometimes very cool, waters. (Note that most geologic ages describing the OL-92 chronology have been rounded to the nearest 5 or 10 ka.) A shallow-but-freshwater lake may have been the result of accelerated sedimentation during an earlier (>800 ka) glaciation in the Sierra Nevada, choking the basin with sediment nearly to its spillway level. Between 645 ka and 450 ka, the lake was probably even shallower, depositing beds of coarse to fine sand, but overflowing periodically allowing its water to remain fresh. Between 450 ka and 5 ka, Owens Lake was mostly deep, alternating between spilling and being closed part of the time. It deposited silt and clay on its floor, yet underwent detectable variations in salinity caused by climate changes; this part of the record is the most easily interpreted and constitutes the main basis for comparing this paleoclimatic record with other long records. From 5 ka to A.D. 1913, when the Owens River was diverted into an aqueduct, Owens Lake was shallow (~2 m to ~15 m), moderately saline (~5% to <15% salts), and depositing oolites. After 1913, the lake desiccated.

Smith, G. I., Bischoff, J. L., and Bradbury, J. P., 1997, Synthesis of the paleoclimatic record from Owens Lake core OL-92, *in* Smith, G. I., and Bischoff, J. L., eds., An 800,000-Year Paleoclimatic Record from Core OL-92, Owens Lake, Southeast California: Boulder, Colorado, Geological Society of America Special Paper 317.

Comparison of the Owens Lake water-depth record with that of Searles Lake, two-basins downstream during much of late Pleistocene time, shows that they underwent similar responses to climate, but sedimentation changes documenting those responses commenced thousands of years apart, apparently because changes in precipitation volumes occurred gradually. Owens Lake, at the base of high mountains, was the first to reflect increasing amounts of regional precipitation; Searles, in a more arid environment, was the first to reflect decreasing amounts of precipitation.

Devils Hole, 150 km east of Owens Lake, has a well dated isotopic-temperature record that resembles the Owens Lake-depth record. Marine records of Pleistocene glacial fluctuations, which measure high-latitude ice-sheet volumes and thus both precipitation and temperature at those latitudes, also resemble the Owens Lake history. There are, however, differences between the ages of the maxima and minima of climatic events as reconstructed from the Owens Lake core and similar-appearing inflections in the other two records; the differences range from 0 to 33 k.y. and average ~15 k.y.

The question arises whether the differences between those ages are results of errors in the time-scale used for the Owens Lake record, or were there significant differences in the times when atmospheric climate change began to affect its different elements. The three records compared here are measurements of different elements and combinations of elements in two latitude belts: the deep-sea marine records measure combinations of temperature and precipitation that determined global ice volumes (at mostly high latitudes), the Devils Hole record measures atmospheric temperatures (in its mid-latitude region), and the Owens Lake record measures effective precipitation (in the same mid-latitude region).

INTRODUCTION

We infer that past variations in the size of Owens Lake reflected changes in precipitation amounts in the Owens Lake watershed. This, in turn, probably caused changes in the glacier sizes in the Sierra Nevada inasmuch as the large volumes of ice required to fill its canyons to the high levels where moraines are now preserved required precipitation that was much greater than at present.

The biological, chemical, mineral, and magnetic variations in the sediments serve as proxies for Owens Lake's fluctuations. These records are similar to those climatic changes that occurred globally, but it is an open question whether these global changes occurred "simultaneously"—say, within a few thousand years of each other. Climate change is, by definition, an *atmospheric* phenomenon, and the search for its cause needs to start with our best reconstructions of the ages of atmospheric changes. However, most records of past changes are proxy records that document the geological consequences of climate change, so it is important to study proxy records that responded quickly to atmospheric changes. The Owens Lake core OL-92 appears be one of those records.

Differences between the ages of local changes in the Owens Lake record and the global changes according to the marine records that integrate variations in polar ice-cap masses, should be expected because of the fundamental differences in the nature of each record. The deep-sea records, as determined from variations in $\delta^{18}O$ in the tests of benthic foraminifera (Emiliani, 1955), primarily integrate changes in the volumes of polar ice-caps, but there had to be a lag of unknown duration between the time when a change in atmospheric climate affected the sizes of polar ice sheets enough to detectably modify the isotopic composition of the oceans and its foraminifera. The history of Owens Lake, in contrast, is a record of variations in effective moisture that was largely determined by changes in the amounts of precipitation falling on the east slopes of the southern Sierra Nevada. Local or regional precipitation change could have altered Owens Lake's size and depositional character within a few decades.

An additional consideration is that the ages of climatic responses also may have varied as a function of elevation and latitude. The Owens Lake record is primarily a record of high-elevation precipitation in this middle-latitude area, whereas the deep-sea record is mainly a reflection of temperature and precipitation change in both low- and moderately high-elevation land in high-latitude areas.

It is difficult to envision a *lack* of impact of high-latitude glaciation on the climates at all latitudes, and for these reasons, we use the word and concept of "correlation" to indicate climate changes that we conclude are *interrelated* even if not *synchronous*. Consequently, we do not consider differences in the ages of events recorded by different proxy criteria of climate as necessarily compelling evidence of noncorrelation or errors.

PAST CLIMATE-INDUCED CHANGES IN OWENS LAKE SEDIMENTATION

Climatic significance of lake-depth changes

The lower third of the clastic deposits recovered by the OL-92 core, between ~323 m and ~206 m, has a relatively high percentage of sand and a lithologic variability that allows small fluctuations in lake depth to be inferred (Menking, this volume; Smith, this volume). Most of the sediments in the upper two thirds of that core, between ~206 m and ~5 m, appear to represent deep-water lacustrine deposition. The lithologic variation documented by OL-92 thus suggests that the lower third of the core represents a period of frequently shallow lakes and the upper two-thirds represents a period of mostly deep lakes. A rapid and brief change in lake depth most likely indicates a climate change inasmuch as atmospheric processes can change so rapidly—they do it every year. However, a gradual change in lake depth does not necessarily indicate a gradual change in climate; it can also reflect a change in subsidence or sedimentation rates, changes which ordinarily proceed at a much slower pace.

The coarse lower part of OL-92 has been interpreted as a possible consequence of an earlier glacial period (Smith, this volume). The co-existence of moderately coarse sediments and freshwater fossils that have affinities for water as shallow as 2 m (Bradbury, this volume) requires the elevation of the lake floor at the drill site to have been raised almost to the spillway of Owens Lake. This conclusion is supported by the facts that mollusks are common from 800 ka to 765 ka, suggesting shallow, quite fresh water during the early part of this regime, fossil fish remains are found in samples having ages ranging from 765 ka to 690 ka, and fossil trout and whitefish remains at 725 ka and 700 ka indicate periods of fresh, and almost certainly cooler, lake waters than any other period represented by the OL-92 core (Firby et al., this volume). Freshwater ostracodes also indicate Owens Lake during this period to have been a mostly fresh, probably overflowing lake (Carter, this volume).

The elevation of a tectonic basin's floor is a balance between sediment supply (serving to fill it) and tectonic subsidence (serving to deepen it). The sediment supply is sensitive to climate, and we conclude that the shallow basin at that time could have been a result of a greatly increased lake-sedimentation rate associated with the earlier Sherwin glacial episode (>760 ka) that some evidence indicates occurred between 1,300 ka and 1,000 ka (Smith et al., 1983). It is also possible, however, that the subsidence rate simply decreased.

The sizes of sediment grains deposited in shallow lakes may be indicators of small changes in water depth, and rapid fluctuations in sediment sizes can represent short-term fluctuations in lake depth. However, changes in current directions or velocities can produce similar depositional records. Criteria other than grain size, however, must be used to estimate changes in the volume of inflow and outflow.

Climatic significance of carbonate budget

The plot of weight percent inorganic CO_3 in the sediment against its age (Fig. 1) shows a horizontal line at ~7 wt % CO_3. This represents the theoretical CO_3 content of the sediment that should differentiate between (1) periods when Owens was a closed lake much of the time, retaining and depositing, as carbonates, much or all of the Ca and Mg entering the basin, and (2) periods

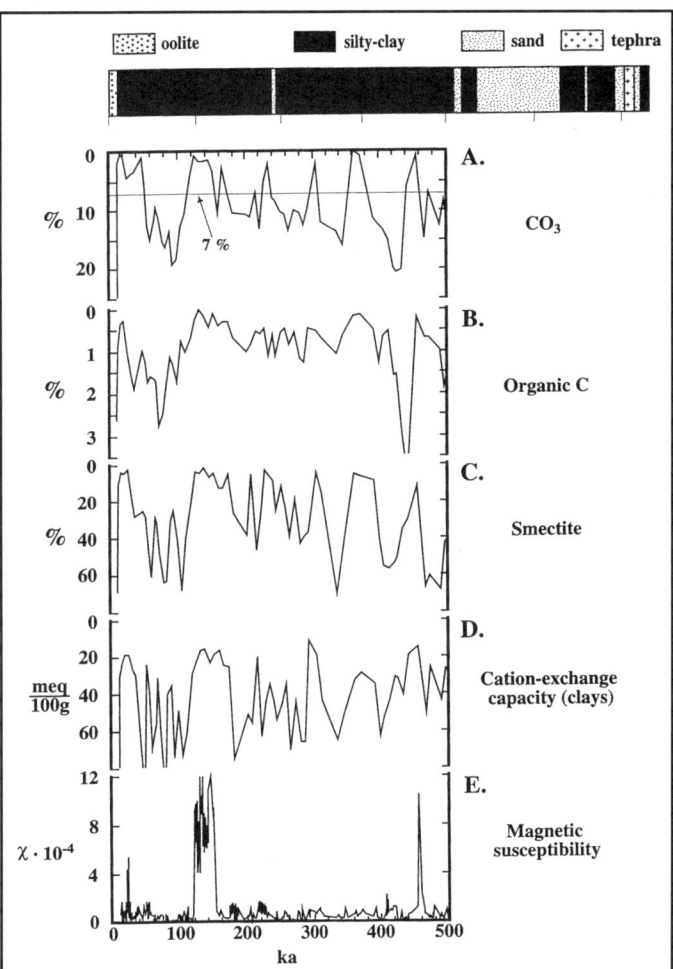

Figure 1. Plots for the period 0 ka to 500 ka, based on chemical, mineralogical, and geophysical data from core OL-92, showing variations with age in (A) weight-percentages CO_3, (B) weight percentages organic carbon, (C) weight percentages smectite, (D) cation-exchange capacities as milliequivalents per 100 g of sample, and (E) magnetic susceptibilities. Note that scales for A through D increase downward, so that peaks = wet periods and troughs = dry periods. Data from Bischoff et al. (this volume, Chapter 4), Menking (this volume), and Glen and Coe (this volume).

when it was most commonly an overflowing lake, exporting most of its Ca, Mg, and CO_3 to downstream basins (Bischoff et al., this volume, Chapter 4). This calculated value is based on the the assumption that the Ca flux of the historic Owens River has been constant for the past 500 k.y. As plotted (Fig. 1), it suggests that Owens Lake had discontinuous periods of overflow for about two-thirds of the time since 500 ka, but this is a statistical expression of overflow versus nonoverflow during the period of ~7,000 yr represented by each analyzed channel sample. Today, annual and monthly flow variations of the Owens River combine to produce volume fluctuations exceeding an order of magnitude (Hollett et al., 1991, table 2; Lee, 1912, tables 46 and 47). Therefore, whenever Owens Lake stood at a level near its spillway, the CO_3 percentage could have been between ~5 wt % and ~10 wt % because overflow occurred during some part of most years. Higher CO$_3$ percentages represent times when lake level stood below its spillway level more of the time; lower percentages represent times of more—at times, year-round—overflow.

For example, the 65 k.y. between ca. 50 ka and ca. 115 ka, during which the CO$_3$ data indicate frequently closed conditions (Fig. 1), is supported by some of the data obtained from the ostracode fauna (Carter, this volume) and diatom flora (Bradbury, this volume) extracted from OL-92 (Fig. 2) which indicate saline benthic diatoms and saline ostracodes during the last half of this period. However, both the diatoms and ostracodes also indicate a number of horizons within this interval that were characterized by fresh water. That 65 k.y. of basin closure also conflicts with interpretations of sediments from downstream Searles Lake (Smith, 1979; Smith, 1984; Bischoff et al., 1985; Jannik et al., 1991) that suggest that a deep lake, fed primarily by overflow from Owens Lake, existed in Searles Basin during most of that period. Searles, possibly as deep as 300 m at that time, probably had an evaporation rate >1 m/yr (Smith and Street-Perrott, 1983), so that substantial overflow from Owens had to reach it more frequently than every 300 yr to prevent desiccation.

Climatic significance of chemical, mineralogical, biological, and geophysical changes

During more than two-thirds of the time span represented by OL-92, changes in Owens Lake's inflow and outflow volumes are reflected by chemical, mineralogical, biological, and geophysical criteria, not reconstructed water depths (Figs. 1 and 2). In those figures, note that some vertical scales are inverted, so that peaks represent what we interpret as "cool" or "wet" (and "glacial"?) periods, and troughs represent "warm" or "dry" (and "interglacial"?) periods. The most sensitive of the chemical and mineralogical criteria (Fig. 1) are (1) the percentages of inorganic carbonate (CO$_3$) and organic carbon (organic C) (Bischoff et al., this volume, Chapter 4), (2) the illite-smectite ratios (Menking, this volume) and the confirmatory cation exchange capacity (CEC) determinations that reflects illite's low (and smectite's high) capacities (Bischoff et al., this volume, Chapter 4), and (3) the sediment's magnetic susceptibility which reflects changes in the species or abundance of magnetic minerals (Glen and Coe, this volume).

The chemical criteria, when plotted together over short time scales, show strong concordance during periods of high inflow and overflow. Minimal concentrations of inorganic CO$_3$ and organic C in the same segments of core are interpreted as indicators of clastic sedimentation in a very fresh, well-oxygenated lake whose waters had such a short residence time that evaporative concentration was minor and most organic C was oxidized and thus not preserved. These components, however, are not well correlated in the zones between those minimums (Figs. 3, 4, and 5).

Minima in CEC and smectite percentages, and the accompanying maxima in illite percentages, also coincide temporally with the minimums in inorganic CO$_3$ and organic C. These are

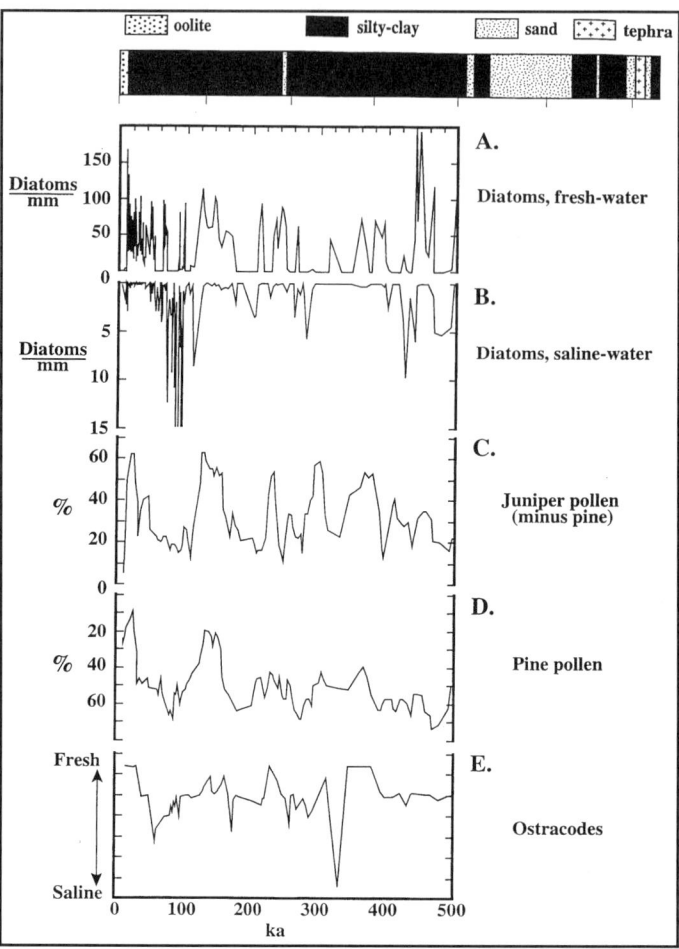

Figure 2. Plots for period 0 ka to 500 ka, based on biological data from core OL-92, showing variation with age in (A) freshwater diatom abundances (individuals/mm of microscope traverse), (B) saline-water diatom abundances, (C) juniper-pollen relative percentages (pine pollen excluded), (D) pine-pollen relative percentages, and (E) ostracode salinity indexes. Note that scales for B, D, and E increase downward, so that peaks = wet or cool periods and troughs = dry or warm periods. Data from Bradbury (this volume), Litwin et al. (this volume), and Carter (this volume).

interpreted as evidence of different weathering regimes (Menking, this volume). We infer that increased regional precipitation generated or expanded existing glaciers in the Sierra Nevada, and that the augmented runoff volumes and glacier activity accelerated mechanical erosion and produced increased volumes of comminuted illite/mica which provided little CEC, and clay- and silt-size glacial flour (mostly quartz and feldspar) which provided none. Lower temperatures accompanying these glacial periods would have further decreased the rates of chemical weathering and smectite production.

The changes in the magnetic properties, specifically the magnetic susceptibility (Fig. 1), are discussed by Glen and Coe (this volume). Notable increases in susceptibility are measured in three segments that other criteria indicate to be periods of high runoff (centered at 460 ka, 135 ka, and 20 ka); decreases

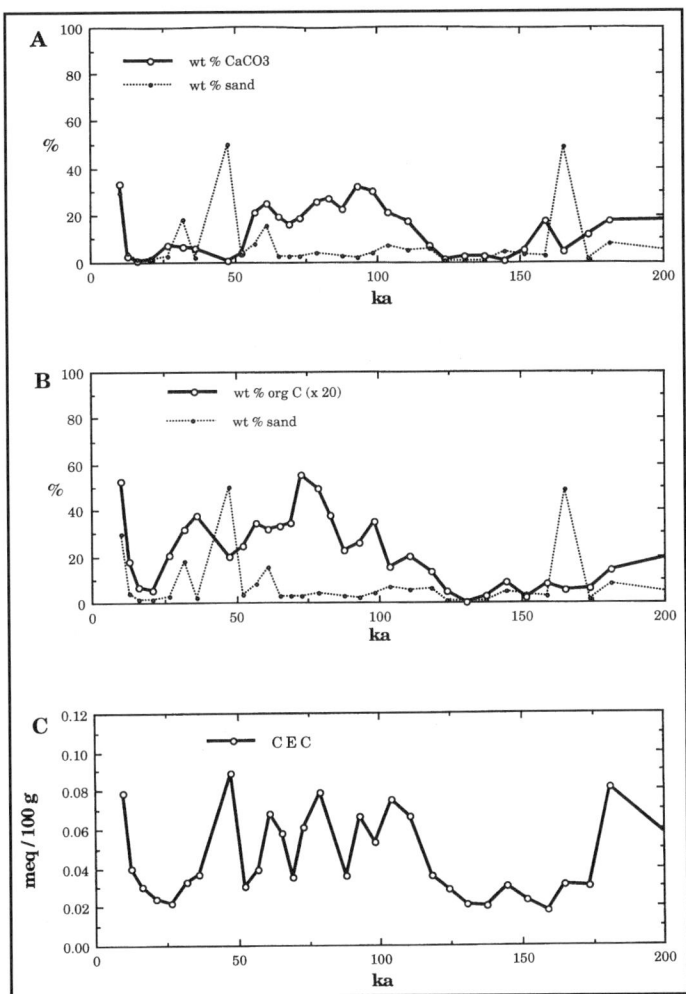

Figure 3. Plot of data from OL-92 core for period 0 ka to 200 ka, showing variation with age in (A) weight percentages $CaCO_3$, (B) weight percentages organic carbon, and (C) cation-exchange capacities (CEC) in milliequivalents of Cs per 100 g of sample. Sand percentages also plotted in A and B to identify samples that have their weight percentages of both acid-soluble $CaCO_3 + MgCO_3$ and organic carbon significantly diluted; the CEC values are relative only to the clay-size fraction of each sample.

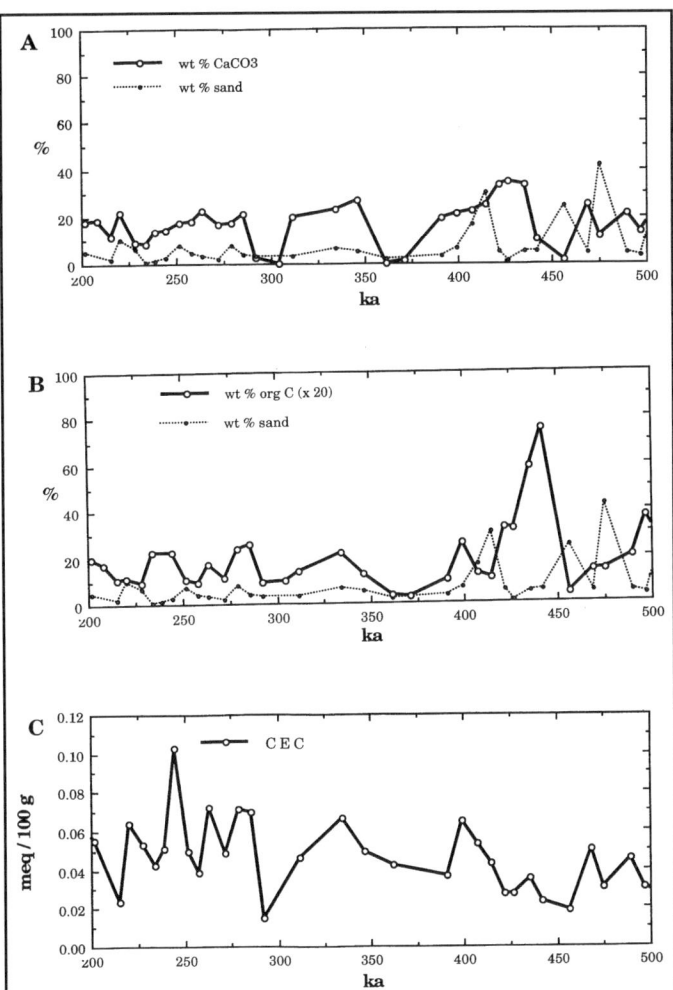

Figure 4. Plot same as Figure 3 but for the period 200 ka to 500 ka.

in susceptibility are found in periods of low runoff (as well as a few periods considered to have had high runoff).

Figure 2 shows cyclic fluctuations in biological criteria that resemble the patterns shown in Figure 1. Abundances of fresh- and saline-water diatoms, and pollen indicative of juniper and pine, show reciprocal patterns (Bradbury, this volume; Litwin et al., this volume). The salinity tolerances of the ostracodes (Carter, this volume) also show similar responses, especially since 200 ka. Variations in the numbers of mollusca and fish (Firby et al., this volume) indicate changes in the habitat provided by the lake. For example, only the horizons estimated to be 725 ka and 700 ka (prior to the period plotted in Fig. 2) contain remains of trout and whitefish that are now found only in high-mountain streams and lakes, documenting exceptionally fresh and cool lake waters at those times.

The OL-92 record thus reflects cyclic changes in the climate of the area surrounding the Owens River, most prominently the ~100 k.y. cycles that are so evident in the marine records that document cycles of high-latitude glaciation and interglaciation (Imbrie et al., 1984). The 20 k.y. and 40 k.y. cyclic periods also detected in marine records are not evident in this core.

Late Pleistocene climate changes indicated by Core OL-92

To interpret the OL-92 record, we plotted the age-depth relation (Bischoff et al., this volume, Chapter 8) along the abscissa, and the analyzed amounts of inorganic CO_3, organic C, and CEC (Bischoff et al., this volume, Chapter 4) along the ordinate (Figs. 3, 4, and 5). These three criteria were chosen in part because their values are based on analyses of channel samples rather than point samples (Bischoff et al., this volume, Chapter 4) and thus average out brief fluctuations in lake chemistry.

The carbonate and organic C plots (Figs. 3, 4, and 5) also show the accompanying percentages of sand containing little to

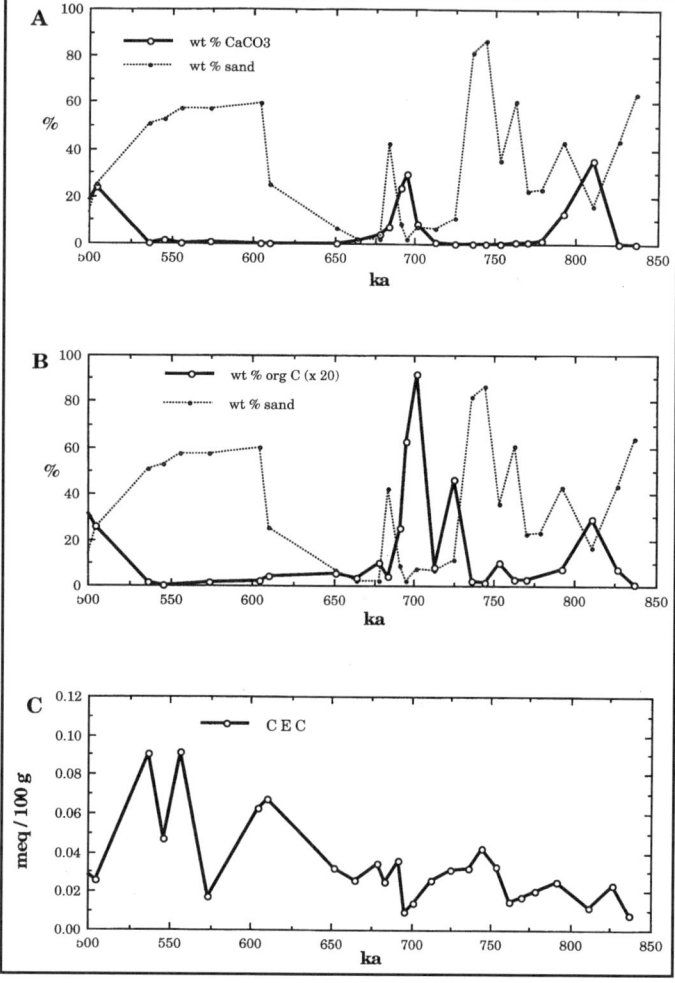

Figure 5. Plot same as Figure 3 but for the period 500 ka to 850 ka.

none of those components, so that large amounts of sand reduce the reported weight percentages of CO_3 and organic C. The CEC values are relative only to the weight percentage of the clay-size fraction.

Two or more adjoining samples characterized by low values for $CaCO_3$, organic C, and CEC are found at levels representing deposition between 15 ka and 20 ka and between 125 ka and 150 ka (Fig. 3). The minimum in the $CaCO_3$ and organic C curves at 50 ka has questionable significance because of the accompanying high sand content, and interpreting it as a wet period would also conflict with the CEC value which shows that sample to be high in smectite, a criterion of aridity.

Comparable diagrams for the intervals 200 ka to 500 ka (Fig. 4) and 500 ka to 850 ka (Fig. 5) show coherent troughs at 300 ka, 370 ka, and 460 ka, indicating periods of high runoff. Because the sand percentage is low, the trough at 715 ka may also represent a high-runoff period, but the values for CEC do not strongly support this interpretation. In fact, throughout the history plotted in Figure 5, the percentages of sand and of both $CaCO_3$ and organic C are nearly mirror images of each other, making meaningful interpretations of those data nearly impossible. The CEC values in Figure 5, however, reveal an interesting trend; they are generally low between 650 ka and 850 ka, although they erratically increase with time, suggesting that a period of high runoff characterized the lake when the oldest sediments in core OL-92 were deposited, and that runoff decreased slowly, although erratically, until 650 ka.

Holocene climate changes indicated by Core OL-92 and other studies

In the late 1800s, before settlers in Owens Valley began to divert a significant part of the Owens River for crop irrigation and domestic needs, its flow was sufficient to stabilize Owens Lake at a maximum depth near 15 m and maintain its salinity between about 6 wt % and 9 wt % dissolved solids (Gale, 1914). Prior to desiccation, the lake-floor sediments at the drill site consisted of oolites, presumably products of a lake that was alkaline enough to precipitate all the dissolved calcium that entered as carbonate, and shallow enough for wind energy to create bottom currents strong enough to agitate the carbonate granules on its floor.

Radiocarbon dates (Bischoff et al., this volume, Chapter 8) reveal an erosional hiatus that removed sediments from the youngest part of the pre-oolite depositional record, stopping at a horizon ca. 8.3 ka in age. We infer that, during early or middle Holocene time, evaporation began to exceed lake inflow and lowered its surface to a level where wind-driven currents caused sublacustrine erosion of ~3.3 m of sediments at the OL-92 core site, carrying them to other parts of the basin and causing a ~3.2 k.y. discontinuity in the core (Bischoff et al., this volume, Chapter 8). Oolite deposition commenced at 5.1 ka (middle Holocene), either because the lake depth increased and/or its alkalinity reached levels that accelerated oolite-deposition rates enough to prevent further erosion.

Little Lake, 40 km south of Owens Lake, occupies the channel through which overflow from Owens Lake passed; the small lake now exists because an alluvial fan dammed its southern end after overflow from Owens Lake ceased. Radiocarbon dates on the basal sediments in a core from that lake are 5.0 ka (Mehringer and Sheppard, 1978), very close to the 5.1 ka age of the initial deposition of oolites in Owens Lake. These relations provide additional evidence that overflow from Owens Lake has not occupied this channel since about 5 ka.

The age on the basal oolites from Owens Lake is also similar to the youngest radiocarbon dates from tree trunks now submerged 5 m below the spillway level of Lake Tahoe, 360 km northwest of Owens Lake and at ~1,899 m elevation. The trees document a period when intense aridity caused that lake's level to fall at least that much, allowing tree seeds to germinate on the newly exposed lake floor; the resulting forest died ~1,500 yr later when subsequent precipitation increases caused the lake level to rise above their roots. A series of ^{14}C dates on the submerged trunks are interpreted by Furgurson (1992) as representing drought in the Tahoe Basin that lasted from about 6.5 ka to 5.0 ka.

The coincidence of the dates from Lake Tahoe, Owens Lake, and Little Lake suggests that a severe regional drought is what caused Owens Lake to begin its recession. The return to "normal" precipitation at about 5 ka only allowed Owens to return to a fraction of its former size.

It is notable that in core OL-92, we recovered no evidence older than 5 ka of near desiccation—oolites, salts, or salt pseudomorphs. Four types of evidence support this conclusion:

1. Desiccation (with or without salts) could have characterized Owens Lake for short periods, but these processes would have been preceded and followed by periods of higher salinity, possibly including a period of gaylussite crystallization that would have precipitated from high-pH waters at salinities ≥15% (Bury and Redd, 1933; Bischoff et al., this volume, Chapter 4). Gaylussite crystals, or calcite pseudomorphs after them, are distinctive and easily identified (Gale, 1914; Smith and Haines, 1964; Smith et al., 1987).

2. Diatoms indicative of saline conditions in *both* planktic and benthic environments were found in only five narrow zones in OL-92 (Bradbury, 1993), but except for one horizon, they came from beds of clay and silt that probably indicate relatively deep water.

3. An oolite horizon like that near the modern Owens Lake surface would probably have been very hard and identified from its drilling characteristics; also, its spherical white oolites would have been conspicuous if they were mixed with "slumped material" or rested on the tops of well-compacted cores from greater depths.

4. Salt beds thicker than a few decimeters are most commonly only partially dissolved by influxes of fresh water, as shown by observation of this process during 1970 and 1971 at Owens Lake (Friedman et al., 1976; Smith et al., 1987), and by the evidence of minimal solution of seven thin salt beds by freshwater Pleistocene lakes in Searles Valley which followed their crystallization during deposition of the Lower Salt (Smith, 1979).

The lack of any of these indicators of near-desiccation leads us to conclude that the last half of the Holocene in this region has been more arid than during any other period since 800 ka.

COMPARISON OF CORE OL-92 RECORD WITH OTHER PALEOHYDROLOGIC RECORDS

Searles Lake, California

Water from Owens Lake overflowed into China and Searles Lakes many times during the Pleistocene. When Owens was not overflowing because of a dry period, the smaller and lower-elevation catchment areas that drain directly into China and then Searles Lakes (4,645 km^2, 2,575 m maximum elevation), if also affected by the same dry period, would not have supplied enough water to offset evaporation from their extensive surfaces. Seemingly, even a small perennial lake in Searles Basin required enough overflow from Owens Lake to first fill China Lake until it spilled into Searles. However, exposed and subsurface evidence from Searles Valley indicates that the Overburden Mud, which rests on the Upper Salt (Fig. 6), was deposited by a moderately saline lake that was ~40 m deep and had an area of ~270 km^2; its deposits contained a piece of wood having a ^{14}C age of ca. 3.5 ka (Stuiver, 1964; Smith, 1979). Considering the previously cited evidence from Owens Lake and Little Lake (Mehringer and Sheppard, 1978), leading to the conclusion that water has not flowed out of Owens Lake since 5 ka, the subsequent lake in Searles Valley suggests that for a period of time centered near 3.5 ka, either a period of monsoonal climate brought abnormal amounts of summertime(?) moisture from the southwest to an area southeast of the Sierra Nevada, or that the ^{14}C age was too young because contaminated by younger carbon.

Changes in the approximate depths and lake chemistry of Searles Lake are recorded back to 3.2 Ma by the lacustrine sediments and salts in core KM-3 (Smith et al., 1983). The ages of two horizons in the Searles record needed for comparing the 0 ka to 800 ka histories of Owens and Searles Lakes, however, are not closely dated. The younger of these horizons is the contact (Fig. 6) between Unit A+B and Unit C (Smith et al., 1983). The U-series dates from Unit A+B, however, allow downward extrapolation of the lowest four finite dates, and two "infinite" dates (>330 ka) from deeper horizons within the same unit suggest the age of that contact is 350 ka (Bischoff et al., 1985). The age of the second boundary, between Unit C and the underlying Unit D+E (Fig. 6), is known to be slightly younger than 670 ka, the age of the Lava Creek ash (Izett et al., 1992; Smith, 1984). We use an age of 665 ka.

Generalized versions of the histories of Owens and Searles Lakes (Fig. 6), based on the lithologic character of sediments recovered by core OL-92 and core KM-3, indicate that there were four separable climatic "regimes" reflected in both cores by sediments deposited since 810 ka. Two of those regimes lie within the sand-rich sediments in the lower third of the OL-92 core; the third regime accounts for most of the upper two-thirds of the core. Climatic trends leading to decreased regional runoff should have been recorded earlier in Searles Lake than in Owens Lake because downstream lakes were the first to stop receiving water from Owens as aridity increased. Conversely, trends leading to increased regional runoff should have been recorded later in Searles because Owens was the first to receive those increased volumes of water. If the ages assigned to the regime boundaries in these two cores are approximately correct, the depositional record confirms this relation (Fig. 6).

Regime I. In OL-92, fine-grained deposits represent much of the time assigned to Regime I, although several beds of sand (and >5 m of tephra) interrupted the sequence (Smith, this volume). We interpret this to mean that perennial lake waters, deep enough to overflow, prevailed most of the time between 810 ka and 645 ka. The mollusks, fish, diatoms, and ostracodes, discussed earlier, support this conclusion.

In Core KM-3 from Searles Lake, greenish to brownish, clastic, and probably perennial-lake, clastic sediments that account for ~80 vol % of Unit D+E (Fig. 6) that is correlated with Regime I. The color variations of these sediments suggest

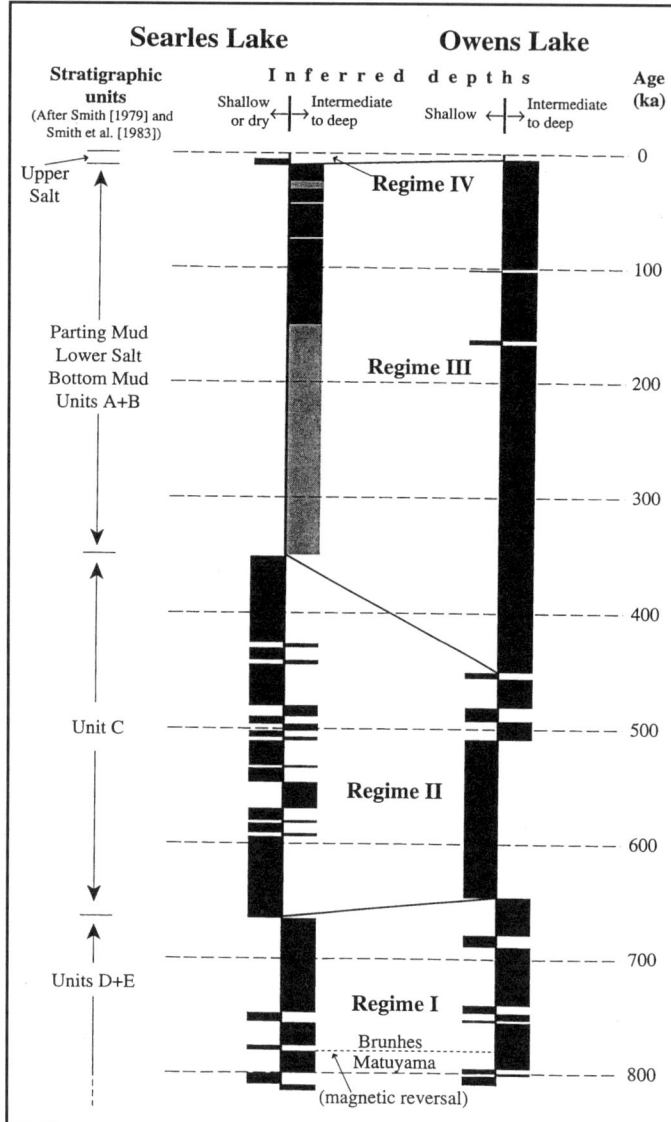

Figure 6. Variations in the inferred depths of water in Searles and Owens Lakes between 500 ka and middle Holocene time (5 ka). Left half of Searles Lake record shows names and age ranges of stratigraphic units discussed in text. Right half of the Searles Lake record plots inferred water depths (table 2 and pl. 1 in Smith et al., 1983). To the left of the vertical axis that bisects the Searles Lake record, black areas represent zones characterized by numerous multimineralic salt beds, indicative of periods when the lake approached or reached desiccation (Smith, 1979, p. 91–95). To right of that vertical axis, black areas represent zones of marl, clay, or silt indicative of wet periods and deep perennial lakes; screened area represents period when Searles Lake was a shallow but perennial lake that that periodically developed salinities between about 3% and 15% but almost never approached desiccation, thus requiring some overflow from Owens Lake (Smith, 1979, p. 86, 108). In the Owens Lake record, black areas plotted to the left of the vertical axis represent zones dominated by fine to very coarse sand, indicating shallow water; black areas to the right of that axis represent zones dominated by silt and clay, indicating deep water (Smith, this volume). Ages for the Searles record are adapted from Stuiver and Smith (1979), Bischoff et al. (1985), using methods described in text. Ages for the Owens Lake record follow Bischoff et al. (this volume, Chapter 8). Position of the Matuyama-Brunhes magnetic reversal (ca. 780 ka) in core OL-92 shown near base of diagram (Glen and Coe, this volume); its position in Searles Lake core from Liddicoat et al., 1980.

that fluctuating and occasionally small lakes existed most of the time between 810 ka and 665 ka (Smith et al., 1983), requiring overflow from Owens Lake in at least moderate amounts. This depositional mode was interrupted by three short(?) periods of salt deposition.

Both OL-92 and KM-3 thus indicate that Regime I was a period mostly characterized by runoff that allowed perennial-lake sediments to be deposited in Searles Lake whenever enough overflow from Owens Lake reached it. There were, however, intervening drier periods when shallower-water, sand-size sediments were deposited in Owens Lake and salts were deposited in Searles Lake.

Regime II. The period between 645 ka and 450 ka was more arid. According to the lithologies in OL-92 that represent this interval, Owens Lake was depositing many beds of fine to very coarse sand that suggest shallower lakes than those that characterized Regime I, probably too shallow to overflow much of the time, reducing the amount of water reaching Searles Lake. Sporadically preserved fossil assemblages in this part of the section, however, mostly indicate fresh lake water, implying periodic overflow that transported salts out of the low-volume basin. Mollusks and fish are common only in the more fine-grained segment between 505 ka and 450 ka (Fig. 6; Firby et al., this volume). Diatoms and ostracodes, also sparsely preserved in sections of coarse sand, indicate sporadic episodes of fresh water during the period between 645 ka and 505 ka, and more frequent episodes of fresh water during periods between 505 ka and 450 ka (Bradbury, this volume; Carter, this volume).

Between 665 ka and 350 ka, Searles Lake deposited salts much of the time, mostly as beds containing two or more salt-mineral species, which is indicative of desiccation or near-desiccation (Smith, 1979). Beds composed of perennial-lake sediments constitute only ~10 vol % of the unit, although a few of these beds are as thick as 1 m, possibly representing a few thousand years. These beds document the intermittent perennial lakes in that basin as implied by the intermittent low salinities in Owens Lake as indicated by some of its fossils during the correlative regime.

The annual flux of Cl from the modern Owens River is calculated to be ~5.8 × 10^6 kg/yr (Table 3 in Bischoff et al., this volume, Chapter 4). The calculated amount of Cl in the Searles Lake unit equivalent to Regime II (Unit C; Fig. 6) is 4.2 × 10^{12} kg (Smith et al., 1983; G. I. Smith, 1995, unpublished data). This quantity would have required ~725 k.y. of modern Owens River inflow, 2.3 times the ~315 k.y. depositional period indicated by the boundaries assigned to that regime. Most of the Cl is likely to have come from springs associated with the Long Valley caldera (Fig. 1 in Smith and Bischoff, this volume) that erupted ~100 k.y. prior to deposition of the base of these deposits, and our calculation thus suggests that the Cl flux in the Owens River at that time was several times the present flux. This conclusion agrees with that based on study of the interstitial waters in OL-92 (Friedman et al., this volume).

Regime III. Owens at 450 ka, and Searles at 350 ka, evolved into perennial lakes. Between 350 ka and 10 ka, Searles Lake stood almost continuously at intermediate to high levels. However, between 32 ka and 24 ka, six thin salt beds indicative of shallow saline lakes plus one bed indicative of brief desiccation at 28 ka, were deposited (Stuiver and Smith, 1979). During the period represented by Regime III, therefore, nearly continuous overflow from Owens Lake was required to offset evaporation from the surfaces of China and Searles Lakes downstream from it. But the variations in Owens Lake sediment chemistry (Bischoff et al., this volume, Chapter 4) show that cyclic changes in inflow and outflow volumes did take place, but they rarely reduced lake depth to levels where coarse sediments were deposited at the site of the OL-92 core, presumably because by then, the rate of tectonic subsidence of Owens Lake's floor equaled or exceeded the rate of deposition.

Two coarse-sand layers are present in the Owens Lake sediments deposited during Regime III, at 165 ka and 100 ka (Fig. 6). If those two beds represent periods of shallow water and nonoverflow, they had to be brief because China and Searles Lakes probably had lake-evaporation rates of >1 m/yr (Smith and Street-Perrott, 1983), so that when overflow from Owens Lake ceased, <300 yr would be required for a full Searles Lake to evaporate to dryness, leaving a multimineralic salt layer in the subsurface record. Evidence of such an event is not found, and the sand beds seem more likely to represent extreme flood events or turbidity currents.

Regime IV. The most recent climatic regime, representing part or all of the Holocene, is recorded by sediments in both cores that indicate a period of aridity that was unprecedented in the preceding 800 k.y. period. Searles Lake began deposition of ~10 m of salts at 10 ka, and Owens Lake began deposition of oolites at 5 ka. Although periods of desiccation are recorded in Searles Lake during Regimes II and III, oolites, salts, or other lithologies indicating comparable aridity in the Owens Lake area are not found in the equivalent-age horizons of the OL-92 core.

Another measure of the amounts of water reaching Searles Lake from Owens Lake during each of the above four regimes is found in the concentration of Ca in the correlative sediments of KM-3 (pl. 2 in Smith et al., 1983). The Ca content of Owens River water (today, ~24 mg/L) was concentrated by evaporation from Owens and China Lakes only ~2 to ~3 times (table 10-2 in Smith and Street-Perrott, 1983). When it mixed with the high-pH Searles Lake water, the Ca precipitated as aragonite, calcite, and dolomite (Smith, 1979). The deposits in Searles Lake assigned to Regime I (Fig. 6) average ~5% Ca, those assigned to Regime II average ~1% Ca, those assigned to Regime III vary between ~6% and ~12% Ca, and Regime IV sediments contain from 0% to ~0.5% (pl. 2 in Smith et al., 1983).

Differences in the apparent ages of the climate transitions between Regimes I, II, III, and IV, as reflected by the sediments from Owens and Searles Lakes, do not mean that atmospheric climate changes were not uniform over a large area. They mean that the time that sediments of each basin began to record a climate change depended (1) on each basin's position in the chain of lakes relative to the main source of runoff in their combined drainage areas and (2) on the direction of climate change that was underway. For example, the onset of arid Regime II in the Searles Lake was first recorded by 665 ka, ~20 k.y. before a comparable onset was recorded in Owens Lake sediments, because the incremental decrease in runoff eventually stopped overflow into Searles Lake yet remained sufficient to supply perennial water bodies in Owens Lake for almost 20 k.y. after. Conversely, the return to the wetter period of Regime III was first recorded permanently in Owens Lake by 450 ka, whereas Searles Lake did not begin to receive enough water from Owens Lake overflow to establish a perennial lake until ~100 k.y. later, at 350 ka (Fig. 6).

Intuitively, such long climatic transitions seem unlikely. However, in each transition, the sequence of basin responses is correct for the direction of climatic change that was in progress, the climatic interpretations of the Searles Lake stratigraphy seems unambiguous, and the age interpretations of the regime boundaries seem unlikely to be so incorrect that they would invalidate the above conclusions. A basic constraint on these interpretations is that *correct* paleoclimatic histories of Owens and Searles Lakes have to be compatible. (One could not, for example, have "correctly interpreted" histories that show Owens Lake receding below its overflow level during a period when Searles Lake was expanding.) By elimination, only very slow *rates* of regional change in precipitation and runoff seem adequate to produce the relations plotted here (Fig. 6). The apparent magnitudes of these differences also illustrate the risk involved in correlating continental paleoclimatic records solely on the basis of age.

In contrast, the the Pleistocene-Holocene climatic transition was much more rapid. In the KM-3 and OL-92 cores, the transition to aridity at 10 ka at Searles Lake required only ~5 k.y. to be recorded also in Owens Lake (Fig. 6, Regime III/IV). Both boundaries are well constrained by ^{14}C ages.

Panamint Lake, California

The diatom stratigraphy of Owens Lake suggests a potential correlation with shoreline tufa deposits on the margins of Panamint Lake. That lake was three basins downstream from Owens Lake during periods of near-maximum overflow (Smith and Bischoff, this volume). A distinctive form of the freshwater planktonic diatom *Cyclotella bodanica* is common in shoreline tufa deposits in Panamint Valley that are tentatively dated as 95 ka to 55 ka by U-series methods, but those dates show considerable scatter (Fitzpatrick and Bischoff, 1993). This diatom species, in a possibly reworked assemblage, occurs prominently in the Owens Lake samples assigned ages between 150 ka and

140 ka (Bradbury, this volume). The exisence of this uncommon diatom in both basins may indicate that the two basins were connected, suggesting that this period was one of near-maximum flow of the Owens River (Figs. 1, 2, and 3). The same diatom species was also detected in samples from Owens Lake having an age of 700 ka, when trout and whitefish also migrated into the lake, indicating exceptionally fresh water and cool temperatures (Firby et al., this volume), possibly confirming that Owens and Panamint were also connected at that time as suggested by Jannik et al. (1991). These associations appear to support the conclusion that this diatom characterized very wet, cool, and quite possibly glacial, periods.

Death Valley Lake, California

U-series dates and lithologic studies of a core from Death Valley, the last in the Pleistocene series of four lakes downstream from Owens Lake, indicate the existence of a large lake in that basin between 186 ka and 128 ka (Lowenstein, 1994). The chemical data from OL-92 indicate a period of high runoff during the period between 125 ka and 150 ka (Fig. 3), and lithologic data from Searles Lake generally confirm this (fig. 41 in Smith, 1979; Bischoff et al., 1985). However, between 150 ka and 186 ka, the chemical and lithologic data indicate a period of only intermediate runoff in Owens and Searles basins, although the data from diatoms in Owens Lake (Bradbury, this volume), magnetic susceptibility (Glen and Coe, this volume), and "juniper" pollen (Litwin et al., this volume) suggest that period to have been one of increasing moisture.

Lake Lahontan, Nevada

Lake Lahontan, a large (22,400 km^2) pluvial lake that existed during the late Pleistocene in northwestern Nevada, is composed of seven subbasins. In the late 1800s, before irrigation was extensive, perennial water bodies existed in four of these subbasins, Pyramid-Winnemucca, Walker, Honey, and Humbolt-Carson. They received water directly from the then perennially flowing Truckee, Walker, Susan, Humbolt, and Carson Rivers (Russell, 1885). In view of the evidence from the Owens Lake core that the past 5 k.y. in the western Great Basin has been more arid than during any other period since 800 ka, it is possible that these historic lakes in Lahontan's subbasins existed continuously throughout the last half of the Pleistocene.

Lake Lahontan has a history of expanded lakes extending back >100 ka, but radiometric age control of that history is best between 10 ka and 22 ka. On the basis of a large number of ^{14}C dates, Lahontan has been shown to have stood at an intermediate level (~1,265 m elevation) from 22 ka to 16 ka; after a brief recession it then rose and stood at its maximum level (~1,335 m) from 15 ka to 13.6 ka. After falling to a level below ~1,180 m, it rose again briefly to 1,225 m from 11 ka to 10 ka, then fell to near its modern level until between 4 ka and 2 ka when a brief resurgence brought its level to ~1,185 m (Benson and Thompson, 1987; Benson et al., 1995). A number of ^{230}Th ages show that intermediate-level lake stands also occurred in these basins prior to the period documented by reliable ^{14}C ages (Lao and Benson, 1988), but they do not as yet provide a coherent history.

The intermediate-level stand in the Lake Lahontan record between 22 ka and ca. 16 ka is about the same age as that of the period of maximum runoff according to the Owens Lake record between 15 ka to 25 ka (Fig. 3). Lake Lahontan's maximum stand (15 ka to 13.6 ka) and its brief resurgence (11 ka to 10 ka) occurred while Owens Lake's runoff volume apparently was decreasing, but the Searles Lake record shows that Owens again increased its overflow into Searles between 13.5 ka and 10.5 ka (G. I. Smith, 1995, unpublished data).

Devils Hole, Nevada

Owens Lake and Devils Hole, about 150 km east of Owens, recorded different elements of climate. The Owens Lake study reconstructs variations in earth-surface precipitation and runoff amounts whereas the Devils Hole record (DH-11) reconstructs variations in atmospheric temperatures. The DH-11 record, which extends from 566 ka to 60 ka, is based on variations in δ^{18}O values in calcite deposited by ground water derived from precipitation falling ~60 km to the north-northeast (Winograd et al., 1988, 1992; Ludwig et al., 1992; Szabo et al., 1994). The Devils Hole recharge area is ~350 km inland from the Pacific Ocean, and the isotopic character of precipitation originating there is modified in proportion to the overland length of each incoming storm trajectory and the elevation of the terrain underlying that trajectory (Smith et al., 1979; Friedman et al., 1992); that terrain includes the southern part of the Sierra Nevada and several other high mountain ranges in the southwestern Great Basin. These combine to decrease the absolute humidity of the incoming air mass reaching the Devils Hole recharge area, and thus determine the elevation and temperature at which it condenses.

Comparison of the OL-92 paleoclimatic record with the δ^{18}O record from calcite in Devils Hole shows marked similarities in shape, with both reflecting most strongly the ~100 k.y. cycles (Figs. 7A and 7B) that seem to dominate the marine record. However, the ages of the correlated peaks and troughs indicating climate reversals differ by an average ~16 k.y. (Table 1). Periods of maximum inflow into Owens Lake indicated in Table 1 by peaks "A" through "E" *preceded* times of lowest temperatures according to the Devils Hole isotopic record; the periods of maximum inflow indicated by peaks "F" through "K" *followed* the maximum cold periods in the Devils Hole record. Similarly, the periods of minimal inflow to Owens indicated by troughs "a" through "f" also *preceded* periods of relative warmth according to the Devils Hole study, and periods of minimal inflow in Owens indicated by troughs "g" through "j" also *followed* periods of relative warmth according to the Devils Hole record.

The record from from Devils Hole is probably the most

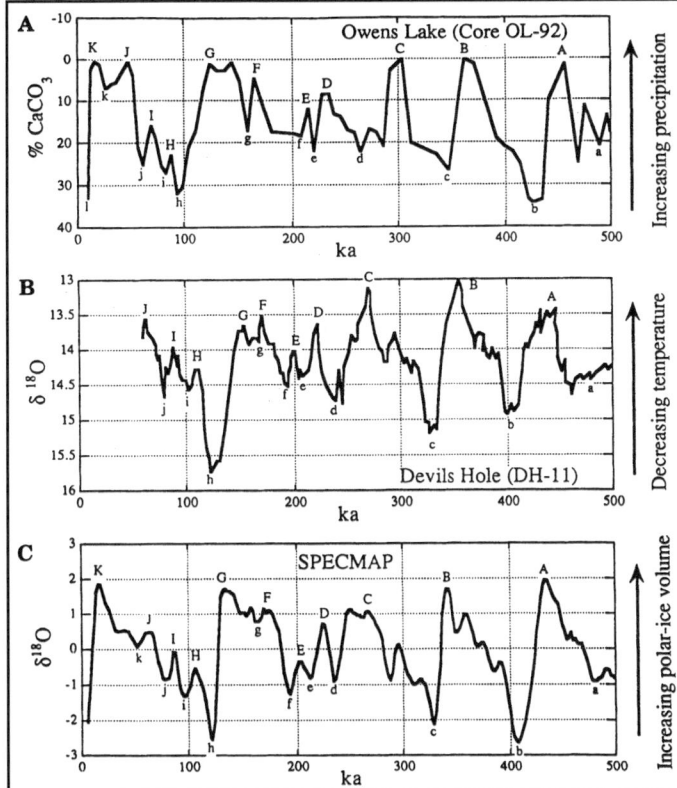

Figure 7. Paleoclimatic diagrams for the interval 0 ka to 500 ka, showing (A) the $CaCO_3$ percentages in the OL-92 core from Owens Lake (Figs. 3A and 4A; vertical scale is inverted), (B) the isotopic record of paleotemperatures from the DH-11 record from Devils Hole, Nevada (Winograd et al., 1992), and (C) the SPECMAP curve indicating fluctuations in global ice volume (Imbrie et al., 1984; Bassinot et al., 1994). As in Figures 1 and 2, curves are plotted so that wet, cold, or glacial periods are peaks, and dry, warm, or interglacial periods are troughs. Inflections that we infer represent related changes in the elements of climate recorded by the three studies are labeled with the same upper-case letters (peaks) or lower-case letters (troughs).

vicinity of Tule Lake were fresh and productive during both summer and winter at those times (Bradbury, 1992; Smith and Bischoff, this volume).

During glacial episodes, the southward movement of westerly storm tracks (Kutzbach and Guetter, 1986) increased winter precipitation in the Owens-Searles system while making Tule Lake drier and colder. In the Tule Lake record (Fig. 8), some glacial climates are suggested by high percentages of *Fragilaria* species, indicating fresh but shallow and closed-marsh environments (Bradbury, 1992). This is particularly evident for what is interpreted as glacial stage 6 at Tule Lake, where large percentages of *Fragilaria* characterize this record between 120 and 170 ka. However, glacial stage 2 at Tule Lake, centered near 20 ka, was apparently so cold and dry that *Fragilaria* was not especially productive, and Pliocene diatoms, reworked from local outcrops in the basin by wind and storms, record harsh glacial environments. In contrast, increased precipitation at Owens Lake during periods centered at 20 ka and 135 ka is shown by high concentrations and large percentages of freshwater planktic diatoms (Fig. 2; Bradbury, this volume).

At Tule Lake, interglacial climates were characterized by increased numbers of freshwater planktic diatoms, especially species of *Aulacoseira* that bloom during the summer and fall (Bradbury, 1992). At Owens Lake, the diatom record for the last interglacial is dominated by saline diatoms (Fig. 2) whereas the late Holocene lake in this basin was so saline that oolites were formed (Smith, this volume) and diatoms were seldom preserved.

OWENS LAKE HISTORY AND SIERRA NEVADA GLACIATION

The existence of Pleistocene glaciers in the Sierra Nevada has been known for more than a century (McGee, 1885; Russell, 1889). More modern studies (Blackwelder, 1931; Putnam, 1950; Sharp and Birman, 1963; Dalrymple, 1964; Sharp, 1968, 1972; Burke and Birkeland, 1979; Gillespie, 1982; Clark and Gillespie, 1994) further subdivided the Pleistocene units, and reported one or more minor Holocene(?) stages. The Pleistocene stages may be grouped and described as follows:

Group I. *Sherwin (and McGee?) stages*: Oldest(?) Pleistocene tills of the Sherwin glaciation are widespread but their moraine morphology is destroyed, and their large plutonic fragments are mostly decomposed; weathered till is covered by the 760-ka Bishop ash.

Group II. *Tahoe and Mono Basin stages*: Older late Pleistocene moraines have subdued crests, their deposits are commonly more extensive than those of the younger glaciations, and their boulders are distinctly weathered but preserved.

Group III. *Recess Peak, Tioga, and Tenaya stages*: Youngest late Pleistocene moraines have sharp crests, their deposits are mostly less extensive than older moraines, and exposed boulders are relatively unweathered.

We suggest that the minima in CO_3 and organic C percentages between 15 ka and 20 ka in the OL-92 record (Fig. 3) are expressions of the Group III glaciations, and the minima

reliably dated paleoclimatic record so far recovered. The Devils Hole, SPECMAP (^{18}O, deep-sea samples), and Vostok (ice cores, Antarctica) records are remarkably similar (fig. 3 and 4 in Winograd et al., 1992). The pattern of shifts from correlated events in the OL-92 core, which first preceded and then followed the correlated inflections in the DH-11 record, suggests a systematic error in the OL-92 age-depth curve, as discussed in a later section.

Tule Lake, California

A comparison of diatom records from Owens Lake (36.4°N) and Tule Lake (41.9°N) for the past 180 k.y. illustrates the different character of glacial and interglacial climates from areas almost 700 km apart. Relatively saline shallow lakes characterized the Owens basin during Holocene and possibly during some Pleistocene interglacial periods, whereas lakes in the

TABLE 1. AGE COMPARISONS OF LABELED "PEAKS" AND "TROUGHS" IN FIGURES 7A AND 7B*

Label	Peak Ages		Difference†	Label	Trough Ages		Difference†
	DH-11 (ka)	OL-92 (ka)	(k.y.)		DH-11 (ka)	OL-92 (ka)	(k.y.)
...	...	...	...	l	...	10	...
K	...	16	...	k	...	32	...
J	62	48	+14	j	79	61	+18
I	89	69	+20	i	102	84	+18
H	110	89	+21	h	121	93	+28
G	129	125	+4	g	160	159	+1
F	171	166	+5	f	194	208	-14
E	200	215	-15	e	204	220	-16
D	222	236	-14	d	239	264	-25
C	270	303	-33	c	326	346	-20
B	355	362	-7	b	400	426	-26
A	448	456	-8	a	480	490	-10

*"Peaks" in these two figures represent cold periods in the Devils Hole record (DH-11) and wet periods in the Owens Lake record (OL-92), both mid-latitude sites; "troughs" represent periods having opposite characteristics. Ages estimated from Figure 7; probable accuracy ±3 k.y.
†DH-11 age minus OL-92 age. Positive age differences mean that reversals in OL-92 curve, which reflect precipitation changes, occurred after the reversals in DH-11 curve, which reflect temperature changes. Negative age differences mean the opposite.

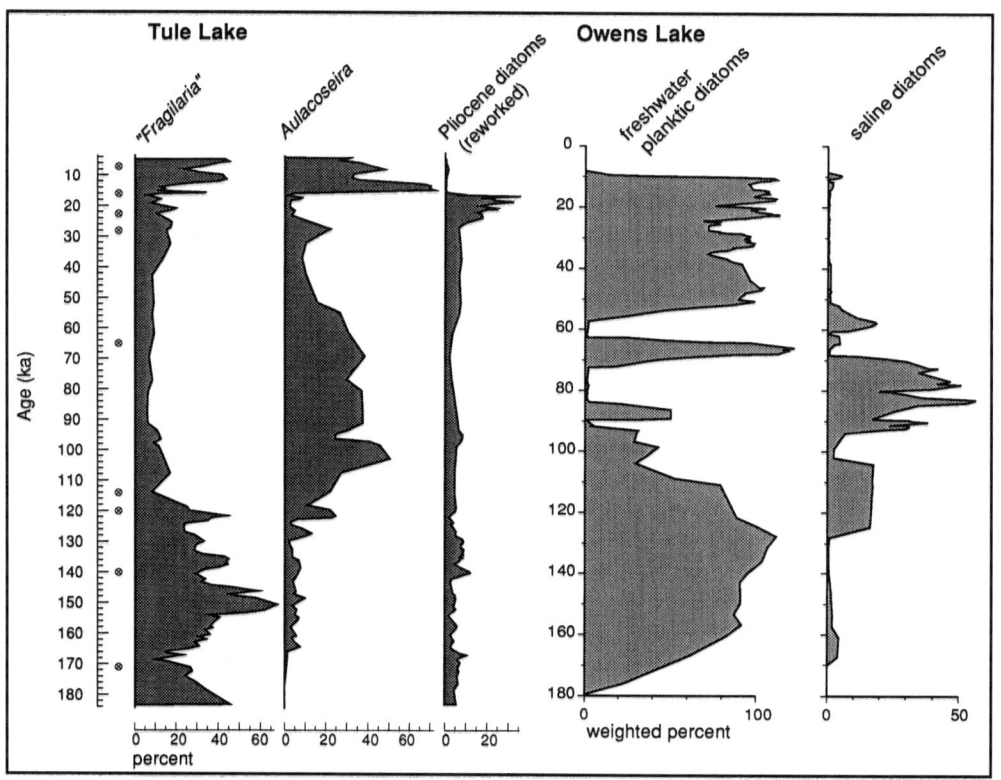

Figure 8. Comparison of planktic freshwater and saline diatoms from Owens Lake core OL-92 with selected diatom profiles from Tule Lake, California, for the past 180 k.y. (Bradbury, 1991). Data have been smoothed by 5-sample averaging, and diatom percentages from Owens Lake have been weighted graphically to illustrate the effect of high and low diatom concentrations (Bradbury, this volume). The Tule Lake chronology is modified from Rieck et al. (1992), and the Owens chronology is from Bischoff et al. (this volume, Chapter 8).

between 125 ka and 150 ka in that figure are expressions of the Group II glaciations. The less-pronounced troughs near 50 ka, 70 ka, and 90 ka in those plots might represent less-pronounced glacial re-advances in the Sierras, but they are not reflected uniformly by the several criteria of runoff volumes, and we conclude that if they do represent advances, they were minor and accompanied by less intense increases in precipitation than accompanied the Group II and III glaciations.

On the basis of the moraine record exposed in the Sierra Nevada, the period between the Group I and the Group II glaciations is commonly inferred to have been free of glaciers. However, evidence indicating three (or four?) possible glaciations during this interval is found in the OL-92 core, as shown in Figures 4 and 5 which together cover the period 200 ka to 800+ ka. The percentages of sand in each sample are also plotted in those figures to allow differentiation between troughs in the curves caused by increased percentages of sand rather than the chemical signature of high precipitation, runoff, and possible glaciation. Troughs at 300 ka, 360 ka, and 670 ka do not coincide with high percentages of sand, and thus appear to have the characteristics of wet and possibly glacial episodes. Another trough at 455 ka approaches the baseline of the diagram, but even though associated with a 25 wt % peak in the sand plot, the low analyzed values of all three chemical criteria suggest a wet interval.

The waning stages of the Group I glaciation, most of which preceded deposition of the Bishop ash at 760 ka, may be equivalent to sediments near the base of OL-92 (Fig. 5), but the evidence is inconclusive. As noted earlier, the troughs in these curves are close to being mirror images of the sand curves (Fig. 5A and 5B), and we place little reliance on them as indications of glacial environments. However, the CEC values (Fig. 5C), which are not affected by sand percentages, have mostly low values prior to 650 ka, and may accurately reflect the waning stages of an earlier wet period and glaciation.

OWENS LAKE HISTORY VERSUS GLOBAL GLACIATION: 0 TO 500 KA

Correlation of the OL-92 record with global glaciation, as derived from the stable-isotope records obtained by analyses of foraminifera from deep-sea cores, is here based primarily on the similarities in the ages and senses of major and minor reversals in their climatically sensitive records (Figs. 7A and 7C). The points of correlation we emphasize here are those that indicate "reversals" in climate trends—peaks and troughs in those figures—because we suspect that the events that may be of most importance to paleoclimatic studies are the ages at which existing atmospheric-climate regimes, representing in many instances long-lasting equilibriums, *begin* to be replaced by new atmospheric-climate regimes.

Correlations between climatic reversals

Most of the periods of high runoff in OL-92 (Figs. 1 and 2) can be correlated with those of the isotopically heavy $\delta^{18}O$ (glacial) cycles in the marine record back to about 500 ka, and periods of low runoff can be correlated with isotopically light (interglacial) cycles. Sediments from OL-92 older than 500 ka have percentages of sand that are too high for many of our criteria to be reliable.

The main peaks in the OL-92 record (Fig. 7A) are correlated with identically lettered peaks in the SPECMAP pattern (Fig. 7C) as well as in the Devils Hole record (Fig. 7B). Similarly, the main troughs in Figures 7A and 7C are correlated. Differences between the ages of the correlated peaks and troughs of OL-92 and SPECMAP (Table 2) vary from 0 k.y. to 33 k.y., with a mean of 16 k.y.

Correlations with glacial "terminations"

Many studies of paleoclimatic records focus on the ages of "terminations," the ages of points in the marine record that are half way between the first indication of the end of a high-latitude glacial maximum and the first indication that interglacial conditions were fully established (Broecker and van Donk, 1970). The *age* of this termination, however, identifies neither the age when the "old" climatic regime began to falter nor the age when the "new" regime became firmly established. More importantly, the shape of these isotopically recognized terminations may have been substantially altered by the complex behavior of ice sheets during climatic change which depends not only on climate but also the nonlinear physical properties of ice as well as each glacier bed's slope, geometry, and topography (Kamb, 1964). Glacier responses to short- and long-term changes in climate have been estimated to require 3–30 years for a typical mountain glacier, to 5,000 years for the Antarctic ice sheet (Nye, 1960). Recent studies have also revealed that deglaciation at the end of the Pleistocene was characterized by periods of massive iceberg discharges, episodic discharges of meltwater, and rapid water- and air-temperature fluctuations (Bond et al., 1993; Lindstrom and MacAyeal, 1993; Oerlemans, 1993; Broecker, 1994). These processes may well have had a nonlinear relationships relative to atmospheric climates, meaning that changes in the isotopic composition of seawater resulting from the uneven introduction of runoff and melted glacier ice might have inaccurately reflected changes in atmospheric climate.

In the OL-92 core, Terminations I, II, IV, and V, seem best correlated with increases in the inorganic CO_3 content, the organic C content, and the CEC determinations (Figs. 3 and 4), reflecting change from overflowing to nearly closed (or closed) lake condition. Marine isotope records and Milankovich theory estimate the age of Termination II to be 128 ka (Williams et al., 1988). The Owens Lake age-depth curve depicts a change from low- to high-CO_3 percentages during the period 125 ka to 93 ka (Fig. 3A; Fig. 7A, points "G" to "h"). We believe this change is correlative with Termination II. Furthermore, we suggest that this 32-k.y.-long period, with its midpoint at about 109 ka, may be a more accurate reflection of the actual pace of atmospheric climate change. The ~19 k.y. difference between the midpoints of the marine termination and the Owens Lake termination may also reflect the time lag between changes in the high-latitude ice volumes and the mid-latitude manifestation of the climate change that influenced the geochemistry and sedimentation of Owens Lake. While it may also reflect imprecision in this part of the Owens Lake time-depth curve (Bischoff et al., this volume, Chapter 8), similar-size differences characterize the entire core (Tables 1 and 2).

Termination III in the marine record, at the Stage 7/8 boundary, occurs at 251 ka (Shackleton and Opdyke, 1976). It was also sought in the Owens Lake record (Fig. 4), but it shows only small changes near that time in the inorganic CO_3 and organic C percentages and the CEC values—unless one correlates Termination III with the small, abrupt increase in all three components at 290 ka, 39 k.y. earlier. Similarly, the diatom (Bradbury, this volume) and ostracode (Carter, this volume) records from OL-92 show no evidence of a major shift from wet to dry conditions, although the pollen record (which responds more to temperature change than most of the other indicators of climate in OL-92) shows a short period of change (Fig. 5 in Litwin et al., this volume). It is also significant that the Stage 7/8 boundary is also not evident in the well-dated LDW-6 core record from downstream Searles Lake (Bischoff et al., 1985), nor are any climatic transitions reflected by the sediments of that core at the times assigned to the boundaries between marine Stages 6/7 or 8/9.

Marine Terminations IV and V occur at 347 ka and 440 ka (Shackleton and Opdyke, 1976). The chemical criteria from OL-92 (Figs. 4A and 4B) that indicate decreases in runoff reaching Owens Lake—theoretically, the hydrologic equivalent of terminations—at 355 ka and 440 ka appear to mean that climate changes accompanying these events that affected high-latitude ice sheets were also recorded by the Owens Lake record. The lack of detectable Searles Lake-level responses to Terminations IV and V shows that they did not include *sufficient* precipitation change to be also reflected in Searles Lake's hydrologic balance.

DISCUSSION

The similarities between the three climatic histories plotted in Figure 7 suggest to us that their variations reflect related, global-scale climatic events since 500 ka. Important questions remain, however, concerning the credibility and the causes of the apparent age discrepancies between the Owens Lake record and the other two records plotted in Figure 7. However, we consider the likelihood of major errors to be small: The uncertainties in the laboratory data used to calculate the age-depth relations are small (Bischoff et al., this volume, Chapter 8); grain-density variations and average dry-sediment weights between dated horizons varied by only ~2%; and the ages of the 11 paleomagnetic excursions used to evaluate the age-depth curve have uncertainties that range from 4 k.y. to 31 k.y. and average 13 k.y. (Glen and Coe, this volume). Whereas these paleomagnetic error bars are based on a variety of evidence, and whereas errors having these magnitudes could incorrectly "validate" the calculated age-depth relations used here, it appears improbable that all except 1 of the 11 paleomagnetic error bars bracket that horizon's age as calculated from sediment and pore-water measurements—and the end of the one error bar that is "off" extends to within ~1 k.y. of the calculated mass-accumulation age (Bischoff et al., this volume, Chapter 8, Fig. 5). We conclude, therefore, that the ages assigned to OL-92 are approximately correct, and that the discrepancies between the ages of its peaks and troughs when compared with the two other records (Fig. 7) are probably real.

When the "difference" values from Tables 1 and 2 are tabulated, comparing the discrepancies between the ages of peaks and

TABLE 2. AGE COMPARISONS OF LABELED "PEAKS" AND "TROUGHS" IN FIGURES 7A AND 7C*

Label	Peak Ages		Difference†	Label	Trough Ages		Difference†
	SPECMAP (ka)	OL-92 (ka)	(k.y.)		SPECMAP (ka)	OL-92 (ka)	(k.y.)
L	…	…	…	l	6	10	-4
K	16	16	0	k	52	32	+20
J	64	48	+16	j	78	61	+17
I	86	69	+17	i	96	84	+12
H	106	89	+17	h	122	93	+29
G	134	125	+9	g	171	159	+12
F	170	166	+4	f	194	208	-14
E	203	215	-12	e	212	220	-8
D	224	236	-12	d	234	264	-30
C	266	303	-37	c	328	346	-18
B	340	362	-22	b	408	426	-18
A	434	456	-22	a	480	490	-10

*"Peaks" in these two figures represent high-latitude glaciation in the SPECMAP record and mid-latitude wet periods in the Owens Lake record; "troughs" represent periods having opposite characteristics. Ages for SPECMAP are from Bassinot et al., 1994; ages for OL-92 are estimated from Figure 7, probable accuracy ±3 k.y.
†SPECMAP age minus OL-92 age. Positive age differences mean that reversals in OL-92 curve, which reflect precipitation changes, occurred after reversals in SPECMAP curve, which reflect high-latitude ice volumes. Negative age differences mean the opposite.

TABLE 3. APPARENT DIFFERENCES IN AGES OF ADJOINING "PEAKS" AND "TROUGHS" IN OL-92 AND DH-11 RECORDS*

Peak-Trough Pair	Older Age (ka)	Younger Age (ka)	Interval (k.y.)	Δ Sed Rate† (k.y.)	Δ/k.y.§
a to A	480	448	32	2	0.06
A to b	448	400	48	-18	-0.38
b to B	400	355	45	19	0.42
B to c	355	326	29	-13	-0.45
c to C	326	270	56	-13	-0.23
C to d	270	239	31	8	0.26
d to D	239	222	17	11	0.65
D to e	222	204	18	-2	-0.11
e to E	204	200	4	1	0.25
E to f	200	194	6	1	0.17
f to F	194	171	23	19	0.83
F to g	171	160	11	-4	-0.36
g to G	160	129	41	3	0.09
G to h	129	121	8	24	3.0
h to H	121	110	11	-7	-0.64
H to i	110	102	8	-3	-0.38
i to I	102	89	13	2	0.15
I to j	89	79	10	-2	-0.20
j to J	79	62	17	-4	-0.24

*Data from Figures 7A and 7B, probable accuracy ±3 k.y.
†Change in apparent sedimentation rate in OL-92; see text for conventions determining sign.
§Change in apparent sediment rate per 1,000 yr.

troughs in the OL-92 record and the correlated inflections in the DH-11 and the SPECMAP records (Fig. 7), a pattern appears. Using the Devils Hole (DH-11) record as the primary basis for comparison, because its ages are determined by radiometric measurements, the youngest five peaks and four troughs in the DH-11 record (Table 1; peaks "F" through "J," and troughs "g" through "j") pre-date the correlated climatic responses in Owens Lake by 1–28 k.y. The older peaks and troughs post-date the correlated response in Owens Lake by 7–33 k.y.

A similar pattern is revealed by comparison of the OL-92 record with SPECMAP (Table 2). Comparison of ages of correlated peaks and troughs in the DH-11 and SPECMAP records finds their age differences to be mostly <5 k.y. and a more random pattern of small leads and lags.

Another consideration that leads us to accept the calculated ages is that the age differences between *adjoining* points of correlation—the peak-trough pairs in the OL-92 and DH-11 diagrams (Figs. 7A and 7B)—are too irregular to be a result of a systematic miscalculation of the age-depth relationship in core OL-92. Again, considering the ages of the labeled peaks and troughs of the DH-11 record (Fig. 7B, Table 1) to be "correct," and those of the OL-92 record (Fig. 7A, Table 1) to be "questionable," we find algebraic differences between the ages of *adjoining* peak-trough pairs of as much as 24 k.y. (peak "G" versus trough "h," an interval of 8 k.y.), and as little as 1 k.y. (peak "E" versus troughs "e" and "f," intervals of 4 k.y. and 6 k.y.). The other adjoining-peak-trough differences range from 2 k.y. to 19 k.y.

To estimate the apparent sedimentation-rate change per unit of time, we took the age differences (in k.y.) between the ages of the correlated peaks and troughs of the DH-11 and OL-92 curves, and determined the direction of change in that value as sedimentation progressed. For example, using the "C"–"d" pair, the DH-11 versus OL-92 "discrepancy" of –33 k.y. between the two "C" peaks decreased to –25 k.y. between the two slightly younger "c" troughs, suggesting that an increase in the sedimentation rate during deposition of core OL-92 reduced by 8 k.y. the age difference between the correlated troughs (Table 1). Where the difference indicated an apparent increase in the OL-92 sedimentation rate, it was listed with a "+" sign, if a decrease, it was listed with a "–" sign. Designating that age difference as Δ_a, and, determining the duration of the peak-to-trough transition from the ages of the peaks and troughs in DH-11 (Δ_d), we calculated

$$\Delta_a / \Delta_d = R_c,$$

where R_c = rate of apparent change in the relative sedimentation rate per 1,000 yr in OL-92.

The values of R_c were then tabulated (Table 3), showing 10 points positive and 9 negative, with each sign almost evenly divided between the upper and lower halves of that table. Their algebraic sum is +2.89 because, as noted above, the overall trend was for the apparent sedimentation rate to increase. The most positive value is +3.00/k.y., indicating an increase in apparent sedimentation rate of ~300% above its norm within a brief (8 k.y.) period. The next highest rate is +0.83, indicating an apparent sedimentation rate increase of ~83% above its norm during a longer (23 k.y.) period. The most negative value is –0.64/k.y., indicating an apparent sedimentation rate decrease of ~64% below its norm during that period (11 k.y.). Omitting the most positive value, these indicate a variation in relative sedimentation rates of almost 150%.

The points were then plotted as a scatter chart and four types of best-fit lines calculated. They produced a variety of subhorizontal lines whose *coefficient of determination* (r^2) values ≤0.05, indicating no correlation and showing that any calculated slope is meaningless. These data show that apparent sedimentation rates of sediments in the OL-92 core vary randomly over an improbably large range. Variations of these magnitudes are not observed in the radiometrically and paleomagnetically dated parts of the sediments in the Owens or Searles Basins (Liddicoat et al., 1980; Bischoff et al., 1985; Bischoff et al., this volume, Chapter 8; Glen and Coe, this volume), leading us to conclude that the correlated events in each record (Fig. 7) actually occurred at different times.

The SPECMAP ages for peaks and troughs (Table 2) are similar to those for DH-11, and numerical comparison of the SPECMAP record with the OL-92 record leads to a similar conclusion.

Nevertheless, the overall similarities between the three records (Fig. 7) reflect a very similar sequence of climatic changes. But those changes had to be transmitted around the globe via changes in air-mass trajectories, ocean-current patterns, global-surface albedos, air and sea-surface temperatures, and other controls of the climate that caused the waxing and waning of high-latitude ice sheets. It is difficult to estimate the time that was required to transmit each of those controls to other areas, but almost certainly they differed. Tables 1 and 2 suggest that the average time may have been ~16 k.y.

An example of how differences in the recorded ages of climate change can occur is provided by the earlier-described differences of several tens of thousands of years between the ages of the inferred climatic responses by the once-connected Owens and Searles Lakes, even though the lakes in both basins were almost solely dependent on precipitation and runoff from the same area. Accordingly, we suggest that when reconstructing local responses to Pleistocene climate change, especially when those responses occurred in different areas and/or reflected different elements of climate, age variations of these magnitudes should be expected.

Sedimentation rates in Owens Lake possibly did vary, especially during times when glaciation in the Sierra Nevada was increasing or decreasing. However, the constant mass-accumulation rate and the age-depth relations in the ^{14}C-dated part of OL-92 that represented the transition from the last Pleistocene glaciation to the Holocene show little evidence of this (Bischoff et al., this volume, Chapter 8). In fact, they imply that changes in the clastic sedimentation rate were offset by other changes, possibly in the precipitation rate of Ca- and Mg-carbonate minerals whose percentages range from <1% to >45%. A decrease in Owens Lake's surface area could have also minimized any decrease in its sedimentation rate by depositing its sediment over a smaller area, but the Searles Lake record (Fig. 6) shows that for ~70% of the period plotted in Figure 7 Owens Lake had to be full, because water from it was reaching Searles Lake.

Our conclusion at present is that pluvial and interpluvial periods in the western Great Basin (and elsewhere?) did at times precede, follow, or coincide with the cold- and warm temperature fluctuations responsible for the Devils Hole record and (in part) for global ice-volume fluctuations. Regardless of the degree of synchroneity found between the Devils Hole, SPECMAP, and Vostok records (Winograd et al., 1992), which largely or entirely reflect atmospheric temperatures, the Owens Lake record of precipitation and runoff apparently did not follow those chronologies as closely as convention might suggest. Confirmation of this possibility is found when part of the Searles Lake history is compared with the global glacial chronology.

In the Searles Lake record, downstream from Owens Lake during much of the Pleistocene (Smith, 1984), non-linear—even reciprocal—relations between that lake's history and global glaciation are recorded. As noted earlier, no changes in sedimentation character took place at the radiometrically determined boundaries between marine Stages 9/8, 8/7, and 7/6 that lie within Unit A+B (Fig. 6) in a core from that lake (Bischoff et al., 1985). The composition of the 45-m-thick Unit A+B required a narrow set of hydrologic conditions. Lake salinity rarely exceeded ~15% during the times it precipitated ~30 beds of mostly monomineralic salts that represent about 25 vol % of this unit; higher salinities would have precipitated additional species of salts. These salts are separated by beds of deeper-water sediments that average ~1 m thick, probably representing average periods of ~3 k.y. (Smith and Pratt, 1957, 67.3 m to 119.1 m; Smith, 1979; Smith et al., 1983; Smith et al., 1987). Repeatedly reverting to such a limited salinity range during periods of monomineralic saline-bed deposition required ~30 lowered, but almost identical, volumes of overflow from Owens Lake, an indication of minimal change in the regional hydrologic budget through isotope Stages 6, 7, 8, and 9.

In contrast, at the time of the Stage 6/5 boundary (Termination II, glacial-to-interglacial), Searles Lake expanded to become a deep lake that maintained its size for ~90% of the following ~100 k.y., yet at the stage 2/1 boundary (Termination I, also glacial-to-interglacial), Searles Lake desiccated (Smith, 1984). These three different responses to global climate change show that precipitation changes in the southwestern Great Basin did not always respond to atmospheric temperature changes in the same way—or at all.

Perhaps this lack of simple relations between these middle-latitude lakes and global cooling and glaciation should not be surprising. Pluvial lakes in low-latitude regions were expanding during many of the same periods when lakes in middle-latitude western North America were shrinking, and vice versa (Street and Grove, 1979). Clearly, "glacial" periods were not world-wide "wet" periods. The latitude of the drainage area supplying water to Owens Lake and its downstream lakes (36° to 38°N) is only a few degrees north of the present boundaries between these low- and middle-latitude regions (~30°N and S), and the precipitation element of past climates in this near-border region, during both "glacial" and "interglacial" periods, might well have had a mix of the precipitation characteristics of these two zones.

REFERENCES CITED

Bassinot, F. C., Labeyrie, L. D., Vincent, E., Quidelleur, X., Shackleton, N. J., and Lancelot, Y., 1994, The astronomical theory of climate and the age of the Brunhes-Matuyama magnetic reversal: Earth and Planetary Science Letters, v. 126, p. 91–108.

Benson, L. V., and Thompson, R. S., 1987, Lake-level variations in the Lahontan Basin for the past 50,000 years: Quaternary Research, v. 28, p. 69–85.

Benson, L. V., Kashgarian, M., and Rubin, M., 1995, Carbonate deposition, Pyramid Lake subbasin, Nevada: 2. Lake levels and polar jet stream positions reconstructed from radiocarbon ages and elevations of carbonates (tufas) deposited in the Lahontan Basin: Palaeogeography, Palaeoclimatology, and Palaeoecology, v. 117, p. 1–30.

Bischoff, J. L., Rosenbauer, R. J., and Smith, G. I., 1985, Uranium-series dating of sediments from Searles Lake: Differences between continental and marine climate records: Science, v. 227, p. 1222–1224.

Blackwelder, E., 1931, Pleistocene glaciation in the Sierra Nevada and Basin Ranges: Geological Society of America Bulletin, v. 42, p. 865–922.

Bond, G., Broecker, W., Johnsen, S., McManus, J., Labeyrie, L., Jouzel, J., and Bonani, G., 1993, Correlations between climate records from North Atlantic sediments and Greenland ice: Nature, v. 365, p. 143–147.

Bradbury, J. P., 1991, The late Cenozoic diatom stratigraphy of Tule Lake,

Siskiyou County, California: Journal of Paleolimnology, v. 6, p. 205–255.

Bradbury, J. P., 1992, Late Cenozoic lacustrine and climatic environments at Tule Lake, northern Great Basin, USA: Climate Dynamics, v. 6, p. 275–285.

Bradbury, J. P., 1993, Diatoms in sediments, in Smith, G. I., and Bischoff, J. L., eds., Core OL-92 from Owens Lake, southeast California: U.S. Geological Survey Open-File Report 93-683, p. 261–302.

Broecker, W. S., 1994, Massive iceberg discharges as triggers for global climate change: Nature, v. 372, p. 421–424.

Broecker, W. S., and van Donk, 1970, Insolation changes, ice volumes, and the O^{18} record in deep-sea cores: Reviews of Geophysics and Space Physics, v. 8, p. 169–198.

Burke, R. M., and Birkeland, P. W., 1979, Reevaluation of multiparameter relative dating techniques and their application to the glacial sequence along the eastern escarpment of the Sierra Nevada, California: Quaternary Research, v. 11, p. 21–51.

Bury, C. R., and Redd, R., 1933, The system sodium carbonate-calcium carbonate-water: Chemical Society Journal [London], p. 1160–1162.

Clark, D. H., and Gillespie, A. R., 1994, A new interpretation for late-glacial and Holocene glaciation in the Sierra Nevada, California, and its implications for regional paleoclimate reconstructions: Geological Society of America Abstracts with Programs, 1994 Annual Meeting, Seattle, Washington, p. A-447.

Dalrymple, G. B., 1964, Potassium-argon dates of three Pleistocene interglacial basalt flows from the Sierra Nevada, California: Geological Society of America Bulletin, v. 75, p. 753–757.

Emiliani, C., 1955, Pleistocene temperatures: Journal of Geology, v. 63, p. 538–578.

Fitzpatrick, J. A., and Bischoff, J. L., 1993, Uranium-series dates on the sediments of the high shoreline of Panamint Valley, California: U.S. Geological Survey Open File Report 93-232, 15 p.

Friedman, I., Smith, G. I., and Hardcastle, K. G., 1976, Studies of Quaternary saline lakes—II. Isotopic and compositional changes during desiccation of brines in Owens Lake, California, 1969–1971: Geochimica et Cosmochimica Acta, v. 40, p. 501–511.

Friedman, I., Smith, G. I., Gleason, J. D., Warden, A., and Harris, J. M., 1992, Stable isotope composition of waters in southeast California: Part 1, Modern precipitation: Journal of Geophysical Research—Atmospheres, v. 97, no. D5, p. 5795–5812.

Furgurson, E. B., 1992, Lake Tahoe, playing for high stakes: National Geographic, v. 181, no. 3, p. 113–132.

Gale, H. S., 1914, Salines in the Owens, Searles, and Panamint Basins, southeastern California: U.S. Geological Survey Bulletin 580-L, p. 251–323.

Gillespie, A. R., 1982, Quaternary glaciation and tectonism in the southeastern Sierra Nevada, Inyo County, California [Ph.D. thesis]: Pasadena, California Institute of Technology, 695 p.

Hollett, K. J., Danskin, W. R., McCaffrey, W. F., and Walti, C. L., 1991, Geology and water resources of Owens Valley, California, in Hydrology and soil-plant relations in Owens Valley, California: U.S. Geological Survey Water-Supply Paper 2370, p. 1–77.

Imbrie, J., Hays, J. D., Martinson, D. G., McIntyre, A., Mix, A. C., Morley, J. J., Pisias, N. G., Prell, W. L., and Shackleton, N. J, 1984, The orbital theory of Pleistocene climate: Support from a revised chronology of the marine $\delta^{18}O$ record, in Berger, A., Imbrie, J., Hays, J., Kukla, G., and Saltaman, B., eds., Milankovich and climate: Boston, D. Reidel Publishing Co., p. 269–305.

Izett, G. A., Pierce, K. L., Naeser, N. D., and Jaworowski, C., 1992, Isotopic dating of Lava Creek B tephra in terrace deposits along the Wind River, Wyoming—Implications for post 0.6 Ma uplift of the Yellowstone hotspot: Geological Society of America Abstracts with Programs, v. 24, no. 7, p. A102.

Jannik, N. O., Phillips, F. M., Smith, G. I., and Elmore, D., 1991, A ^{36}Cl chronology of lacustrine sedimentation in the Pleistocene Owens River system: Geological Society of America Bulletin, v. 103, p. 1146–1159.

Kamb, B., 1964, Glacier geophysics: Science, v. 146, p. 353–365.

Kutzbach, J. E., and Guetter, P. J., 1986, The influence of changing orbital parameters and surface boundary conditions of climate simulations for the past 18,000 years: Journal of the Atmospheric Sciences, v. 43, no. 16, p. 1726–1759.

Lao, Y., and Benson, L. V., 1988, Uranium-series age estimates and paleoclimatic significance of Pleistocene tufas from the Lahontan Basin, California and Nevada: Quaternary Research, v. 30, p. 165–176.

Lee, C. H., 1912, An intensive study of the water resources of a part of Owens Valley, California: U.S. Geological Survey Water-Supply Paper 294, 135.

Liddicoat, J. C., Opdyke, N. D., and Smith, G. I., 1980, Palaeomagnetic polarity in a 930-m core from Searles Valley, California: Nature, v. 286, no. 5768, p. 22–25.

Lindstrom, D. R., and MacAyeal, D. R., 1993, Death of an ice sheet: Nature, v. 365, p. 214.

Lowenstein, T. K., 1994, Death Valley salt core: 200,000 year record of closed-basin subenvironments and climate [abs.]: Geological Society of America, Abstracts with Programs, 1994 Annual Meeting, Seattle, Washington, p. A-169.

Ludwig, K. R., Simmons, K. R., Szabo, B. J., Winograd, I. J., Landwehr, J. M., Riggs, A. C., and Hoffman, R. J., 1992, Mass-spectrometric $^{230}Th/^{238}U$ dating of the Devils Hole calcite vein: Science, v. 258, p. 284–287.

McGee, W. J., 1885, On the meridional deflection of ice streams: American Journal of Science, v. 29, 3d ser., p. 386–392.

Mehringer, P. J., and Sheppard, J. C., 1978, Holocene history of Little Lake, Mojave Desert, California, in Davis, E. L., ed., The ancient Californians, Rancholabrean hunters in the Mojave Lakes country: Los Angeles, Natural History Museum of Los Angeles County, Science Series 29, p. 153–166.

Nye, J. F., 1960, The responses of glaciers and ice-sheets to seasonal and climatic changes: Proceedings of the Royal Society of London, v. 256, p. 559–588.

Oerlemans, J., 1993, Evaluating the role of climate cooling in iceberg production and the Heinrich events: Nature, v. 364, p. 783–786.

Putnam, W. C., 1950, Moraine and shoreline relationships at Mono Lake, California: Geological Society of America Bulletin, v. 61, p. 115–122.

Rieck, H. J., Sarna-Wojcicki, A. M., Meyer, C. E., and Adam, D. P., 1992, Magnetostratigraphy and tephrochronology of an upper Pliocene to Holocene record in lake sediments at Tulelake, northern California: Geological Society of America Bulletin, v. 104, p. 409–428.

Russell, I. C., 1885, Geologic history of Lake Lahontan, a Quaternary lake of northwestern Nevada: U.S. Geological Survey Monograph 11, 288 p.

Russell, I. C., 1889, Quaternary history of Mono Valley, California: Eighth Annual Report of the United States Geological Survey, J. W. Powell, Director, 1886-7, Part Ib, p. 261–394.

Shackleton, N. J., and Opdyke, N. D., 1976, Oxygen-isotope and paleomagnetic stratigraphy of Pacific core V28-239, late Pliocene to latest Pleistocene, in Cline, R. M., and Hays, J. D., eds., Investigation of late Quaternary paleoceanography and paleoclimatology: Boulder, Colorado, Geological Society of America Memoir 145, p. 449–464.

Sharp, R. P., 1968, Sherwin till–Bishop Tuff geological relationships, Sierra Nevada, California: Geological Society of America Bulletin, v. 79, p. 351–364.

Sharp, R. P., 1972, Pleistocene glaciation, Bridgeport Basin, California: Geological Society of America Bulletin, v. 83, p. 2233–2260.

Sharp, R. P., and Birman, J. H., 1963, Additions to classical sequence of Pleistocene glaciations, Sierra Nevada, California: Geological Society of America Bulletin, v. 74, p. 1079–1086.

Smith, G. I., 1979, Subsurface stratigraphy and geochemistry of late Quaternary evaporites, Searles Lake, California: U.S. Geological Survey Professional Paper 1043, 130 p.

Smith, G. I., 1984, Paleohydrologic regimes in the southwestern Great Basin, 0–3.2 my ago, compared with other long records of "global" climate: Quaternary Research, v. 22, p. 1–17.

Smith, G. I., 1993, Field log of Core OL-92, in Smith, G. I., and Bischoff, J. L., eds., Core OL-92 from Owens Lake, southeast California: U.S. Geological Survey Open-File Report 93-683, p. 4–57.

Smith, G. I., Barczak, V. J., Moulton, G. F., and Liddicoat, J. C., 1983, Core

Smith, G. I., Friedman, I., Klieforth, H., and Hardcastle, K., 1979, Areal distribution of deuterium in eastern California precipitation: Journal of Applied Meteorology, v. 18, no. 2, p. 172–188.

Smith, G. I., Friedman, I., and McLaughlin, R. J., 1987, Studies of Quaternary saline lakes—III: Mineral, chemical, and isotopic evidence of salt solution and crystallization processes in Owens Lake, California, 1969–1971: Geochimica et Cosmochimica Acta, v. 51, p. 811–827.

Smith, G. I., and Haines, D. V., 1964, Character and distribution of nonclastic minerals in the Searles Lake evaporite deposit, California: U.S. Geological Survey Bulletin 1181-P, 58 p.

Smith, G. I., and Pratt, W. P., 1957, Core logs from Owens, China, Searles, and Panamint basins, California, U.S. Geological Survey Bulletin 1045-A, 62 p.

Smith, G. I., and Street-Perrott, F. A., 1983, Pluvial lakes of the western United States, *in* Wright, H. E., Jr., ed., Late-Quaternary Environments of the United States: Minneapolis, University of Minnesota Press, p. 190–212.

Street, F. A., and Grove, A. T., 1979, Global maps of lake-level fluctuations since 30,000 yr B.P.: Quaternary Research, v. 12, p. 83–118.

Stuiver, M., 1964, Carbon isotopic distribution and correlated chronology of Searles Lake sediments: American Journal of Science, v. 262, p. 377–392.

Stuiver, M., and Smith, G. I., 1979, Radiocarbon ages of stratigraphic units, *in* Subsurface stratigraphy and geochemistry of late Quaternary evaporites, Searles Lake, California: U.S. Geological Survey Professional Paper 1043, p. 68–78.

Szabo, B. J., Kolesar, P. T., Riggs, A. C., Winograd, I. J., and Ludwig, K. R., 1994, Paleoclimatic influences from a 120,000-yr calcite record of water-table fluctuations in Browns Room of Devils Hole, Nevada: Quaternary Research, v. 41, p. 59–69.

Williams, D. F., Thunell, R. C., Tappa, C., Rio, D., and Raffi, I., 1988, Chronology of the Pleistocene oxygen isotope record: 0–1.88 m.y. B.P.: Palaeogeography, Palaeoclimatology, Palaeoecology, v. 64, p. 221–240.

Winograd, I. J., Szabo, B. J., Coplen, T. B., and Riggs, A. C., 1988, A 250,000 year climatic record from Great Basin vein calcite: Science, v. 242, p. 1275–1280.

Winograd, I. J., Coplen, T. B., Lanwehr, J. M., Riggs, A. C., Ludwig, K. R., Szabo, B. J., Kolesar, P. T., and Revesz, K. M., 1992, Continuous 500,000-year climate record from vein calcite in Devils Hole, Nevada: Science, v. 258, p. 255–260.

MANUSCRIPT ACCEPTED BY THE SOCIETY JUNE 17, 1996

Index

[Italic page numbers indicate major references]

A

Abert Lake, Oregon, 33
    potassium-feldspar in, 33
Abies
    concolor, 130
    magnifica, 130
Achnanthes sp., 101
Adobe Valley, 3
age, 56
age-depth curve, *94*, *96*, *97*, 147, 155
age-depth model, 37, 45, 100, 128
age-depth relation, 25, 34, 71, *91*, 156, 157
Alabama Hills, California, 110
    midden records, 132, 136, 137, 138
Alamo River, 61
albite, 26, 30
Aleutian Low, 99, 101
alkalinity, 37, 43
All American Canal, 61
alluvium, Mt. Jefferson, 85
Amargosa Desert, Nevada, 127
Ambrosia
    dumosa, 130
    spp., 130
ammonia (NH_3), 13, 23
Amnicola sp., 122, 123, 124
Amphora
    coffaeiformia, 101
    ovalis, 101
    perpusilla, 101
andesite, 26
Anodonta sp., 122, 123, 124, 125
Anomoeoneis costata, 101
anoxic conditions, 21, 117
Antarctic ice sheet, 155
aragonite, 18, 39
aridity, 151. See also drought, precipitation
Artemia, 100
Artemisia
    spinescens, 130
    tridentata, 130, 135
    spp., 130, 136
Artemisia pollen, 110
ash
    air-fall, 21, 23
    Bishop, 6, 7, 9, 19, 23, 49, 64, 71, 79, 91, 94, 99, 103, 106, 128, 139
    Dibekulewe, 7, 19, 71, 79, 80, 83, 88, 97
    Glass Mountain, 80, 83, *88*
    Lava Creek, 71, 88, 89, 149
    Orange, 88
    Rockland, 83, 88
    Thermal Canyon, 79, 81, 88
    Walker Lake, 79
Asterionella formosa, 101, 108, 109
Atriplex
    confertifolia, 128
    polycarpa, 128

Aulacoseira
    ambigua, 101
    granulata, 101
    islandica, 101
    solida, 101, 103, 106
    subarctica, 101, 106
    spp., 101, 103, 106, 109, 111, 153

B

bedding, 21, 23
    graded, 22
    rhythmic, 9, 23
Bend, Oregon, 88
Benton Valley, 3
bicarbonate ions, 103
big sagebrush, 130
biotite, 18
bioturbation, 9, 13, 21
bioturbation structures, 22, 23
Bishop, California, 5, 81, 86, 130
Bishop ash, 6, 7, 9, 71, 94, 99
    age, 130
    deposition of, 79, 87, 155
    depth, 128
    eruption of, 19, 49, 64, 79, 87, 103, 106
    mass-accumulation rate, 91
    redated, 86
    reworked, 7, 23, 103
    stratigraphic relation to Matuyama-Brunhes boundary, 85
    See also Bishop Tuff
Bishop tephra, reworked, *87*
Bishop Tuff, deposition of, 20, 92. See also Bishop ash
bitterbush, 135
blackbrush scrub, 130
Blake inclination anomaly, 71
Borrego Badlands, California, 83
brine shrimp, 100. See also *Artemia*
bristlecone pine, 94, 130
Brunhes-Matuyama paleomagnetic reversal, 91. See also Matuyama-Brunhes polarity reversal
Brunhes Normal Polarity, 85
burkeite, 18
burrowing organisms, 22
bursage, 130

C

calcite, 18, 39, 94, 152
calcium, 44, 145
calcium carbonate ($CaCO_3$), 6, 13, 19, 33, 37, 39, 40, 42, *44*, 45, 46, 55, 71, 73, 74, 98, 118, 146, 147, 148
    as indicator of closed lake conditions, 43
    in ostracodes, 118
calcium carbonate budget, 43, 98
Caloneis, sp., 101

Campylodiscus
    clypeus, 101
    sp., 101
Candona
    caudata, 113, 117, 118
    sp., 113, 117, 118
carbon, organic, 6, 34, 37, 38, 55, 146, 147, 148, 153, 155, 156
carbonate (CO_3), 19, 20, 25, 34, 38, 39, 91, 118, 124, 145, 148, 153
    inorganic, 153, 155
carbonate budget, climatic significance of, 145
carbonate content
    correlation with glacial terminations, 155
    as reflection of lake level variation, 34
Carex sp., 130
Carson River, Nevada, 152
Cascade Range, 88
cation-exchange capacity (CEC), 37, 38, 41, 44, 45, 146, 147, 148
    correlation with glacial terminations, 155
Catostomus fumeiventris, 122. See also sucker
cedar, 45
Chaetoceros mulleri spores, 101
Chalfant Valley, 3
Chamaebatiaria, sp., 136
chemical sediments, 6, 18
chenopods/amaranths, 127, 130
China basin, 99
China Lake, 6, 149
    clay minerals in, 26
chloride, 49, 50, 56, 64, 103
    Yellowstone hydrothermal system, 64
chlorine, 19
chlorinity, 49, 55, 64
chlorite, 26, 29, 32, 33
clam, 123
clastic dikes, 9, 21, 23
clastic materials, 23, 25
clastic sediments, 6, 13, 19
clay, 13, 14, 21, 22, 56, 81, 92, 100, 146
    diatomaceous, 83
    lacustrine, 9, 25
clay fraction, cation-exchange capacity of, 37
clay mineralogical analyses, 28
    methods, 28
clay mineralogy, 73
clay minerals, 6, 25, 26, 33, 94. See also specific minerals
Clear Lake, California, 109, 137, 140
climate, 25, 74, *100*
    changes in, 20, 25, 37, 73, 89, *99*, 137, 139, 143, *146*
    global, 25
    link between North American and North Atlantic, 76

climate (continued)
  past cycles, 6
  relation to diatoms, 111
  relation to lake depth, *144*
  relation to lake sedimentation, *144*
  relation to magnetic susceptibility, 73, 146
climatic signals, in clay mineralogy, 25
climatic trends, by geochemical evidence, 127
closed lake conditions, 42, 43
Coast Ranges, California, 137
Cocconeis placentula, 101
Coleogyne ramosissima, 130
color, 9, 13, 75
color veins, 21, 23
Colorado Plateau, pollen, 132
Columbia River, Washington, 64
components, of lake water, 19
coral reefs, Oahu, 138
coring procedures, 6
Coso Range, 2
Cupressaceae juniper pollen, 109
cycles, of open/closed lake conditions, 44, 45
Cyclostephanos spp., 101, 103, 108, 109
Cyclotella
  bodanica, 101, 103, 151
  caspia, 101
  meneghiniana, 101, 103, 106, 108
  ocellata, 101, 103, 109
  quillensis, 101
Cymbella sp., 101
Cyperaceae, 130
Cytherissa lacustris, 113, 116, 117, 118
Cytheromorpha sp., 113

D

Dansgaard-Oeschger events, 76
Dead Sea, Israel, 140
Death Valley, 3, 6, 9
Death Valley basin, 99
Death Valley Lake, California, *152*
demagnetization, 68, 70, 71, 85
desiccation, 19, 37, 55, 56, 61, 64, 92, 143, 148, 149, 150, 151
deuterium, 50, 60
deuterium-hydrogen ratios, 49, 50, 56
Devils Hole, Nevada, 7, 110, 127, 132, 137, 139, 144, *152*, 155, 158
diatoms, 4, 7, 11, 45, 75, *99*, 143, 146, 149, 150, 152, 153, 156
  ecology, *101*
  correlation with glaciation, *109*
  correlation with precipitation, *109*
  freshwater, 34, 44, 45, 99, 101, 103, 106, 109, 147, 151
  as measure of glacial-type climates, 111
  saline, 34, 45, 101, 106, 146, 147, 149
Dibekulewe ash bed, 7, 19, 71, 79, 80, 83, 88, 97

Diploxylon, 130
discontinuities, 9
dissolved-ion composition, 113
Distchlis spicata, 130
dolomite, 39
Draba sp., 130
dropstones, 14
drought
  Great Basin, 94
  Sierra Nevada, 94
  Tahoe basin, 94, 148
Dry Falls, Washington, 64

E

earthquakes, 21
Emperor inclination anomaly, 71
environment of deposition, 12
Ephedra
  nevadensis, 130
  viridis, 130, 135
  spp., 130
Epilobium sp., 130
Epithemia sp., 101
Ericamieria cooperi, 130
Eriogonum fasciculatum, 130
erosion, 22
evaporation, 5
evaporation rate, 5
evaporite minerals, 33, 100. *See also* salt

F

faults, 2, 9, 22, 23
feldspar, 18, 25, 26, 29, 34, 94, 146
feldspar, correlation with chlorite, 33
feldspar, correlation with illite, 33
ferric oxide (Fe_2O_3), 74
Festuca sp., 130
fish, 143, 147, 149, 150
  cold water, 106
  fossil, 7, 100, 145
  See also specific fish
Fish Lake Valley, 81
Fish Slough, 83
flora, variations in, 7
fluctuations, of lake level, 1
foraminifera, 110, 144, 155
fossil assemblages, 20
fossils, aquatic, 4, 11
foxtail pine, 130
Fragilaria
  capucina, 101
  crotonensis, 101, 103, 108
  vaucheriae, 101
  sp., 101, 103, 106, 153
Funza, Colombia, 140

G

Gara Dibas wadi, northwestern Sahara, 60
gastropods, 122, 125

gaylussite, 7, 18, 19, 20, 37, 39, 42, 45, 55, 149
  Mono Lake, 43
geologic setting, of drilling project, 1, 9
geothermal system, Long Valley, 44
Gila (Siphatelos) bicolor, 122. *See also* tui chub
glacial abrasion, 26, 32
glacial advances, Sierran, 37
glacial comminution, 143
glacial Lake Missoula, 26
glacial periods, 74
  identification of, 33, 34
glaciations, 155
  correlation with diatoms, *109*
  cyclic, 25, 35
  global, 44
Glass Mountain, California, 81
  ash beds, 80, 83, *88*
  tephra eruptions, 103
Glass Mountain–Long Valley Caldera, California, 81
glass shards, 81, 83, 87
  volcanic, 79, 80
global climate changes, 25
global glaciations, 44
Gomphonema, sp., 101
goosefoot, 130
Great Basin
  drought, 94
  lakes, 100, 103
  midden records, 127, 128, 132
  ostracodes, 117
  pollen, 132
Great Basin sagebrush, 135
Great Lakes, 103
grain density measurements, 94
grain size, 25
  lacustrine, 26
  as reflection of lake level variation, 34
  relation to climate change, 26, 31
  relation to lake level, 31
grain-size analyses, 14, 17, 27, 30
  methods, 27
  results, 27
grain-size variations, 25, 33
Grand Coulee, Washington, 64
Grande Pile, France, 140
granodiorite, 38, 39
granules, 14, 25
Grayia spinosa, 130
Great Basin
  lakes of, 1
  precipitation, 4
Great Salt Lake, Utah
  salinity, 4
  water level, 4
green joint-fir, 135
Greenland ice cores, 76
Greenland Summit ice core, 76
greigite, 11, 74
Gulf of California, moisture from, 101
gypsum, 37, 39, 42, 45
Gyrosigma sp., 101

H

Haiwee, California, precipitation, 5
halite, 18, 19
Hammil Valley, 3
Hantzschia, sp., 101
Haploxylon, 130
Haystack Mountain, California, midden records, 135, 136
Heinrich events, 76
Helisoma
  (Carinifex) newberryi, 123
  sp., 122, 123, 125
histories, lake, 1, 3, *153*
Honey subbasin, Nevada, 152
hornblende, 18
humates, 75, 91, 92, 93
Humbolt-Carson subbasin, Nevada, 152
Humbolt River, Nevada, 152
Humbolt River Canyon, Nevada, 71
Hydrobia sp., 124
hydrogen sulfide (H_2S), 13, 23
hydrology, 49, *100*
  climatically caused changes, 20
hydrotroilite, 11
Hymenoclea salsola, 130

I

ice, 14
illite, 6, 25, 26, 29, 30, 32, 33, 34, 35, 42, 146
Imperial Valley, California, 61
inclination anomalies, 71
Independence, California, 5
Indian Wells Valley, 3, 9
Inyo County, California, 99
Inyo Mountains, 2, 87, 101, 130
irrigation, 4
isotope cycles, marine, 38
isotopic makeup, 6

J

Jamaica inclination anomaly, 71
Jeffrey pine, 128, 130
Joshua tree, 128, 132
juniper, 45, 110
  pinyon, 110
juniper pollen, 109, 127, 137, 138, 152
Juniperus
  australis, 130
  occidentalis, 130
  osteosperma, 110, 130, 132, 135, 136
  scopulorum, 110, 135, 136

K

Kamikatsura Normal Polarity Sub-chron, 86
kaolinite, 26, 29, 32
kerolite, 40

L

Lahontan paleolake system, 99
Lake Biwa, Japan, 103, 140
Lake Cochise, Arizona, 137
Lake County, California, 109
Lake Lahontan, Nevada, *152*
lake level
  fluctuations in, 20, 87
  inferred from lithologies, 20
  related to carbonate content, 34
  related to grain size, 34
Lake Maracaibo, Venezuela, 26
Lake Michigan, ostracodes, 116
Lake Missoula, glacial, 26
lake size, 4
Lake Tahoe, 94, 148, 149
Lake Tecopa, 81, 85, 88, 89
Lanx sp., 122, 123
Larrea tridentata, 130
Laschamp inclination anomaly, 71
Lassen Park, California, 88
last interglacial, 45, 56
Late Glacial Interstadial, 76
Lava Creek ash, 71, 88, 89, 149
Les Echets, France, 140
Levantine inclination anomaly, 71
limber pine, 130
Limnocythere
  bradburyi, 113, 117, 118
  ceriotuberosa, 113, 116, 117, 118
  friabilis, 113, 116, 117, 118
  itasca, 113
  paraornata, 113
  platyforma, 113
  sappaensis, 113, 116, 117, 118
Little Lake, 64, 148, 149
location, of drilling project, 1
lodgepole pine, 130
Long Valley, California, 88
  eruption in, 64
  geothermal system of, 44
Long Valley Caldera, California, 5, 19, 81
  eruption of, 100, 103, 150
  magmatic activity in, 44
Long Valley Lake, 99, 106
Los Angeles Aqueduct, diversion of water into, 2, 4, 18, 55, 64, 92
Lower Salt, 149

M

mackinawite, 11
macrofossil record, Alabama Hills, 137
magmatic activity, Long Valley caldera, 44
magnesium, 145
magnesium carbonate ($MgCO_3$), 40
magnetic field, direction changes, 77
magnetic reversal, Bishop ash bed, 79
magnetic susceptibility, 67, 106, 111, 143, 146, 152

magnetic susceptibility (continued)
  measurements, 68
  relation to climate, 73
magnetization, remanent, 68
Mammoth Mountain, rhyolite, 83
mass accumulation rate (MAR), 7, 25, *91*, *94*, 99
  model, 96, 98
Matuyama-Brunhes boundary, 99, 100, 106
  stratigraphic relation to Bishop ash, 85
Matuyama-Brunhes polarity reversal, 6, 9, 67, 68, 71, 77, 79, 85, 86, 88, 128
Matuyama Polarity Chron, 100
Matuyama Reversed Polarity Chron, 85
McGee stage, 153
Mecca Hills, California, 83
Melosira
  varians, 101
  spp., 109
methane (CH_4), 13, 23
Mexican Plateau, ostracodes, 117
micas, weathering of, 32
microcline, 26
midden records, Great Basin, 128
Minarette Summit, 5
minnow, 122
models
  age-depth, 37, 45, 100, 128
  of ancestral Owens Lake, 63
  of evaporating lakes in arid environments, 59
  mass accumulation rate, 96, 98
  Rayleigh distillation, 59
Mojave creosote bush scrub, 130, 135
mollusks, fossil, 7, 143, 147, 149, 150
Mono Basin, 42
Mono Basin stage, 153
Mono Craters, eruption of, 80, 83, 85
Mono Lake, California, 3, 26, 83, 88, 99
  gaylussite in, 43
  salinity of, 43
montmorillonite, 26
Mormon joint fir, 130
Mount Diablo Base and Meridian, 6
Mt. Jefferson, Nevada, alluvium, 85
mudcracks, 75

N

Navicula
  pygmaea, 101
  subinflatoides, 101
  sp., 101
Negit Island, 83
  tephra layer, 85
Neotoma middens, 127, 132
New River, 61
Nitzschia
  frustulum, 101
  monoensis, 101
  pusilla, 101
  sp., 101

O

Oahu, Hawaii, coral reefs, 138
Oncorhymchus clarki, 122. *See also* trout
oolites, 6, 7, 9, 18, 20, 23, 27, 33, 34, 75, 92
   aragonite, 18
   calcareous, 55
   deposition, 151
   sedimentation rate, 93
   sizes, 18
Orange ash beds, 88
organic carbon, 6, 34, 37, 38, 55, 146, 147, 148, 153, 155, 156
orthoclase, 26
ostracodes, 4, 11, 34, 45, 100, 123, 146, 147, 149, 150, 156
   Canadian, 116
   fossil, 7, 143
   freshwater, 45, *113*, 144
   Lake Michigan, 116
   New Mexico, 117
   Oregon, 116
   as proxies for past climates, 113
   saline, 45, *113*, 146
Overburden Mud, 149
Owens basin, 99
   volcanic impacts, 109
Owens hydrologic system, *100*
Owens Lake drainage system, 7
Owens Lake playa, 99
Owens River
   discharge of, 42
   diversion of, 2, 4, 18, 23, 37, 43, 143, 148. *See also* Los Angeles Aqueduct
   east fork of, 3
Owens River delta, 56
Owens River drainage area, 2, 5, 99
   precipitation, 4
Owens Valley Aqueduct, 5. *See also* Los Angeles Aqueduct
Owens Valley fault, 22
OWL cores, 75
oxidation, 11, 13, 21
oxygen isotopic composition, 25, 37, 44, 45, 71, 73, *121*, 124, 125, 127, 137, 138, 139, 143, 152

P

Pacific Ocean, moisture from, 101
packrat middens, 110, 127, 132
paleoclimatic reconstruction, modern data relevant to, 5
paleoclimatic record, synthesis of, *143*
paleohydrologic record, or climate change, *99*
paleomagnetic excursion events, 91
paleomagnetism, 67
palynomorph assemblages, 127
Panamint basin, 6, 99
Panamint Lake, California, 151
   clay minerals in, 26
Panamint Valley, 3, 9, 151
Paoha Island, California, 83, 88
pelecypods, 122
Phlox sp., 130
phyllosilicate, 35, 40
phytoplankton, 101, 103
pine, 94, 130, 132, 147
pine pollen, 34, 127, 130, 131
   correlated to oxygen isotope record, 138
Pinnularia sp., 101
piñon pine, 130, 135
Pinus, 45. *See also* pine, pine pollen
Pinus
   albiculis, 130
   balfouriana, 130
   contorta, 130
   flexilis, 130
   jeffreyi, 130
   longaeva, 130
   monophylla, 130, 135
   monticola, 130
   murrayana, 130
pinyon juniper, 110
Pisidium
   compressum, 123
   sp., 122, 123, 124
plagioclase, 18, 30, 32, 33, 35
Poa
   rupicola, 130
   spp., 130
Poaceae, 130
polar ice-cap, 144
pollen, 45, 100, *127*
   concentrations, 130
   deposition rates, 130
   fossil, 7, 130, 132, 143
   juniper, 109, 127
   pine, 34, 127, 130, 131, 138
   profile, 132
   record, 156
pore fluids, 49
   transport of, 58
pore water, 156
   age of, 55
   chemistry of, 54
   composition of, 6
   content, 91, 94
   data, 77
potassium-feldspar, 18, 30, 32, 33, 35
precipitation, 2, 5
   annual, 5
   correlation with diatoms, *109*
   correlation with lake size, 144
   past changes in, 1
   patterns of, 109
   seasonal, 5
   Sierra Nevada, 6, 45, 99
Prosopium sp., 222. *See also* whitefish
Prunus sp., 136
Pseudostaurosira brevistriata, 101
Psorothamnus fremonti, 130
pumice, 87
pumice clasts, 87
pumice lapilli, 81
pumiceous shards, 81
pupfish, 122
Purshia
   tridentata, 130, 135
   sp., 136
Pyramid Lakes, Nevada, 123
Pyramid-Winnemucca subbasin, Nevada, 152

Q

quartz, 18, 25, 26, 29, 30, 32, 33, 34, 94, 146
Querus spp., 130, 137, 139

R

ragweed, 130
rationale, for drilling project, 1
Recess Peak stage, 153
relative humidity, variations in, 3
remanent magnetization, 68
Rhoicosphenia curvata, 101
Rhopalodia constricta, 101
Rhopalodia gibba, 101
Rhopalodia gibberula, 101
rhyolite, Mammoth Mountain, 83
rhythmites, 26
Ribes sp., 136
Rock-Color Chart classification system, 13
Rockland ash bed, 83, 88
Rocky Mountain juniper, 135
runoff, 5
   changes in, 1
Rye Patch Dam Bed, 88

S

sagebrush, pollen, 127, 130
sagebrush scrub zone, 130
Sahara, 60, 61
saline/fresh cycles, 37
saline minerals, 18. *See also* salt
salinity, 7, 20, 33, 34, 37, 39, 43, 45, 49, 55, 63, 64, 94, 99, 101, 113, *118*, 121, 143, 147, 149, 150, 158
   closed lake, 42, 43
   Great Salt Lake, 4
   Mono Lake, 43
salinity-depth profile, 55
Salix spp., 130
salt, 149, 158
   anthropogenic, 6, 9
   oolitic, 92
   Searles Lake, 100
salt flats, 3
salt pseudomorphs, 149
Salton Sea, California, 60, 61, 83
San Joaquin Valley, California, 137
sand, 9, 13, 14, 20, 25, 56, 92, 143, 144, 148
sand pods, 9, 21, 23
Sarcobatus vermiculatus, 130

Searles Basin, 6, 75, 99, 146, 149, 152
Searles Lake, California, 20, 37, 44, 64, 77, 111, *149*, 156, 158
  clay minerals in, 26
  climate record, 7, 99
  lithologic record, 152
  potassium-feldspar in, 33
  salt deposition in, 76, 100
  sediments in, 26, 33, 56
  water depth, 144
Searles Lake core, 25, 26
Searles Valley, 3, 9, 149
Sebkha el Melah, Algerian Sahara, 61
sedges, 130
sediment, 156
  analyses, 38, 45
  deposition of, 20
  geochemistry, relation to climate change, 37
  sorting, 27
sediment size analyses, 14
sedimentary structures, 21
sedimentation, climate-induced changes in, *144*
sedimentation rate, 10, 71, 77, 145
  oolites, 93
sepiolite, 40
Shepherdia argentea, 130
Sherwin glaciation, 20, 103, 145, 153
Sierra Nevada, 1, 20
  climatic conditions in, 25, 111, 139
  drought, 94
  erosion of, 108
  glaciation, 20, 26, 32, 35, 37, 106, 108, 109, 139, 143, 144, 146, 153, 158. *See also* Sierran glacial advances
  granite, 108
  influence of elevation on precipitation, 4
  influence on Owens Lake, 4, 143
  juniper, 110
  pollen, 132
  precipitation, 6, 45, 99
  snow, 5, 100, 101
  uplift, 5
  weathering in, 34
  woodland taxa, 127
Sierran escarpment, 130
Sierran glacial advances, 37
Sierran glaciers, expansion of, 103
Sierran lakes, 7
Sierran montane forest, 128
silicate minerals, weathering of, 32
silt, 9, 13, 14, 21, 22, 25, 56, 81, 100, 143, 146
silver buffaloberry, 130
smectite, 6, 25, 26, 29, 30, 32, 33, 34, 35, 42, 45, 71, 74, 118, 146, 148
Soap Lake, Washington, 64
sodium, 19
Staurosira construens, 101
Staurosirella
  leptostauron, 101

Staurosirella (continued)
  pinnata, 101
Stephanodiscus
  carconensis, 109
  niagarae, 75, 106, 108, 109
  oregonicus, 106, 108, 109
  sp., 101, 106, 108
stevensite, 40
Suaeda spp., 130
sucker, 7, 122, 123
sulfate ions, 103
Surirella
  hoefleri, 101
  ovalis, 101
  striatula, 101
  sp., 101
Susan River, Nevada, 152
Synedra
  acus, 101, 108
  mazamaensis, 101
  rumpens, 101
  ulna, 101

T

Tahoe basin, drought in, 94
Tahoe glacial moraines, 109
Tahoe stage, 153
Taylor Canyon, tuff, 81
temperature, 2, 7, 113
  air, 14
  variations in, 3
Tenaya stage, 153
tephra, 14, 19
tephra layers, 9, 21, 79, 100
  ages of, 88
  Borrego Badlands, 83
  Mecca Hills, 83
  Negit Island, 85
  Walker Lake, 85, 97
terminations, glacial, 34, 44, 45, 46, 74, 113, 117, 119, 137, 155
Tetradymia
  axillaris, 130
  sp., 136
Thermal Canyon, ash of, 79, 81, 88
time-depth curve. *See* age-depth curve
time-depth relation. *See* age-depth relation
Tioga moraines, 109
Tioga stage, 153
Toba eruption, Sumatra, 109
Tolara Lake, California, 137
Toquima Range, Nevada, 85
total organic carbon (TOC), 38, 39, 45
tree rings, 94
tree trunks, submerged, 94
trona, 18
trout, 7, 106, 122, 123, 145, 147, 148
Truckee River, 152
Tryonia sp., 122, 123, 124
Tsuga mertensiana, 130
tuff, Taylor Canyon, 81

tui chub, 7, 122
Tule Lake, California, *153*
  core, 85, 140
  diatom record, 108
turbidity-current structures, 9, 22

U

Upper Salt, 149
Utah juniper, 132, 135, 136

V

Valvata
  sincera, 123
  sp., 122, 123
vegetational distribution, *128*
vegetational trends, 127, 139
volcanic cones, 2
volcanic flows, 2
volcanic glass shards, 79, 80
volcanic impacts, 109
Volcanic Tableland, California, 81, 83, 85, 86, 136
Vostok ice core, Antarctica, 153, 158

W

Walker Lake, Nevada, 80, 83, 88, 93, 103
  ash beds in, 79
  diatoms, 108
  tephra layers, 83, 97
Walker River, Nevada, 152
Walker subbasin, Nevada, 152
water content, in core, 54
weathering, 143
  of micas, 32
  in Sierra Nevada, 34
  of silicate minerals, 32
western white pine, 130
White Mountains, 2, 87, 94, 101
White Mountains, junipers, 110
whitebark pine, 130
whitefish, 7, 106, 122, 123, 145, 147
Wilcox playa, Arizona, 137
Wilson Creek beds, 83
wind velocity, variations in, 3
woodland taxa, 127

Y

Yellowstone hydrothermal system, 64
Yellowstone National Park, Wyoming, 88
Younger Dryas Stadial, 76, 77
Yucca brevifolia, 132, 135

Z

zeolites, 33
zircon, 39

```
QE 696 .A14 1997

An 800,000-year
  paleoclimatic record from
```

Typeset and printed in U.S.A. by Johnson Printing, Boulder, Colorado